MW00837698

Emergence, Complexity and Computation

Volume 43

The Emergence, Complexity and Computation (ECC) series publishes new developments, advancements and selected topics in the fields of complexity, computation and emergence. The series focuses on all aspects of reality-based computation approaches from an interdisciplinary point of view especially from applied sciences, biology, physics, or chemistry. It presents new ideas and interdisciplinary insight on the mutual intersection of subareas of computation, complexity and emergence and its impact and limits to any computing based on physical limits (thermodynamic and quantum limits, Bremermann's limit, Seth Lloyd limits…) as well as algorithmic limits (Gödel's proof and its impact on calculation, algorithmic complexity, the Chaitin's Omega number and Kolmogorov complexity, non-traditional calculations like Turing machine process and its consequences,…) and limitations arising in artificial intelligence. The topics are (but not limited to) membrane computing, DNA computing, immune computing, quantum computing, swarm computing, analogic computing, chaos computing and computing on the edge of chaos, computational aspects of dynamics of complex systems (systems with self-organization, multiagent systems, cellular automata, artificial life,…), emergence of complex systems and its computational aspects, and agent based computation. The main aim of this series is to discuss the above mentioned topics from an interdisciplinary point of view and present new ideas coming from mutual intersection of classical as well as modern methods of computation. Within the scope of the series are monographs, lecture notes, selected contributions from specialized conferences and workshops, special contribution from international experts.

Indexed by zbMATH.

More information about this series at https://link.springer.com/bookseries/10624

Yaroslav D. Sergeyev · Renato De Leone
Editors

Numerical Infinities and Infinitesimals in Optimization

Editors
Yaroslav D. Sergeyev
Dipartimento di Ingegneria
Informatica, Modellistica, Elettronica e
Sistemistica
University of Calabria
Rende, Italy

Lobachevsky University
Nizhny Novgorod, Russia

Renato De Leone
Scuola di Scienze e Tecnologie
Università degli Studi di Camerino
Camerino, Italy

ISSN 2194-7287 ISSN 2194-7295 (electronic)
Emergence, Complexity and Computation
ISBN 978-3-030-93641-9 ISBN 978-3-030-93642-6 (eBook)
https://doi.org/10.1007/978-3-030-93642-6

This Springer imprint is published by the registered company Springer Nature Switzerland AG
The registered company address is: Gewerbestrasse 11, 6330 Cham, Switzerland

The editors thank their families for their love, support, and really infinite patience during the years of preparation of this book on ① infinity.

To Adriana and Ekaterina

To Chiara, Dmitriy, and Robert

Preface

This book is dedicated to two topics that, up to the recent times, were believed incompatible. In fact, infinities and infinitesimals for centuries were considered within highly theoretical research areas whereas optimization is a part of applied mathematics and deals with algorithms and codes implemented on computers. The possibility to join these two topics under the same cover is a consequence of the fact that infinities and infinitesimals considered here are not traditional symbolic entities but numerical ones, namely, they can be represented on a new supercomputer patented in several countries and called *the Infinity Computer*. This computer is able to execute numerical (i.e., floating-point) operations with numbers that can have different infinite, finite, and infinitesimal parts represented in the positional numeral system with an infinite base called *grossone* expressed by the numeral ①. The numerical character of these infinite and infinitesimal numbers is one of the key differences with respect to traditional theories of infinity and infinitesimals. This novel point of view opens completely new horizons not only in pure mathematics but also in computer science and applied mathematics.

During the last years there was an increasing interest in this new computational paradigm. In 2013, 2016, and 2019 there were international conferences "NUMTA—Numerical Computations: Theory and Algorithms" where the Infinity Computer and its applications were among the main topics. Several special issues in leading scientific journals have been published. The last NUMTA conference in 2019 had more than 200 participants from 30 countries. Three special issues and proceedings in volumes 11973 and 11974 of the Springer book series "Lecture Notes in Computer Science" have been published.

The interest of the international scientific community to this paradigm is also due to the fact that, for the first time, infinities and infinitesimals were involved not only in scientific fields related to pure mathematics such as foundations of mathematics, logic, and philosophy but also in computer science and applied mathematics. Moreover, there exist several software simulators of the Infinity Computer that are actively used by researchers in their work with numerical methods. Nowadays among the fields of application of the new computational paradigm we find: numerical differentiation and numerical solution of ordinary differential equations, game theory, probability theory, cellular automata, Euclidean and hyperbolic geometry, percolation, fractals, infinite series and the Riemann zeta function, the first Hilbert problem, Turing machines, teaching, etc. A huge bulk of results using the Infinity Computer paradigm in optimization (linear, non-linear, global, and multi-criteria) is collected in this volume co-authored by 22 leading experts working in these fields. Thus, the reader will have a unique opportunity to find in the same book the state-of-the-art knowledge and practices in optimization using the Infinity Computing methodology.

The first two chapters of the book written by the editors have an introductory character and describe the Infinity Computer methodology (Chapter "A New Computational Paradigm Using Grossone-Based Numerical Infinities and Infinitesimals", written by Yaroslav D. Sergeyev) and some of the most important results in unconstrained and constrained optimization (Chapter "Nonlinear Optimization: A Brief Overview", written by Renato De Leone). Chapter "The Role of *grossone* in Nonlinear Programming and Exact Penalty Methods", also written by Renato De Leone, is dedicated to exact penalty methods for solving constrained optimization problems. Using penalty functions, the original constrained optimization problem can be transformed in an unconstrained one. The chapter shows how ① can be utilized in constructing exact continuously differentiable penalty functions for the case of only equality constraints, the general case of equality and inequality constraints, and quadratics problems. It is also discussed how these new penalty functions allow one to recover the exact solution of the unconstrained problem from the optimal solution of the unconstrained problem. Moreover, Lagrangian duals associated to the constraints can also be automatically obtained thanks to the ①-based penalty functions.

Chapter "Krylov-Subspace Methods for Quadratic Hypersurfaces: a Grossone–based Perspective" is written by Giovanni Fasano who studies the role of ① to deal with two renowned Krylov-subspace methods for symmetric (possibly indefinite) linear systems. First, the author explores a relationship between the Conjugate Gradient method and the Lanczos process, along with their role of yielding tridiagonal matrices which retain large information on the original linear system matrix. Then, he shows that coupling

① with the Conjugate Gradient method provides clear theoretical improvements. Furthermore, reformulating the iteration of this algorithm using this ①-based framework allows the author to encompass also a certain number of Krylov-subspace methods relying on conjugacy among vectors. The last generalization remarkably justifies the use of a ①-based reformulation of the Conjugate Gradient method to solve also indefinite linear systems. Finally, pairing this method with the algebra of grossone easily provides relevant geometric properties of quadratic hypersurfaces.

Marco Cococcioni, Alessandro Cudazzo, Massimo Pappalardo, and Yaroslav D. Sergeyev show in Chapter "Multi-objective Lexicographic Mixed-Integer Linear Programming: An Infinity Computer Approach" how a lexicographic multi-objective linear programming problem can be transformed into an equivalent single-objective problem by using the grossone methodology. Then, the authors provide a simplex-like algorithm, called GrossSimplex, able to solve the original problem using a single run of the algorithm. In the second part of the chapter, the authors tackle the mixed-integer lexicographic multi-objective linear programming problem and solve it in an exact way, by using a ①-version of the Branch-and-Bound scheme. After proving the theoretical correctness of the associated pruning rules and terminating conditions, they present a few experimental results executed on an Infinity Computer simulator.

In Chapter "The Use of Infinities and Infinitesimals for Sparse Classification Problems", Renato De Leone, Nadaniela Egidi, and Lorella Fatone discuss the use of grossone in determining sparse solutions for special classes of optimization problems. In fact, in various optimization and regression problems, and in solving overdetermined systems of linear equations it is often necessary to determine a sparse solution, that is a solution with as many as possible zero components. The authors show how continuously differentiable concave approximations of the pseudoâ€“norm can be constructed using ① and discuss properties of some new approximations. Finally, the authors present some applications in elastic net regularization and Sparse Support Vector Machine.

Chapter "The Grossone-Based Diagonal Bundle Method" written by Manlio Gaudioso, Giovanni Giallombardo, and Marat S. Mukhametzhanov discusses a fruitful impact of the Infinity Computing paradigm on the practical solution of convex nonsmooth optimization problems. The authors consider a class of methods based on a variable metric approach: the use of the Infinity Computing techniques allows them to numerically deal with quantities which can take arbitrarily small or large values, as a consequence of nonsmoothness. In particular, by choosing a diagonal matrix with positive entries as a metric, the authors modify the well-known Diagonal Bundle algorithm by means of matrix updates based on the Infinity Computing paradigm and provide computational results obtained on a set of benchmark test problems.

Chapter "On the Use of Grossone Methodology for Handling Priorities in Multi-objective Evolutionary Optimization" written by Leonardo Lai, Lorenzo Fiaschi, Marco Cococcioni, and Kalyanmoy Deb describes a new class of problems, called mixed Pareto-lexicographic multi-objective optimization problems, a suitable model for scenarios where some objectives have priority over some others. Two relevant subclasses of this problem are considered: priority chains and priority levels. It is shown that the Infinity Computing methodology allows one to handle priorities efficiently. It is remarkable that this technique can be easily embedded in most of the existing evolutionary algorithms, without altering their core logic. Three algorithms are described and tested on benchmark problems, including some real-world problems. The experiments show that the algorithms using the Infinity Computing methodology are able to produce more solutions and of higher quality.

Chapter "Exact Numerical Differentiation on the Infinity Computer and Applications in Global Optimization" written by Maria Chiara Nasso and Yaroslav D. Sergeyev shows how exact numerical differentiation can be executed on the Infinity Computer and efficiently applied in Lipschitz global optimization for computing derivatives. Optimization algorithms using smooth piece-wise quadratic support functions to approximate the global minimum are also discussed and their convergence conditions are provided. It is shown that all the methods can be implemented both in the traditional floating-point arithmetic and in the novel Infinity Computing framework. Numerical experiments confirm that methods using analytic derivatives and derivatives computed numerically on the Infinity Computer exhibit the identical behavior.

Chapter "Comparing Linear and Spherical Separation Using Grossone-Based Numerical Infinities in Classification Problems" authored by Annabella Astorino and Antonio Fuduli investigates the role played by the linear and spherical separations in binary supervised learning and in Multiple Instance Learning (MIL), in connection with the use of the grossone-based numerical infinities. While in classical binary supervised learning the objective is to separate two sets of samples, a binary MIL problem consists in separating two different type of sets (positive and negative), each of them constituted by a finite number of samples. The authors focus on the possibility to construct binary spherical classifiers characterized by an infinitely far center adopting the Infinity Computing methodology. They show that this approach allows them to obtain a good performance in terms of average testing correctness and to manage very easily numerical computations without any tuning of the "big M" parameter.

In Chapter "Computing Optimal Decision Strategies Using the Infinity Computer: The Case of Non-Archimedean Zero-Sum Games", Marco Cococcioni, Lorenzo Fiaschi, and Luca Lambertini investigate non-Archimedean zero-sum games allowing payoffs to be infinite, finite, and infinitesimal. Since any zero-sum game is associated to a linear programming problem, the search for Nash equilibria of non-Archimedean games requires optimization of a non-Archimedean linear programming problem whose peculiarity is to have the constraints matrix populated by both infinite and infinitesimal numbers. The authors implement and test a grossone-based version of the Simplex algorithm called Gross-Matrix-Simplex method and stress the advantages of numerical computations with respect to symbolic ones. Some possible applications related to such games are also discussed in the chapter.

Chapter "Modeling Infinite Games on Finite Graphs using Numerical Infinities" written by Louis D'Alotto studies infinite games played on finite graphs. A new model of infinite games played on finite graphs using the grossone paradigm is presented. It is shown that the new grossone-based model provides certain advantages such as allowing for draws, which are common in board games, and a more accurate and decisive method for determining the winner. Numerous examples and illustrations are provided.

Chapter "Adopting the Infinity Computing in Simulink for Scientific Computing" authored by Alberto Falcone, Alfredo Garro, Marat S. Mukhametzhanov, and Yaroslav D. Sergeyev describe a software simulator of the Infinity Computer written in the visual programming environment Simulink. Since this environment is adopted in many industrial and research domains for addressing important real-world issues (in particular, handling problems in control theory and dynamics), the appearance of the Infinity Computer Simulink-based solution is very important.

Chapter "Addressing Ill-Conditioning in Global Optimization Using a Software Implementation of the Infinity Computer" addressing some aspects of ill-conditioning in global optimization is written by Marat S. Mukhametzhanov and Dmitri E. Kvasov. One of the desirable properties of global optimization methods is *strong homogeneity* meaning that a method produces the same sequences of points at which the objective function is evaluated both for the original function and a scaled one. It is shown in the chapter that even if a method possesses this property theoretically, numerically very small and large scaling constants can lead to ill-conditioning of the scaled problem. In these cases the usage of numerical infinities and infinitesimals can avoid ill-conditioning produced by scaling. Numerical experiments illustrating theoretical results are reported.

In conclusion, it is obligatory to mention that all the chapters have been peer-reviewed by two reviewers each and by the editors. It is our pleasant duty to thank the reviewers for their meticulous work. The editors of the present volume thank the editors of the Springer series "Emergence, Complexity and Computation" Andrew Adamatzky, Guanrong Chen, and Ivan Zelinka as well as the Springer Editorial Director Thomas Ditzinger for their friendly support. Special thanks go to Marco Cococcioni and Lorenzo Fiaschi who not only co-authored several chapters but also greatly helped the editors with the Latex compilation of the book.

The editors hope that this book presenting the power of the Infinity Computing methodology in optimization will become a source of further interesting developments in this field and not only.

Rende, Italy — Yaroslav D. Sergeyev
Camerino, Italy — Renato De Leone

Contents

Contributors

Annabella Astorino Institute for High Performance Computing and Networking (ICAR), National Research Council, Rende, Italy

Marco Cococcioni Department of Information Engineering, University of Pisa, Pisa, Italy

Alessandro Cudazzo Dipartimento di Informatica, (University of Pisa), Pisa, Italy

Louis D'Alotto York College, The City University of New York, Jamaica, Queens, NY, USA;
The Graduate Center, The City University of New York, New York, NY, USA

Renato De Leone School of Science and Technology, University of Camerino, Camerino (MC), Italy

Kalyanmoy Deb Department of Electrical and Computer Engineering, Michigan State University, East Lansing, MI, USA

Nadaniela Egidi School of Science and Technology, University of Camerino, Camerino, MC, Italy

Alberto Falcone Department of Informatics, Modeling, Electronics and Systems Engineering (DIMES), University of Calabria, Rende, Italy

Giovanni Fasano Department of Management, University Ca' Foscari of Venice, Venice, Italy

Lorella Fatone School of Science and Technology, University of Camerino, Camerino, MC, Italy

Lorenzo Fiaschi Department of Information Engineering, University of Pisa, Pisa, Italy

Antonio Fuduli Department of Mathematics and Computer Science, University of Calabria, Rende, Italy

Alfredo Garro Department of Informatics, Modeling, Electronics and Systems Engineering (DIMES), University of Calabria, Rende, Italy

Manlio Gaudioso Università della Calabria, Rende, Italy

Giovanni Giallombardo Università della Calabria, Rende, Italy

Dmitri E. Kvasov University of Calabria, Rende, Italy;
Lobachevsky University of Nizhny Novgorod, Nizhny Novgorod, Russia

Leonardo Lai Department of Information Engineering, University of Pisa, Pisa, Italy

Luca Lambertini Department of Economics, University of Bologna, Bologna, Italy

Marat S. Mukhametzhanov Università della Calabria, Rende, Italy

Maria Chiara Nasso University of Calabria, Rende, Italy

Massimo Pappalardo Dipartimento di Informatica, (University of Pisa), Pisa, Italy

Yaroslav D. Sergeyev University of Calabria, Rende, Italy;
Lobachevsky University of Nizhny Novgorod, Nizhny Novgorod, Russia

Theoretical Background

A New Computational Paradigm Using Grossone-Based Numerical Infinities and Infinitesimals

Yaroslav D. Sergeyev

Abstract This Chapter surveys a recent computational methodology allowing one to work with infinities and infinitesimals numerically on a supercomputer called *the Infinity Computer* that has been patented in several countries. This methodology applies the principle *The whole is greater than the part* to all numbers (finite, infinite, and infinitesimal) and to all sets and processes (finite and infinite). It is shown that, from the theoretical point of view, the methodology allows one to consider infinite and infinitesimal quantities more accurately w.r.t. traditional approaches such as Cantor's cardinals and non-standard analysis of Robinson. On the other hand, the methodology has a pronounced numerical character that gives an opportunity to construct algorithms of a completely new type and run them on the Infinity Computer (there already exist numerous applications not only in optimization but also in numerical differentiation, ODEs, game theory, etc.). In this Chapter, some key aspects of the methodology are described and several examples are given. Due to the breadth of the subject, the main attention is limited to methodological and practical aspects required to grasp applications in optimization contained in the other Chapters of the book. The material is presented in an accessible form that does not require any additional knowledge exceeding the first year university course of mathematical analysis.

Y. D. Sergeyev (✉)
University of Calabria, Rende (CS), Italy
e-mail: yaro@dimes.unical.it

Lobachevsky State University, Nizhny Novgorod, Russia

Y. D. Sergeyev and R. De Leone (eds.), *Numerical Infinities and Infinitesimals in Optimization*, Emergence, Complexity and Computation 43,
https://doi.org/10.1007/978-3-030-93642-6_1

1 Introduction

The point of view on infinitesimals and infinities accepted nowadays (see, e.g., [5, 14, 17, 34, 43, 52]) takes its origins from the fundamental ideas of Leibniz, Newton, and Cantor (see [10, 48, 56]). More recently, in the late fifties of the 20th century, Robinson (see [63]) has introduced his symbolic non-standard analysis approach showing that non-archimedean ordered field extensions of the reals contained numbers that could serve as infinitesimals and their reciprocals as infinities. However, since he used in his approach Cantor's mathematical tools and terminology (cardinal numbers, countable sets, continuum, one-to-one correspondence, etc.), his methodology has incorporated advantages and disadvantages of Cantor's ideas. Among their disadvantages we find numerous paradoxes that go against our every day experience. Is it true that they are inevitable? Is it possible to propose an alternative viewpoint that would allow us to avoid some of them and to work with the infinite in a more precise and, especially, numerical way?

In this Chapter, we describe a recent methodology (see a comprehensive survey in [75] and a popular presentation in [67]) that, on the one hand, allows us to avoid numerous classical paradoxes related to infinity and infinitesimals and, on the other hand, gives the possibility to solve numerically[1] problems involving quantities of this kind (see patents [70]). In effect, this book collects results where this methodology has been applied in optimization. Before we start a technical consideration let us mention that numerous papers studying consistency of the new methodology and its connections to the historical panorama of ideas dealing with infinities and infinitesimals have been published (see [29, 50, 51, 53, 55, 71, 76, 81]). In particular, in [76] it is stressed that it is not related to non-standard analysis. However, the methodology that will be discussed here is not a contraposition to the traditional views. In contrast, it can be viewed as an applied evolution of the intuition of Cantor and Robinson regarding the existence of different infinities and infinitesimals. Traditional approaches and the methodology described here do not contradict one another, they can be considered just as different tools having different accuracies and used for observations of certain mathematical objects dealing with infinities and infinitesimals.

---

[1] Recall that numerical computations operate with floating point numbers on a computer where results of arithmetical operations are approximated in order to store them in the computer memory. In their turn, symbolic computations are the exact manipulations with mathematical expressions containing variables that have not any given value and are thus manipulated as symbols. For instance, non-standard analysis is symbolic (see [63]), it is not possible to assign any value to non-standard infinities and infinitesimals.

The new methodology has been successfully applied in several areas of pure and applied mathematics and computer science (more than 60 papers published in international scientific journals can be found at the dedicated web page [37]). We provide here just a few examples of areas where this methodology is useful. First of all, we mention applications in local, global, and multiple criteria optimization covered in this book (see [4, 11–13, 20–24, 30, 46, 79, 85]). Then, we can indicate game theory and probability (see, e.g., [9, 19, 27, 28, 58, 61, 62]), hyperbolic geometry and percolation (see [41, 42, 54]), fractals (see [3, 7, 8, 68, 69, 72, 74]), infinite series (see [65, 66, 75, 84]), Turing machines, cellular automata, ordering, and supertasks (see [18, 60, 62, 73, 77, 78]), numerical differentiation and numerical solution of ordinary differential equations (see [1, 25, 26, 39, 80]), etc. In addition, successful applications of this methodology in teaching mathematics should be mentioned (see [2, 38, 40]). The dedicated web page [36], developed at the University of East Anglia, UK, contains, among other things, a comprehensive teaching manual and a nice animation related to the Hilbert's paradox of the Grand Hotel.

This Chapter is structured as follows. First, in Sect. 2 some traditional numeral systems used to express finite and infinite quantities are studied and compared. Then, in Sect. 3 the main principles of the new computational methodology are described whereas Sect. 4 discusses a new way of counting implementing these principles. After that, in Sect. 5 it is shown how different finite, infinite, and infinitesimal numbers can be represented in a unique framework allowing one to execute numerical operations on the Infinity Computer. Section 6 considers some classical paradoxes of infinity in the traditional and the new frameworks. Section 7 concludes the Chapter that contains numerous examples and a continuous comparison with traditional views on infinity is performed.

2 Numeral Systems Used to Express Finite and Infinite Quantities

A *numeral* is a group of symbols (which can also be one symbol only) that represents a *number* that is a concept. For example, symbols '5', 'five', 'IIIII', and 'V', are different numerals, but they all represent the same number. A *numeral system* is defined as a set of certain basic numerals, rules used to write down more complex numerals using the basic ones, and algorithms that are executed to perform arithmetical operations with these numerals. Thus, numbers can be considered as objects of an observation that are represented (observed) by instruments of the observation, i.e., by numerals and, more general, by numeral systems.

In our everyday activities with finite numbers the *same* finite numerals are used for *all* purposes where we need to express the quantity we are interested in. In contrast, when we need to work with infinities and/or infinitesimals, *different* numerals are used in *different* situations. For instance, we use the symbol ∞ in mathematical analysis, symbol ω for working with Cantor's ordinals, symbols $\aleph_0, \aleph_1, ...$ for dealing with Cantor's infinite cardinals, etc. Moreover, different arithmetics with different rules should be used to operate with infinities in different situations. Let us give some examples:

- There exist undetermined operations ($\infty - \infty$, $\frac{\infty}{\infty}$, etc.) that are absent when we work with finite numbers.
- Arithmetic with ∞ does not satisfy Euclid's Common Notion no. 5 saying 'The whole is greater than the part' that holds when we work with finite quantities. In fact, it follows that, e.g., $\infty - 1 = \infty$ whereas clearly for any finite x it follows $x - 1 < x$.
- Addition and multiplication with Cantor's ordinals are not commutative (e.g., $1 + \omega = \omega$ and $\omega + 1 > \omega$), moreover, there does not exist an ordinal γ such that $\gamma + 1 = \omega$, etc.

It is well known that traditional computers work numerically only with finite numbers and situations where the usage of infinite or infinitesimal quantities is required are studied mainly theoretically (see [10, 16, 31, 33, 35, 48, 49, 63, 82] and references given therein). The fact that numerical computations with infinities and infinitesimals have not been implemented so far on computers can be explained by difficulties mentioned above. There exist also practical troubles that preclude an implementation of numerical computations with infinities and infinitesimals. For example, it is not clear how to store an infinite quantity in a finite computer memory.

In order to understand how one can change his/her view on infinity, let us consider in a historical perspective some numeral systems used to express finite numbers. Through the centuries, numeral systems (see, e.g., [16, 17]) became more and more sophisticated allowing people to express more and more numbers. Since new numeral systems appear very rarely, in each concrete historical period people tend to think that *any* number can be expressed by the current numeral system and the importance of numeral systems for mathematics is very often underestimated (especially by pure mathematicians who often write phrases of the kind 'Let us take any x from the interval [0, 1]' and do not specify how this operation of taking *can be executed*). However, if we think about, we can immediately see limitations that various numeral systems induce. It will become clear that due to practical reasons different numeral systems can express different limited sets of numbers and they can be more or less suitable for executing arithmetical operations.

In fact, people in different historical periods used different numeral systems to count and these systems: (a) can be more or less suitable for counting; (b) can express different sets of numbers. For instance, Roman numeral system is not able to express zero and negative numbers and such expressions as II – VII or X – X are indeterminate forms in this numeral system. As a result, before the appearance of positional numeral systems and the invention of zero mathematicians were not able to create theorems involving zero and negative numbers and to execute computations with them. The positional numeral system not only has allowed people to execute new operations but has led to new theoretical results, as well. Thus, numeral systems not only limit us in practical computations, they induce boundaries on theoretical results, as well.

It should be stressed that the powerful positional numeral system also has its limitations. For example, nobody is able to write down a numeral in the decimal positional system having 10^{100} digits (see a discussion on feasible numbers in [57, 64]). In fact, suppose that one is able to write down one digit in one nanosecond. Then, it will take 10^{91} s to record all 10^{100} digits. Since in one year there are $31.556.926 \approx 3.2 \cdot 10^7$ s, 10^{91} s are approximately $3.2 \cdot 10^{83}$ years. This is a sufficiently long time since it is supposed that the age of the universe is approximately $1.382 \cdot 10^{10}$ years.

As we have seen above, Roman numeral system is weaker than the positional one. However, it is not the weakest numeral system. There exist very poor numeral systems allowing their users to express very few numbers and one of them is illuminating for our story. This numeral system is used by a tribe, Pirahã, living in Amazonia nowadays. A study published in *Science* in 2004 (see [32]) describes that these people use an extremely simple numeral system for counting: one, two, many. For Pirahã, all quantities larger than two are just 'many' and such operations as 2+2 and 2+1 give the same result, i.e., 'many'. Using their weak numeral system Pirahã are not able to see, for instance, numbers 3, 4, and 5, to execute arithmetical operations with them, and, in general, to say anything about these numbers because in their language there are neither words nor concepts for that.

It is worthy of mention that the result 'many' is not wrong. It is just *inaccurate*. Analogously, when we observe a garden with 547 trees, then both phrases: 'There are 547 trees in the garden' and 'There are many trees in the garden' are correct. However, the accuracy of the former phrase is higher than the accuracy of the latter one. Thus, the introduction of a numeral system having numerals for expressing numbers 3 and 4 leads to a higher accuracy of computations and allows one to distinguish results of operations 2+1 and 2+2.

The poverty of the numeral system of Pirahã leads also to the following results

$$\text{`many'} + 1 = \text{`many'}, \quad \text{`many'} + 2 = \text{`many'}, \quad \text{`many'} + \text{`many'} = \text{`many'} \tag{1}$$

that are crucial for changing our outlook on infinity. In fact, by changing in these relations 'many' with ∞ we get relations used to work with infinity in the traditional calculus

$$\infty + 1 = \infty, \qquad \infty + 2 = \infty, \qquad \infty + \infty = \infty. \tag{2}$$

Analogously, if we consider Cantor's cardinals (where, as usual, numeral $\aleph_0$ is used for cardinality of countable sets and numeral $\mathfrak{c}$ for cardinality of the continuum, see, e.g., [83] for their definitions and related discussions) we have similar relations

$$\aleph_0 + 1 = \aleph_0, \qquad \aleph_0 + 2 = \aleph_0, \qquad \aleph_0 + \aleph_0 = \aleph_0, \tag{3}$$

$$\mathfrak{c} + 1 = \mathfrak{c}, \qquad \mathfrak{c} + 2 = \mathfrak{c}, \qquad \mathfrak{c} + \mathfrak{c} = \mathfrak{c}. \tag{4}$$

It should be mentioned that the astonishing numeral system of Pirahã is not an isolated example of this way of counting. In [15], more than 20 languages having numerals only for small numbers are mentioned. For example, the same counting system, one, two, many, is used by the Warlpiri people, aborigines living in the Northern Territory of Australia (see [6]). The Pitjantjatjara people living in the Central Australian desert use numerals one, two, three, big mob (see [47]) where 'big mob' works as 'many'. It makes sense to remind also another Amazonian tribe—Mundurukú (see [59]) who fail in exact arithmetic with numbers larger than 5 but are able to compare and add large approximate numbers that are far beyond their naming range. Particularly, they use the words 'some, not many' and 'many, really many' to distinguish two types of large numbers. Their arithmetic with 'some, not many' and 'many, really many' reminds the rules Cantor uses to work with $\aleph_0$ and $\mathfrak{c}$, respectively. In fact, it is sufficient to compare

$$\text{`some, not many'} + \text{`many, really many'} = \text{`many, really many'} \tag{5}$$

with

$$\aleph_0 + \mathfrak{c} = \mathfrak{c} \tag{6}$$

to see this similarity.

Let us compare now the weak numeral systems involved in (1), (5) and numeral systems used to work with infinity. We have already seen that relations (1) are results of the weakness of the numeral system employed. Moreover, the usage of a stronger numeral system shows that it is possible to pass from records 1+2 = 'many' and 2+2 = 'many' providing for two different expressions the same result, i.e., 'many', to more precise answers 1+2 = 3 and 2+2 = 4 and to see that $3 \neq 4$. In these examples we have the same objects—small finite numbers—but results of computations we execute are different in dependence of the instrument—numeral system—used to represent numbers. Substitution of the numeral 'many' by a variety of numerals representing numbers 3, 4, etc. allows us both to avoid relations of the type (1), (5) and to increase the accuracy of computations.

Relations (2)–(4), (6) manifest a complete analogy with (1), (5). Canonically, symbols ∞, $\aleph_0$, and $\mathfrak{c}$ are *identified* with concrete mathematical objects and (2)–(4), (6) are considered as intrinsic properties of these infinite objects (see e.g., [63, 83]). However, the analogy with (1), (5) suggests that relations (2)–(4), (6) do not reflect the *nature* of infinite objects. They are just a result of weak numeral systems used to express infinite quantities. As (1), (5) show the lack of numerals in numeral systems of Pirahã, Warlpiri, Pitjantjatjara, and Mundurukú for expressing different finite quantities, relations (2)–(4), (6) show shortage of numerals in mathematical analysis and in set theory for expressing different infinite numbers. Another hint leading to the same conclusion is the situation with indeterminate forms of the kind III-V in Roman numerals that have been excluded from the practice of computations after introducing positional numeral systems.

Thus, the analysis made above allows us to formulate the following key observation that changes our perception of infinity:

Our difficulty in working with infinity is not a consequence of the *nature* of infinity but is a result of *weak numeral systems* having too little numerals to express the multitude of infinite numbers.

The way of reasoning where the object of the study is separated from the tool used by the investigator is very common in natural sciences where researchers use tools to describe the object of their study and the used instruments influence the results of the observations and determine their accuracy. The same happens in mathematics studying natural phenomena, numbers, objects that can be constructed by using numbers, sets, etc. Numeral systems used to express numbers are among the instruments of observation used by mathematicians. As we have illustrated above, the usage of powerful numeral systems gives the possibility to obtain more precise results in mathematics in the same way as the usage of a good microscope gives the possibility of obtaining more precise results in Physics. Traditional numeral systems have been developed to express finite quantities and they simply have no sufficiently high number of numerals to express different infinities (and infinitesimals).

3 Three Methodological Postulates

In this section, we describe the methodology that will be adopted hereinafter. In contrast to the modern mathematical fashion that tries to make all axiomatic systems more and more precise (decreasing so degrees of freedom of the studied part of mathematics), we would like to be more flexible and just to define a set of general rules describing how practical computations should be executed leaving as much space as possible for further changes and developments dictated by practice of the introduced mathematical language. Notice also that historically human beings started to count without any help of axiomatic systems (and nowadays children learn to count by practice, without studying, e.g., Peano axioms for natural numbers). The spirit of this Chapter is the same, namely, it shows how to count and gives examples where this way of counting is useful and more precise with respect to traditional ways of counting dealing with infinity. We start by introducing three Postulates that will fix our methodological positions with respect to infinite and infinitesimal quantities and mathematics, in general. Then, in the subsequent section, there will be described practical rules required to execute computations implementing the ideas expressed in the Postulates.

After this preamble, let us start with the first Postulate. Usually, when mathematicians deal with infinite objects (sets or processes) it is supposed that human beings are able to execute certain operations infinitely many times. For example, in a fixed numeral system it is possible to write down a numeral with *any* number of digits. However, this supposition is an abstraction because we live in a finite world and all human beings and/or computers are doomed to finish operations they have started. In the methodology to be introduced, this abstraction is not used and the following Postulate is adopted.

Methodological Postulate 1. *We postulate existence of infinite and infinitesimal objects but accept that human beings and machines are able to execute only a finite number of operations.*

Thus, we accept that we shall never be able to give a complete (in any sense) description of infinite objects due to our finite capabilities. In particular, this means that we accept that we are able to write down only a finite number of symbols to express numbers. This Postulate has a practical meaning because we need to write down numbers and to see results of operations we execute.

The second Postulate is adopted following the way of reasoning used in natural sciences where researchers use tools to describe the object of their study and the used instruments influence the results of the observations. When a physicist uses a weak lens A and sees two black dots in his/her microscope he/she does not say: "The object of the observation *is* two black dots". The physicist is obliged to say: "The lens used in the microscope allows us to see two black dots and it is not possible to say anything more about the nature of the object of the observation until we replace the instrument—the lens or the microscope itself—with a more precise one". Suppose that he/she changes the lens and uses a stronger lens B and is able to observe that the object of the observation is viewed as ten (smaller) black dots. Thus, we have two different answers: (i) the object is viewed as two dots if the lens A is used; (ii) the object is viewed as ten dots by applying the lens B. Which of the answers is correct? Both. Both answers are correct but with the *different accuracies* that depend on the lens used for the observation. The answers are not in opposition one to another, they both describe the reality (or whatever is behind the microscope) correctly with the precision of the used lens. In both cases our physicist discusses what he/she observes and does not pretend to say what the object *is*.

The same happens in mathematics studying natural phenomena, numbers, and objects that can be constructed by using numbers. Numeral systems used to express numbers are among the instruments of observations used by mathematicians. The usage of powerful numeral systems gives the possibility to obtain more precise results in mathematics in the same way as usage of a good microscope gives the possibility of obtaining more precise results in physics. However, even for the best existing tool the capabilities of this tool will be always limited due to Postulate 1 (we are able to write down only a finite number of symbols when we wish to describe a mathematical object) and due to the following Postulate 2 we shall never tell, what is, for example, a number but we shall just observe it through numerals expressible in a chosen numeral system.

Methodological Postulate 2. *We shall not tell* ***what are*** *the mathematical objects we deal with; we just shall construct more powerful tools that will allow us to improve our capacities to observe and to describe properties of mathematical objects.*

This Postulate is important for our study because it emphasizes that the triad—object of a study, instrument of the study, and a researcher executing the study—introduced in physics in the 20th century exists in mathematics, as well. With this Postulate we stress that mathematical results are not absolute, they depend on mathematical languages used to formulate them, i.e., there always exists an accuracy of the description of a mathematical result, fact, object, etc. imposed by the mathematical language used to formulate this result. For instance, the result of Pirahã $2 + 2 =$ 'many' is not wrong, it is just *inaccurate*. The introduction of a stronger tool (in this case, a numeral system that contains a numeral for a representation of the number four) allows us to have a more precise answer. Postulates 1 and 2 impose us to think always about *the possibility to execute* a mathematical operation by applying a numeral system. They tell us that there always exist situations where we are not able to express the result of an operation (people working with numerical codes on computers understand this fact very well).

Numerals that we use to write down numbers, functions, etc. are among our tools of investigation and, as a result, they strongly influence our capabilities to study mathematical objects. This separation (having an evident physical spirit) of mathematical objects from the tools used for their description is crucial for our study but it is used rarely in contemporary mathematics. In fact, the idea of finding an adequate (absolutely the best) set of axioms for one or another field of mathematics continues to be among the most attractive goals for many contemporary mathematicians whereas physicists perfectly understand the impossibility of any attempt to give a *final* description of physical objects due to limitations of the instruments of study.

Instruments (numeral systems in this case) not only bound our practical capabilities to compute, they can influence *theoretical* results in mathematics, as well. We illustrate this statement by considering the following simple phrase 'Let us consider all $x \in [1, 2]$'. For Pirahã, Warlpiri, Pitjantjatjara, and Mundurukú *all numbers* are just 1 and 2. For people who do not know irrational numbers (or do not accept their existence) *all numbers* are fractions $\frac{p}{q}$ where p and q can be expressed in a numeral system they know. If both p and q can assume values 1 and 2 (as it happens for Pirahã), *all numbers* in this case are: 1, $1 + \frac{1}{2}$, and 2. For persons knowing positional numeral systems *all numbers* are those numbers that can be written in a positional system. Thus, in different historical periods (or in different cultures) the phrase 'Let us consider all $x \in [1, 2]$' has different meanings. As a result, without fixing the numeral system we use to express numbers we cannot fix the numbers we deal with and an ambiguity arises.

Let us now introduce the last Postulate. We have already seen in the previous section that situations of the kind (2)–(4), (6) are results of the weakness of numeral systems used to express infinite quantities. Thus, we should treat infinite and infinitesimal numbers in the same manner as we are used to deal with finite ones, i.e., by applying the Euclid's Common Notion no. 5 'The whole is greater than the part'. In our considerations, its straight reformulation 'The part is less than the whole' will be used for practical reasons that will become clear soon. This principle, in our opinion, very well reflects organization of the world around us but is not incorporated in many traditional theories of the infinite where it is true only for finite numbers. As the analysis made in the previous section suggests, the reason of this discrepancy is in the low accuracy of traditional numeral systems used to work with infinity.

Methodological Postulate 3. *We adopt the principle 'The part is less than the whole' to all numbers (finite, infinite, and infinitesimal) and to all sets and processes (finite and infinite).*

As was already discussed in connection with the numerals of Pirahã, Warlpiri, Pitjantjatjara, and Mundurukú, the traditional point of view on infinity accepting such results as $\infty - 1 = \infty$ should be substituted in such way that subtraction of 1 from an infinite quantity Q can be registered, i.e., it should be $Q - 1 < Q$, as it happens with finite quantities. The Postulate 3 fixes this desire.

It can seem at first glance that Postulate 3 contradicts Cantor's one-to-one correspondence principle for infinite sets. However, as it will be shown later, this is not the case. Instead, the situation is similar to the example from physics described above where we have considered two lenses having different accuracies. Analogously, we have two different lenses having different accuracies used to observe the same infinite objects: Cantor's approach and the new one based on Postulates 1–3.

The adopted Postulates impose also the style of exposition of results in this Chapter: we first introduce new mathematical instruments, then show how to use them in several areas of mathematics giving some examples and introducing each item as soon as it becomes indispensable for the problem under consideration. In order to proceed, we need to consider an example arising in practice where there is the necessity to count extremely large finite quantities. This example will help us to develop a new mathematical intuition required to compute with finite, infinite, and infinitesimal quantities in accordance with Postulates 1–3.

Imagine that the owner of a granary asks us to count how much grain he has inside it. Obviously, it is possible to answer that there are *many* seeds in the granary. This answer is correct but its accuracy is low. In order to obtain a more precise answer it would be necessary to count the grain seed by seed but since the granary is huge, it is not possible to do this due to practical reasons. To overcome this difficulty and to obtain a more precise answer, people take sacks, fill them with seeds, and count the number of sacks. In this situation, it is supposed that: (i) all the seeds have the same measure and all the sacks also; (ii) the number of seeds in each sack is the same and is equal to K_1 but the sack is so big that we are not able to count how many seeds it contains and to establish the value of K_1; (iii) in any case the resulting number K_1 would not be expressible by available numerals.

Then, if the granary is huge and it becomes difficult to count the sacks, then trucks or even big train wagons are used. As it was for the sacks, we suppose that all trucks contain the same number K_2 of sacks, and all train wagons contain the same number K_3 of trucks, however, the numbers K_i, $i = 1, 2, 3$, are so huge that it becomes impossible to determine them. At the end of the counting of this type we obtain a result in the following form: the granary contains 27 wagons, 32 trucks, 56 sacks, and 134 seeds of grain. Note, that if we add, for example, one seed to the granary, we not only can see that the granary has more grain but also *quantify* the increment: from 134 seeds we pass to 135 seeds. If we take out two wagons, we again are able to say how much grain has been subtracted: from 27 wagons we pass to 25 wagons.

In the example it is necessary to count a large quantity that is finite but so huge that it becomes impossible to count it directly by using the elementary unit of measure, let us say u_0—seeds. Therefore, people are forced to behave as if the quantity was infinite. To solve the problem of 'infinite' quantity, new units of measure, u_1—sacks, u_2—trucks, and u_3—wagons, are introduced. The new units have an important feature: all the units u_{i+1} contain a certain number K_i of units u_i but these numbers, K_i, $i = 1, 2, 3$, *are unknown*. Thus, the quantity that it was impossible to express using only the initial unit of measure, u_0, are perfectly expressible in the new units u_i, $i = 1, 2, 3$, that we have introduced in addition to u_0. Notice that, in spite of the fact that the numbers K_i, $i = 1, 2, 3$, are unknown, the accuracy of the obtained answer is equal to one seed.

This key idea of counting by introduction of new units of measure with unknown but fixed values K_i, $i \geq 1$, will be used in what follows to deal with infinite quantities together with the relaxation allowing one to use negative digits in positional numeral systems.

4 A New Way of Counting and the Infinite Unit of Measure ①

The way of counting described in the previous section suggests us how to proceed: it is necessary to extend the idea of the introduction of new units of measure from sets and numbers that are huge but finite to infinite sets and numbers. This can be done by extrapolating from finite to infinite the idea that n is the number of elements of the set $\{1, 2, 3, \ldots, n-1, n\}$.

Thus, the infinite unit of measure is introduced as the number of elements of the set, $\mathbb{N}$, of natural numbers[2] and expressed by the numeral ① called *grossone*. Using the granary example discussed above we can offer the following interpretation: the set $\mathbb{N}$ can be considered as a sack and ① is the number of seeds in the sack. Following our extrapolation, the introduction of ① allows us to write down the set of natural numbers as $\mathbb{N} = \{1, 2, 3, \ldots, ① - 1, ①\}$. Recall that the usage of a numeral indicating the totality of the elements we deal with is not new in Mathematics. It is sufficient to mention the theory of probability (see axioms of Kolmogorov in [45]) where events can be defined in two ways. First, as unions of elementary events; second, as a sample space, Ω, of all possible elementary events (or its parts) from which some elementary events have been excluded (or added in case of parts of Ω). Naturally, the latter way to define events becomes particularly useful when the sample space consists of infinitely many elementary events.

Grossone is introduced by describing its properties postulated by the *Infinite Unit Axiom* (IUA) consisting of three parts: Infinity, Identity, and Divisibility. Similarly, in order to pass from natural to integer numbers a new element—zero—is introduced, a numeral to express it is chosen, and its properties are described. The IUA is added to axioms for real numbers (that are considered in sense of Postulate 2). Thus, it is postulated that associative and commutative properties of multiplication and addition, distributive property of multiplication over addition, existence of inverse elements with respect to addition and multiplication hold for grossone as they do for finite numbers.

Let us introduce the axiom and then give some comments upon it. Notice that in the IUA infinite sets will be described in the traditional form, i.e., without indicating the last element. For instance, the set of natural numbers will be written as $\mathbb{N} = \{1, 2, 3, \ldots\}$ instead of $\mathbb{N} = \{1, 2, 3, \ldots, ① - 1, ①\}$. We emphasize that in both cases we deal with the same mathematical object—the set of natural numbers—that is observed through two different instruments. In the first case traditional numeral systems do not allow us to express infinite numbers whereas the numeral system with grossone offers this possibility. Similarly, Pirahã are not able to see finite natural numbers greater than 2 but these numbers (e.g., 3 and 4) belong to $\mathbb{N}$ and are visible if one uses a more powerful numeral system. A detailed discussion will follow shortly to show how to use grossone to calculate and express the number of elements of certain infinite sets.

[2] Notice that nowadays not only positive integers but also zero is frequently included in $\mathbb{N}$. However, since historically zero has been invented significantly later with respect to positive integers used for counting objects, zero is not include in $\mathbb{N}$ in this Chapter.

The Infinite Unit Axiom. The infinite unit of measure is introduced as the number of elements of the set, $\mathbb{N}$, of natural numbers. It is expressed by the numeral ① called *grossone* and has the following properties:

Infinity. Any finite natural number n is less than grossone, i.e., $n <$ ①.

Identity. The following relations link ① to identity elements 0 and 1

$$0 \cdot ① = ① \cdot 0 = 0, \quad ① - ① = 0, \quad \frac{①}{①} = 1, \quad ①^0 = 1, \quad 1^{①} = 1, \quad 0^{①} = 0. \tag{7}$$

Divisibility. For any finite natural number n sets $\mathbb{N}_{k,n}$, $1 \leq k \leq n$, being the nth parts of the set, $\mathbb{N}$, of natural numbers have the same number of elements indicated by the numeral $\frac{①}{n}$ where

$$\mathbb{N}_{k,n} = \{k, k+n, k+2n, k+3n, \ldots\}, \qquad 1 \leq k \leq n, \qquad \bigcup_{k=1}^{n} \mathbb{N}_{k,n} = \mathbb{N}. \tag{8}$$

Let us comment upon this axiom. Its first part—Infinity—is quite clear. In fact, we want to describe an infinite number, thus, it should be larger than any finite number. The second part of the axiom—Identity—tells us that ① interacts with identity elements 0 and 1 as all other numbers do. In reality, we could even omit this part of the axiom because, due to Postulate 3, all numbers should be treated in the same way and, therefore, at the moment we have stated that grossone is a number, we have fixed the usual properties of numbers, i.e., the properties described in Identity, associative and commutative properties of multiplication and addition, distributive property of multiplication over addition, etc. The third part of the axiom—Divisibility—is the most interesting, since it links infinite numbers to infinite sets (in many traditional theories infinite numbers are introduced algebraically, without any connection to infinite sets) and is based on Postulate 3. Let us first illustrate it by an example.

Example 1 If we take $n = 1$, then it follows that $\mathbb{N}_{1,1} = \mathbb{N}$ and Divisibility says that the set, $\mathbb{N}$, of natural numbers has ① elements. If $n = 2$, we have two sets $\mathbb{N}_{1,2}$ and $\mathbb{N}_{2,2}$, where

$$\begin{aligned} \mathbb{N}_{1,2} &= \{1, \quad 3, \quad 5, \quad 7, \ldots\}, \\ \mathbb{N}_{2,2} &= \{\quad 2, \quad 4, \quad 6, \quad \ldots\} \end{aligned} \tag{9}$$

and they have $\frac{①}{2}$ elements each. Notice that the sets $\mathbb{N}_{1,2}$ and $\mathbb{N}_{2,2}$ have the same number of elements not because they are in a one-to-one correspondence but due to the Divisibility axiom. In fact, we are not able to count the number of elements of the sets $\mathbb{N}$, $\mathbb{N}_{1,2}$, and $\mathbb{N}_{2,2}$ one by one because due to Postulate 1 we are able to execute only a finite number of operations whereas all these sets are infinite. To define their number of elements we use Divisibility and implement Postulate 3 in practice by determine the number of the elements of the parts using the whole.

Then, if $n = 3$, we have three sets

$$\begin{array}{llllll} \mathbb{N}_{1,3} = \{1, & & 4, & & 7, & \dots \}, \\ \mathbb{N}_{2,3} = \{ & 2, & & 5, & & 8, \dots \}, \\ \mathbb{N}_{3,3} = \{ & & 3, & & 6, & 9, \dots \} \end{array} \quad (10)$$

and they have $\frac{①}{3}$ elements each. Note that in formulae (9), (10) we have added extra spaces writing down the elements of the sets $\mathbb{N}_{1,2}$, $\mathbb{N}_{2,2}$, $\mathbb{N}_{1,3}$, $\mathbb{N}_{2,3}$, $\mathbb{N}_{3,3}$ just to emphasize Postulate 3 and to show visually that $\mathbb{N}_{1,2} \cup \mathbb{N}_{2,2} = \mathbb{N}$ and $\mathbb{N}_{1,3} \cup \mathbb{N}_{2,3} \cup \mathbb{N}_{3,3} = \mathbb{N}$. □

In general, to introduce $\frac{①}{n}$ we do not try to count elements $k, k+n, k+2n, k+3n, \dots$ one by one in (8). In fact, we cannot do this due to Postulate 1. By using Postulate 3, we construct the sets $\mathbb{N}_{k,n}$, $1 \le k \le n$, by separating the whole, i.e., the set $\mathbb{N}$, in n parts and we affirm that the number of elements of the nth part of the set, i.e., $\frac{①}{n}$, is n times less than the number of elements of the entire set, i.e., than ①.

As was already mentioned, in terms of our granary example ① can be interpreted as the number of seeds in the sack. In that example, the number K_1 of seeds in each sack was fixed and finite but it was impossible to express it in units u_0, i.e., seeds, by counting seed by seed because we had supposed that sacks were very big and the corresponding number would not be expressible by available numerals. In spite of the fact that K_1, K_2, and K_3 were inexpressible and unknown, by using new units of measure (sacks, trucks, etc.) it was possible to count more easily and to express the required quantities. Now our sack has the infinite but again *fixed* number of seeds. It is fixed because it has a strong link to a concrete set—it is the number of elements of the set of natural numbers. Since this number is inexpressible by existing numeral systems with the same accuracy afforded to measure finite small sets[3], we introduce a new numeral, ①, to express the required quantity.

---

[3] First, this quantity is inexpressible by numerals used to count the number of elements of finite sets because the set $\mathbb{N}$ is infinite. Second, traditional numerals existing to express infinite numbers do not have the required high accuracy (recall that we would like to be able to register the variation of the number of elements of infinite sets even when

Then, we apply Postulate 3 and say that if the sack contains ① seeds, then, even though we are not able to count the number of seeds of the nth part of the sack seed by seed, its nth part contains n times less seeds than the entire sack, i.e., $\frac{①}{n}$ seeds. Notice that the numbers $\frac{①}{n}$ are integer since they have been introduced as numbers of elements of sets $\mathbb{N}_{k,n}$.

The new unit of measure allows us to express a variety of infinite numbers (including those larger than ① that will be considered later) and calculate easily the number of elements of the union, intersection, difference, or product of sets of type $\mathbb{N}_{k,n}$. Due to our accepted methodology, we do it in the same way as these measurements are executed for finite sets. Let us consider two simple examples showing how grossone can be used for this purpose.

Example 2 Let us determine the number of elements of the set $A_{k,n} = \mathbb{N}_{k,n} \backslash \{a, b\}$, $a, b \in \mathbb{N}_{k,n}$, $n \geq 1$. Due to the IUA, the set $\mathbb{N}_{k,n}$ has $\frac{①}{n}$ elements. The set $A_{k,n}$ has been constructed by excluding two elements from $N_{k,n}$. Thus, the set $A_{k,n}$ has $\frac{①}{n} - 2$ elements. The granary interpretation can be also given for the number $\frac{①}{n} - 2$: the number of seeds in the nth part of the sack minus two seeds. For $n = 1$ we have $① - 2$ interpreted as the number of seeds in the sack minus two seeds. □

Divisibility and Example 2 show us that in addition to the usual way of counting, i.e., by adding units, that has been formalized in the traditional mathematics, there exists also the way to count by taking parts of the whole and by subtracting units or parts of the whole. The following example shows a slightly more complex situation (other more sophisticated examples will be given later after the reader gets accustomed to the concept of ①).

Example 3 Let us consider the following two sets

$$B_1 = \{3, 8, 13, 18, 23, 28, 33, 38, 43, 48, 53, 58, 63, 68, 73, 78,$$
$$83, 88, 93, 98, 103, 108, 113, 118, \dots\}, \qquad (11)$$

$$B_2 = \{3, 14, 25, 36, 47, 58, 69, 80, 91, 102, 113, 124, 135, \dots\} \qquad (12)$$

and determine the number of elements in the set

$$B = (B_1 \cap B_2) \cup \{3, 4, 5, 113\}.$$

one element has been excluded or added). For example, by using Cantor's Alephs we say that cardinality of the sets $\mathbb{N}$ and $\mathbb{N} \setminus \{1\}$ is the same—$\aleph_0$. This answer is correct but its accuracy is low. Using Cantor's cardinals we are not able to register the fact that one element was excluded from the set $\mathbb{N}$ even though we ourselves have executed this exclusion. Analogously, we can say that both the sets have many elements. Again, the answer 'many' is correct but its accuracy is also low.

It follows immediately from the IUA that $B_1 = \mathbb{N}_{3,5}$ and $B_2 = \mathbb{N}_{3,11}$. From (11) and (12) we can see that their intersection is

$$B_1 \cap B_2 = \mathbb{N}_{3,5} \cap \mathbb{N}_{3,11} = \{3, 58, 113, \ldots\} = \mathbb{N}_{3,55}.$$

Thus, due to the IUA, the set $\mathbb{N}_{3,55}$ has $\frac{①}{55}$ elements. Finally, since 3 and 113 belong to the set $\mathbb{N}_{3,55}$ and 4 and 5 do not belong to it, the set B has $\frac{①}{55} + 2$ elements. The granary interpretation is the following: $\frac{①}{55} + 2$ is the number of seeds in the 55th part of the sack plus two seeds. □

The IUA introduces ① as the number of elements of the set of natural numbers and, therefore, it is the last natural number. We can also talk about the set of *extended natural numbers* indicated as $\widehat{\mathbb{N}}$ and including $\mathbb{N}$ as a proper subset

$$\widehat{\mathbb{N}} = \{\underbrace{1, 2, \ldots, ① - 1, ①}_{\text{Natural numbers}}, ① + 1, ① + 2, \ldots, 2① - 1, 2①, 2① + 1, \ldots$$

$$①^2 - 1, ①^2, ①^2 + 1, \ldots 3①^{①} - 1, 3①^{①}, 3①^{①} + 1, \ldots\}. \quad (13)$$

The extended natural numbers greater than grossone are also linked to infinite sets of numbers and can be interpreted in the terms of grain. For example, $① + 1$ is the number of elements of a set $B_3 = \mathbb{N} \cup \{a\}$, where a is integer and $a \notin \mathbb{N}$. In the terms of grain, $① + 1$ is the number of seeds in a sack plus one seed.

The extended natural numbers greater than grossone are also linked to sets of numbers and can be interpreted in the terms of grain.

Example 4 Let us determine the number of elements of the set

$$C_m = \{(a_1, a_2, \ldots, a_{m-1}, a_m) : a_i \in \mathbb{N}, 1 \le i \le m\}, \quad 2 \le m \le ①.$$

The elements of C_m are m-tuples of natural numbers. It is known from combinatorial calculus that if we have m positions and each of them can be filled in by one of l symbols, the number of the obtained m-tuples is equal to l^m. In our case, since $\mathbb{N}$ has grossone elements, $l = ①$. Thus, the set C_m has $①^m$ elements. In the particular case, $m = 2$, we obtain that the set

$$C_2 = \{(a_1, a_2) : a_i \in \mathbb{N}, i \in \{1, 2\}\},$$

being the set of couples of natural numbers, has $①^2$ elements. This fact is illustrated below

$$\begin{array}{ccccc}
(1,1), & (1,2), & \dots & (1,①-1), & (1,①), \\
(2,1), & (2,2), & \dots & (2,①-1), & (2,①), \\
\dots & \dots & \dots & \dots & \dots \\
(①-1,1), & (①-1,2), & \dots & (①-1,①-1), & (①-1,①), \\
(①,1), & (①,2), & \dots & (①,①-1), & (①,①).
\end{array}$$

□

Let us consider now how the principle 'The part is less than the whole' included in Postulate 3 and the IUA can be reconciled with traditional views on infinite sets. At first sight it seems that there is a contradiction between the two methodological positions. For instance, traditionally it is said that the one-to-one correspondence can be established between the set, $\mathbb{N}$, of natural numbers and the set, $\mathbb{O}$, of odd numbers. Namely, odd numbers can be put in a one-to-one correspondence with all natural numbers in spite of the fact that $\mathbb{O}$ is a proper subset of $\mathbb{N}$

$$\begin{array}{ll}
\text{odd numbers:} & 1,\ 3,\ 5,\ 7,\ 9,\ 11,\ \dots \\
 & \updownarrow\ \ \updownarrow\ \ \updownarrow\ \ \updownarrow\ \ \updownarrow\ \ \updownarrow \\
\text{natural numbers:} & 1,\ 2,\ 3,\ 4\ 5,\ 6,\ \dots
\end{array} \qquad (14)$$

The traditional conclusion from (14) is that both sets are countable and they have the same cardinality $\aleph_0$.

Let us see now what we can say from the new methodological position, in particular, by using Postulate 2. The separation of the objects of study which are two infinite sets from the instrument used to compare them, i.e., from the bijection, suggests that another conclusion can be derived from (14): *the accuracy of the used instrument is not sufficiently high to see the difference between the sizes of the two sets.*

We have already seen that when one executes the operation of counting, the accuracy of the result depends on the numeral system used for counting. If one asked Pirahã to measure sets consisting of four apples and five apples the answer would be that both sets of apples have many elements. This answer is correct but its precision is low due to the weakness of the numeral system used to measure the sets.

Thus, the introduction of the notion of accuracy for measuring sets is very important and should be applied to infinite sets also. As was already discussed, since for cardinal numbers it follows

$$\aleph_0 + 1 = \aleph_0, \qquad \aleph_0 + 2 = \aleph_0, \qquad \aleph_0 + \aleph_0 = \aleph_0,$$

these relations suggest that the accuracy of the cardinal numeral system of Alephs is not sufficiently high to see the difference with respect to the number of elements of the two sets from (14).

In order to look at the record (14) using the new methodology let us remind that due to the IUA the sets of even and odd numbers have ①/2 elements each and, therefore, ① is even. It is also necessary to recall that numbers that are larger than ① are not natural, they are extended natural numbers. For instance, ① + 1 is odd but not natural, it is the extended natural, see (13). Thus, the last odd natural number is ① − 1. Since the number of elements of the set of odd numbers is equal to $\frac{①}{2}$, we can write down not only the initial (as it is usually done traditionally) but also the final part of (14)

$$\begin{array}{ccccccccccc} 1, & 3, & 5, & 7, & 9, & 11, & \dots & ①-5, & ①-3, & ①-1 \\ \updownarrow & \updownarrow & \updownarrow & \updownarrow & \updownarrow & \updownarrow & & \updownarrow & \updownarrow & \updownarrow \\ 1, & 2, & 3, & 4 & 5, & 6, & \dots & \frac{①}{2}-2, & \frac{①}{2}-1, & \frac{①}{2} \end{array} \qquad (15)$$

concluding so (14) in a complete accordance with the principle 'The part is less than the whole'. Both records, (14) and (15), are correct but (15) is more accurate, since it allows us to observe the final part of the correspondence that is invisible if (14) is used.

In many other cases the ①-based methodology allows us to measure infinite sets better with respect to Cantor's cardinals. We conclude this section by Table 1 that illustrates this fact and collects results proved in [75].

5 Positional Numeral System with the Infinite Base ①

We have already started to write down simple infinite numbers and to execute arithmetical operations with them without concentrating our attention upon a general form we wish to use to write down different infinite and infinitesimal numbers. We also did not describe yet algorithms for executing arithmetical operations with these numbers. Let us explore these two issues systematically.

Different numeral systems have been developed to describe finite numbers. In positional numeral systems, fractional numbers are expressed by the record

$$(a_n a_{n-1} \dots a_1 a_0 . a_{-1} a_{-2} \dots a_{-(q-1)} a_{-q})_b \qquad (16)$$

where numerals a_i, $-q \le i \le n$, are called *digits*, belong to the alphabet $\{0, 1, \dots, b-1\}$, and the dot is used to separate the fractional part from the integer one. Thus, the numeral (16) expresses the quantity obtained by summing up

Table 1 Cardinalities and the number of elements of some infinite sets, see [75]

Description of sets	Cantor's cardinalities	Number of elements
The set of natural numbers $\mathbb{N}$	Countable, $\aleph_0$	①
$\mathbb{N} \setminus \{3, 5, 10, 23, 114\}$	Countable, $\aleph_0$	① − 5
The set of even numbers $\mathbb{E}$ (the set of odd numbers $\mathbb{O}$)	Countable, $\aleph_0$	$\frac{①}{2}$
The set of integers $\mathbb{Z}$	Countable, $\aleph_0$	2①+1
$\mathbb{Z} \setminus \{0\}$	Countable, $\aleph_0$	2①
Squares of natural numbers $\mathbb{G} = \{x : x = n^2, x \in \mathbb{N},\ n \in \mathbb{N}\}$	Countable, $\aleph_0$	$\lfloor\sqrt{①}\rfloor$
Pairs of natural numbers $\mathbb{P} = \{(p, q) : p \in \mathbb{N},\ q \in \mathbb{N}\}$	Countable, $\aleph_0$	$①^2$
The set of numerals $\mathbb{Q}_1 = \{\frac{p}{q} : p \in \mathbb{Z},\ q \in \mathbb{Z},\ q \neq 0\}$	Countable, $\aleph_0$	$4①^2 + 2①$
The set of numerals $\mathbb{Q}_2 = \{0, -\frac{p}{q},\ \frac{p}{q} : p \in \mathbb{N},\ q \in \mathbb{N}\}$	Countable, $\aleph_0$	$2①^2 + 1$
The power set of the set of natural numbers $\mathbb{N}$	Continuum, $\mathfrak{c}$	$2^{①}$
The power set of the set of even numbers $\mathbb{E}$	Continuum, $\mathfrak{c}$	$2^{0.5①}$
The power set of the set of integers $\mathbb{Z}$	Continuum, $\mathfrak{c}$	$2^{2①+1}$
The power set of the set of numerals $\mathbb{Q}_1$	Continuum, $\mathfrak{c}$	$2^{4①^2+2①}$
The power set of the set of numerals $\mathbb{Q}_2$	Continuum, $\mathfrak{c}$	$2^{2①^2+1}$
Numbers $x \in [0, 1)$ expressible in the binary numeral system	Continuum, $\mathfrak{c}$	$2^{①}$
Numbers $x \in [0, 1]$ expressible in the binary numeral system	Continuum, $\mathfrak{c}$	$2^{①} + 1$
Numbers $x \in (0, 1)$ expressible in the decimal numeral system	Continuum, $\mathfrak{c}$	$10^{①} - 1$
Numbers $x \in [0, 2)$ expressible in the decimal numeral system	Continuum, $\mathfrak{c}$	$2 \cdot 10^{①}$

$$a_n b^n + a_{n-1} b^{n-1} + \ldots + a_1 b^1 + a_0 b^0 + a_{-1} b^{-1} + \ldots + a_{-(q-1)} b^{-(q-1)} + a_{-q} b^{-q}. \quad (17)$$

Record (16) uses numerals consisting of one symbol each, i.e., digits $a_i \in \{0, 1, \ldots, b-1\}$, to express how many finite units of the type b^i belong to the number (17). Quantities of finite units b^i are counted separately for each exponent i and all symbols in the alphabet $\{0, 1, \ldots, b-1\}$ express finite numbers.

To express infinite and infinitesimal numbers we shall use records that are similar to (16) and (17) but have some peculiarities. In order to construct a number C in the numeral positional system with the base ①, we subdivide C into groups corresponding to powers of ①:

$$C = c_{p_m} ①^{p_m} + \ldots + c_{p_1} ①^{p_1} + c_{p_0} ①^{p_0} + c_{p_{-1}} ①^{p_{-1}} + \ldots + c_{p_{-k}} ①^{p_{-k}}. \quad (18)$$

Then, the record

$$C = c_{p_m} ①^{p_m} \ldots c_{p_1} ①^{p_1} c_{p_0} ①^{p_0} c_{p_{-1}} ①^{p_{-1}} \ldots c_{p_{-k}} ①^{p_{-k}} \quad (19)$$

represents the number C, where all numerals c_i are called *grossdigits*. They are not equal to zero and belong to a traditional numeral system. Grossdigits express finite positive or negative numbers and show how many corresponding units $①^{p_i}$ should be added or subtracted in order to form the number C. Grossdigits can be expressed by several symbols using positional systems, can be written in the form $\frac{Q}{q}$ where Q and q are integers, or in any other numeral system used to express finite quantities.

Numbers p_i in (19) called *grosspowers* can be finite, infinite, and infinitesimal (the introduction of infinitesimal numbers will be given soon), they are sorted in the decreasing order

$$p_m > p_{m-1} > \ldots > p_1 > p_0 > p_{-1} > \ldots p_{-(k-1)} > p_{-k}$$

with $p_0 = 0$.

In the traditional record (16), there exists a convention that a digit a_i shows how many powers b^i are present in the number and the radix b is not written explicitly. In the record (19), we write $①^{p_i}$ explicitly because in the ①-based positional system the number i in general is not equal to the grosspower p_i. This gives the possibility to write, for example, such a number as $3.6①^{44.5}\,134①^{32.1}$ having grosspowers $p_2 = 44.5$, $p_1 = 32.1$ and grossdigits $c_{44.5} = 3.6$, $c_{32.1} = 134$ without indicating grossdigits equal to zero corresponding to grosspowers smaller than 44.5 and greater than 32.1. Note also that if a grossdigit $c_{p_i} = 1$ then we often write $①^{p_i}$ instead of $1①^{p_i}$.

The term having $p_0 = 0$ represents the finite part of C because, due to (7), we have $c_0①^0 = c_0$. The terms having finite positive grosspowers represent the simplest infinite parts of C. Analogously, terms having negative finite grosspowers represent the simplest infinitesimal parts of C. For instance, the number $①^{-1} = \frac{1}{①}$ is infinitesimal. It is the inverse element with respect to multiplication for ①:

$$①^{-1} \cdot ① = ① \cdot ①^{-1} = 1. \qquad (20)$$

Note that all infinitesimals are different from zero. Particularly, $\frac{1}{①} > 0$ because it is a result of division of two positive numbers. It also has a clear granary interpretation. Namely, if we have a sack containing ① seeds, then one sack divided by the number of seeds in it is equal to one seed. Vice versa, one seed, i.e., $\frac{1}{①}$, multiplied by the number of seeds in the sack, ①, gives one sack of seeds.

All of the numbers introduced above can be grosspowers, as well. This gives a possibility to have various combinations of quantities and to construct terms having a more complex structure.

Example 5 The left-hand expression below shows how to write down numbers in the new numeral system and the right-hand shows how the value of the number is calculated:

$$4.1①^{50.3①}(-7.2)①^{2.1}4.6①^{0}1.8①^{-7.4} = 4.1①^{50.3①} - 7.2①^{2.1} + 4.6①^{0} + 1.8①^{-7.4}.$$

The number above has one infinite part with an infinite grosspower, one infinite part having a finite grosspower, a finite part, and an infinitesimal part. □

Finally, numbers having a finite and infinitesimal parts can be also expressed in the new numeral system, for instance, the number $3.6①^{0}{-}5.2①^{-3.7}11①^{-16.5①+0.3}$ has a finite and two infinitesimal parts, the second of them has the infinite negative grosspower equal to $-16.5① + 0.3$. In case a finite number has no infinitesimal parts it is called *purely finite*. This definition will play an important role in a number of applications described in the subsequent Chapters.

Let us now describe arithmetical operations with the ① -based positional numeral system that can be executed on the Infinity Computer. This is a new supercomputer able to work with numbers (19) numerically. It has been patented in several countries (see [70, 75]). A working software simulator of the Infinity Computer has been implemented and the first application—the Infinity Calculator—has been realized. Figure 1 shows operation of multiplication executed on the Infinity Calculator that works using the Infinity

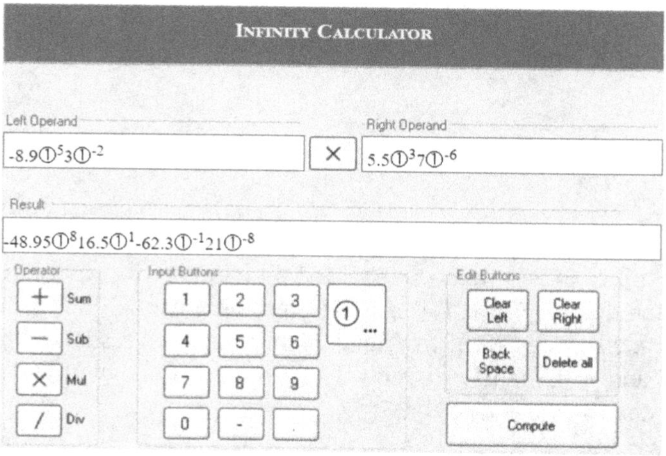

Fig. 1 Operation of multiplication executed on the Infinity Calculator

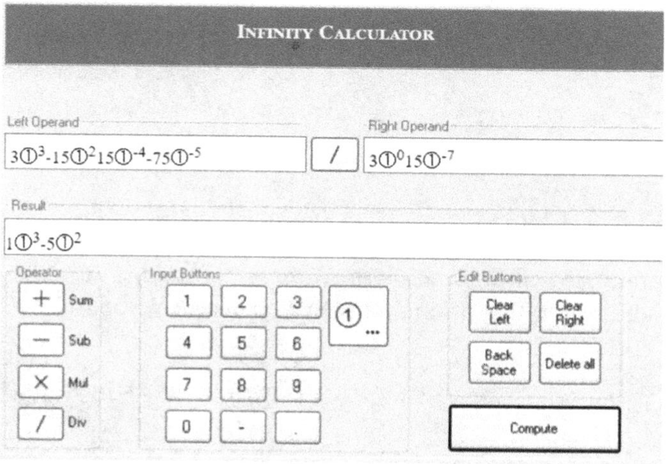

Fig. 2 Operation of division executed on the Infinity Calculator

Computer technology. Both operands have one infinite and one infinitesimal part whereas the result of multiplication has two infinite and two infinitesimal parts. Figure 2 shows division where the first operand has two infinite and two infinitesimal parts and the second operand has one finite and one infinitesimal part. The result is the number having two infinite parts.

Let us give now a general description of operations providing other examples.

We start by discussing the operation of *addition* (*subtraction* is a direct consequence of addition and is thus omitted). Let us consider numbers A, B, and C, where

$$A = \sum_{i=1}^{K} a_{k_i} ①^{k_i}, \qquad B = \sum_{j=1}^{M} b_{m_j} ①^{m_j}, \qquad C = \sum_{i=1}^{L} c_{l_i} ①^{l_i}, \tag{21}$$

and the result $C = A + B$ is constructed by including in it all items $a_{k_i}①^{k_i}$ from A such that $k_i \neq m_j$, $1 \leq j \leq M$, and all items $b_{m_j}①^{m_j}$ from B such that $m_j \neq k_i$, $1 \leq i \leq K$. If in A and B there are items such that $k_i = m_j$, for some i and j, then this grosspower k_i is included in C with the grossdigit $b_{k_i} + a_{k_i}$, i.e., as $(b_{k_i} + a_{k_i})①^{k_i}$.

Example 6 *(Addition)* We consider two infinite numbers A and B, where

$$A = 15①^{33}2①^{7.3}(-2)①^{-4.5}, \quad B = 25①^{33}345.1①^{6}(-3)①^{-2}.$$

Their sum C is calculated as follows:

$$\begin{aligned} C = A + B &= 15①^{33} + 2①^{7.3} + (-2)①^{-4.5} + 25①^{33} + 345.1①^{6} + (-3)①^{-2} \\ &= 40①^{33} + 2①^{7.3} + 345.1①^{6} + (-3)①^{-2} + (-2)①^{-4.5} \\ &= 40①^{33}2①^{7.3}345.1①^{6}(-3)①^{-2}(-2)①^{-4.5}. \end{aligned}$$

The operation of *multiplication* of two numbers A and B in the form (21) returns, as the result, the infinite number C constructed as follows:

$$C = \sum_{j=1}^{M} C_j, \quad C_j = b_{m_j}①^{m_j} \cdot A = \sum_{i=1}^{K} a_{k_i} b_{m_j} ①^{k_i + m_j}, \quad 1 \leq j \leq M. \tag{22}$$

Example 7 *(Multiplication)* We consider two following numbers

$$A = 1①^{14}(-2.5)①^{-3}, \quad B = -1①^{1}2①^{-4}(-5)①^{-5}$$

and calculate the product $C = A \cdot B$. The first partial product C_1 is equal to

$$\begin{aligned} C_1 = (-2.5)①^{-3} \cdot B &= (-2.5)①^{-3}(-①^{1} + 2①^{-4} - 5①^{-5}) \\ &= 2.5①^{-2} - 5①^{-7} + 12.5①^{-8}. \end{aligned}$$

The second partial product, C_2, is computed analogously

$$C_2 = ①^{14} \cdot B = ①^{14}(-①^1 + 2①^{-4} - 5①^{-5})$$
$$= -①^{15} + 2①^{10} - 5①^9.$$

Finally, the product C is equal to

$$C = C_1 + C_2 = -①^{15} 2①^{10} (-5)①^9 2.5①^{-2} (-5)①^{-7} 12.5①^{-8}.$$

In the operation of *division* of a number C by a number B from (21), we obtain a result A and a reminder R (that may equal zero), i.e., $C = A \cdot B + R$. The number A is constructed as follows. The first grossdigit a_{k_K} and the corresponding maximal exponent k_K are established from the equalities

$$a_{k_K} = c_{l_L}/b_{m_M}, \quad k_K = l_L - m_M. \tag{23}$$

Then the first partial reminder R_1 is calculated as

$$R_1 = C - a_{k_K} ①^{k_K} \cdot B. \tag{24}$$

If $R_1 \neq 0$ then the number C is substituted by R_1 and the process is repeated with a complete analogy. The grossdigit $a_{k_{K-i}}$, the corresponding grosspower k_{K-i} and the partial reminder R_{i+1} are computed by formulae (25) and (26) obtained from (23) and (24) as follows: l_L and c_{l_L} are substituted by the highest grosspower n_i and the corresponding grossdigit r_{n_i} of the partial reminder R_i that, in turn, substitutes C:

$$a_{k_{K-i}} = r_{n_i}/b_{m_M}, \quad k_{K-i} = n_i - m_M. \tag{25}$$

$$R_{i+1} = R_i - a_{k_{K-i}} ①^{k_{K-i}} \cdot B, \quad i \geq 1. \tag{26}$$

The process stops when a partial reminder equal to zero is found (this means that the final reminder $R = 0$) or when a required accuracy of the result is reached.

Example 8 *(Division)* Let us divide the number $C = 135①^2 72①^0 9①^{-2}$ by the number $B = 9①^2 3①^0$. For these numbers we have

$$l_L = 2, \quad m_M = 2, \quad c_{l_L} = 135, \quad b_{m_M} = 9.$$

It follows immediately from (23) that $a_{k_K} ①^{k_K} = 15①^0$. The first partial reminder R_1 is calculated as

$$R_1 = 135①^2 72①^0 9①^{-2} - 15①^0 (9①^2 3①^0)$$

$$= 135①^{2}72①^{0}9①^{-2} - 135①^{2}45①^{0} = 27①^{0}9①^{-2}.$$

By a complete analogy we should construct $a_{k_{K-1}}①^{k_{K-1}}$ by rewriting (23) for R_1. By doing so we obtain equalities

$$27 = a_{k_{K-1}} \cdot 9, \quad 0 = k_{K-1} + 2$$

and, as the result, $a_{k_{K-1}}①^{k_{K-1}} = 3①^{-2}$. The second partial reminder is

$$R_2 = R_1 - 3①^{-2} \cdot 9①^{2}3①^{0} = 27①^{0}9①^{-2} - 27①^{0}9①^{-2} = 0.$$

Thus, we can conclude that the reminder $R = R_2 = 0$ and the final result of division is $A = 15①^{0}3①^{-2}$.

Let us now substitute the grossdigit 9 by 7 in C and divide this new number $\tilde{C} = 135①^{2}72①^{0}7①^{-2}$ by the same number $B = 9①^{2}3①^{0}$. This operation gives us the same result $\tilde{A}_2 = A = 15①^{0}3①^{-2}$ (where subscript 2 indicates that two partial reminders have been obtained) but with the reminder $\tilde{R} = \tilde{R}_2 = -2①^{-2}$. Thus, we obtain $\tilde{C} = B \cdot \tilde{A}_2 + \tilde{R}_2$. If we want to continue the procedure of division, we obtain $\tilde{A}_3 = 15①^{0}3①^{-2}\left(-\frac{2}{9}\right)①^{-4}$ with the reminder $\tilde{R}_3 = \frac{2}{3}①^{-4}$. Naturally, it follows $\tilde{C} = B \cdot \tilde{A}_3 + \tilde{R}_3$. The process continues until a partial reminder $\tilde{R}_i = 0$ is found or when a required accuracy of the result will be reached.

6 Some Paradoxes of Infinity Related to Divergent Series

The ①-based methodology allows one to avoid numerous classical paradoxes related to the notion of infinity (for example, Galileo's paradox, Hilbert's paradox of the Grand Hotel, Thomson's lamp paradox, the rectangle paradox of Torricelli, etc., see [66] and the references therein) and provides the first foundations for a mathematical analysis allowing one to work without indeterminate forms and divergences (see [71]). It is interesting that this analysis allows one to operate with functions assuming infinite and infinitesimal values over infinite and infinitesimal domains.

Let us consider just a couple of paradoxes dealing with divergent series first in the traditional fashion and then in the ① -based framework. We shall show that a very simple chain of equalities including addition of an infinite number of summands can lead to a paradoxical result.

Suppose that we have

$$x = 1 + 2 + 4 + 8 + \dots \tag{27}$$

Then we can multiple both parts of this equality by 2:

$$2x = 2 + 4 + 8 + \dots$$

By adding 1 to both parts of the previous formula we obtain

$$2x + 1 = 1 + 2 + 4 + 8 + \dots \tag{28}$$

It can be immediately noticed that the right hand side of (28) is just equal to x and, therefore, it follows

$$2x + 1 = x$$

from which we obtain

$$x = -1$$

and, as a final paradoxical result, the following equality follows

$$1 + 2 + 4 + 8 + \dots = -1. \tag{29}$$

The paradox here is evident: we have summed up an infinite number of positive integers and have obtained as the final result a negative number.

The second paradox considers the well known divergent series of Guido Grandi $S = 1 - 1 + 1 - 1 + 1 - 1 + \dots$ (see [44]). By applying the telescoping rule, i.e., by writing a general element of the series as a difference, we can obtain two different answers using two general elements, 1–1 and –1+1:

$$S = (1 - 1) + (1 - 1) + (1 - 1) + \dots = 0,$$

$$S = 1 + (-1 + 1) + (-1 + 1) + (-1 + 1) + \dots = 1.$$

In the literature there exist many other approaches giving different answers regarding the value of this series (see, e.g., [44]). Some of them use various notions of average (for instance, Cesàro summation assigns the value 0.5 to S, see [44, 84]).

Let us consider now these two paradoxes in the new fashion starting from the definition of x in (27). Thanks to ①, we have different infinite integers and, therefore, we can consider sums having different infinite numbers of summands. Thus, with respect to the new methodology, (27) is not well defined because the number of summands in the sum (27) is not explicitly indicated. Recall that to say just that there are ∞ many summands has the same meaning of the phrase 'There are many summands' (see (1), (2)).

Thus, it is necessary to indicate explicitly an infinite number of addends, k, (obviously, it can be finite, as well). After this (27) becomes

$$x(k) = 1 + 2 + 4 + 8 + \ldots + 2^{k-1}$$

and multiplying both parts by two and adding one to both the right and left sides of the above equality gives us

$$2x(k) + 1 = 1 + 2 + 4 + 8 + \ldots + 2^{k-1} + 2^k.$$

Thus, when we go to substitute, we can see that there remains an addend, 2^k, that is infinite if k is infinite, and that was invisible in the traditional framework:

$$2x(k) + 1 = \underbrace{1 + 2 + 4 + 8 + \ldots + 2^{k-1}}_{x(k)} + 2^k.$$

The substitution gives us the resulting formula

$$x(k) = 2^k - 1$$

that works for both finite and infinite values of k giving different results for different values of k (exactly as it happens for the cases with finite values of k). For instance, $x(①) = 2^{①} - 1$ and $x(3①) = 2^{3①} - 1$. Thus, the paradox (29) does not take place.

Let us consider now Grandi's series. To calculate the required sum, we should indicate explicitly the number of addends, k, in it. Then it follows that

$$S(k) = \underbrace{1 - 1 + 1 - 1 + 1 - 1 + 1 - \ldots}_{k \text{ addends}} = \begin{cases} 0, \text{ if } k = 2n, \\ 1, \text{ if } k = 2n + 1, \end{cases} \quad (30)$$

and it is not important whether k is finite or infinite. For example, $S(①) = 0$ since ① is even. Analogously, $S(① - 1) = 1$ because $① - 1$ is odd.

As it happens in the cases where the number of addends in a sum is finite, the result of summation does not depend on the way the summands are rearranged. In fact, if we know the exact infinite number of addends and the order the signs are alternated is clearly defined, we know also the exact number of positive and negative addends in the sum. Let us illustrate this point by supposing, for instance, that we want to rearrange addends in the sum $S(2①)$ as follows

$$S(2①) = 1 + 1 - 1 + 1 + 1 - 1 + 1 + 1 - 1 + \dots$$

Traditional mathematical tools used to study divergent series give an impression that this rearrangement modifies the result. However, in the ①-based framework we know that this is just a consequence of the weak lens used to observe infinite numbers. In fact, thanks to ① we are able to fix an infinite number of summands. In our example the sum has 2① addends, the number 2① is even and, therefore, it follows from (30) that $S(2①) = 0$. This means also that in the sum there are ① positive and ① negative items. As a result, addition of the groups $1 + 1 - 1$ considered above can continue only until the positive units present in the sum will not finish and then there will be necessary to continue to add only negative summands. More precisely, we have

$$S(2①) = \underbrace{1 + 1 - 1 + 1 + 1 - 1 + \dots + 1 + 1 - 1}_{① \text{ positive and } \frac{①}{2} \text{ negative addends}} \underbrace{-1 - 1 - \dots - 1 - 1}_{\frac{①}{2} \text{ negative addends}} = 0, \tag{31}$$

where the result of the first part in this rearrangement is calculated as $(1 + 1 - 1) \cdot \frac{①}{2} = \frac{①}{2}$ and the result of the second part summing up negative units is equal to $-\frac{①}{2}$ giving so the same final result $S(2①) = 0$. It becomes clear from (31) the origin of the Riemann series theorem. In fact, the second part of (31) containing only negative units is invisible if one works with the traditional numeral ∞.

7 Conclusion

In this Chapter, a recently introduced computational methodology allowing one to work numerically with different infinite and infinitesimal numbers has been described. The exposition was limited to basic rules and ideas that would allow the reader to execute his/her own computations. In order to learn more about this fascinating new opportunity for computing we suggest to refer to a comprehensive technical survey [75] and the popular book [67]. The subsequent Chapters of this book describe a number of striking applications in optimization whereas the list of references includes numerous results in different fields of computer science and pure and applied mathematics (many of them are available at the web page [37]). People interested in teaching this methodology will benefit from the dedicated web page [36] developed at the University of East Anglia, UK.

References

1. Amodio, P., Iavernaro, F., Mazzia, F., Mukhametzhanov, M.S., Sergeyev, Y.D.: A generalized Taylor method of order three for the solution of initial value problems in standard and infinity floating-point arithmetic. Math. Comput. Simul. **141**, 24–39 (2017)
2. Antoniotti, L., Caldarola, F., d'Atri, G., Pellegrini, M.: New approaches to basic calculus: an experimentation via numerical computation. In: Sergeyev, Y.D., Kvasov, D.E., (eds.), Numerical Computations: Theory and Algorithms. NUMTA 2019, LNCS, vol. 11973, pp. 329–342. Springer (2020)
3. Antoniotti, L., Caldarola, F., Maiolo, M.: Infinite numerical computing applied to Hilbert's, Peano's, and Moore's curves. Mediterranean J. Math. **17**, Article Number 99 (2020)
4. Astorino, A., Fuduli, A.: Spherical separation with infinitely far center. Soft. Comput. **24**, 17751–17759 (2020)
5. Bagaria, J., Magidor, M.: Group radicals and strongly compact cardinals. Trans. Am. Math. Soc. **366**(4), 1857–1877 (2014)
6. Butterworth, B., Reeve, R., Reynolds, F., Lloyd, D.: Numerical thought with and without words: evidence from indigenous Australian children. Proc. Natl. Acad. Sci. U.S.A. **105**(35), 13179–13184 (2008)
7. Caldarola, F.: The Sierpinski curve viewed by numerical computations with infinities and infinitesimals. Appl. Math. Comput. **318**, 321–328 (2018)
8. Caldarola, F., Maiolo, M.: On the topological convergence of multi-rule sequences of sets and fractal patterns. Soft. Comput. **24**(23), 17737–17749 (2020)
9. Calude, C.S., Dumitrescu, M.: Infinitesimal probabilities based on grossone. SN Comput. Sci. **1**, Article Number 36 (2020)
10. Cantor, G.: Contributions to the Founding of the Theory of Transfinite Numbers. Dover Publications, New York (1955)
11. Cococcioni, M., Cudazzo, A., Pappalardo, M., Sergeyev, Y.D.: Solving the lexicographic multi-objective mixed-integer linear programming problem using branch-and-bound and grossone methodology. Commun. Nonlinear Sci. Numer. Simul. **84**, 105177 (2020)
12. Cococcioni, M., Fiaschi, L.: The Big-M method with the numerical infinite M. Optim. Lett. **15**, 2455–2468 (2021)
13. Cococcioni, M., Pappalardo, M., Sergeyev, Y.D.: Lexicographic multi-objective linear programming using grossone methodology: theory and algorithm. Appl. Math. Comput. **318**, 298–311 (2018)
14. Colyvan, M.: An Introduction to the Philosophy of Mathematics. Cambridge University Press, Cambridge (2012)
15. Comrie, B.: Numeral bases. In: Dryer, M.S., Haspelmath, M., (eds.), The World Atlas of Language Structures Online. Max Planck Institute for Evolutionary Anthropology, Leipzig (2013). http://wals.info/chapter/131
16. Conway, J.H., Guy, R.K.: The Book of Numbers. Springer, New York (1996)
17. Corry, L.: A Brief History of Numbers. Oxford University Press, Oxford (2015)
18. D'Alotto, L.: Cellular automata using infinite computations. Appl. Math. Comput. **218**(16), 8077–8082 (2012)
19. D'Alotto, L.: Infinite games on finite graphs using grossone. Soft. Comput. **55**, 143–158 (2020)

20. De Cosmis, S., De Leone, R.: The use of grossone in mathematical programming and operations research. Appl. Math. Comput. **218**(16), 8029–8038 (2012)
21. De Leone, R.: Nonlinear programming and grossone: quadratic programming and the role of constraint qualifications. Appl. Math. Comput. **318**, 290–297 (2018)
22. De Leone, R., Egidi, N., Fatone, L.: The use of grossone in elastic net regularization and sparse support vector machines. Soft. Comput. **24**, 17669–17677 (2020)
23. De Leone, R., Fasano, G., Roma, M., Sergeyev, Y.D.: Iterative grossone-based computation of negative curvature directions in large-scale optimization. J. Optim. Theory Appl. **186**, 554–589 (2020)
24. De Leone, R., Fasano, G., Sergeyev, Y.D.: Planar methods and grossone for the conjugate gradient breakdown in nonlinear programming. Comput. Optim. Appl. **71**(1), 73–93 (2018)
25. Falcone, A., Garro, A., Mukhametzhanov, M.S., Sergeyev, Y.D.: Representation of Grossone-based arithmetic in Simulink and applications to scientific computing. Soft. Comput. **24**, 17525–17539 (2020)
26. Falcone, A., Garro, A., Mukhametzhanov, M.S., Sergeyev, Y.D.: A Simulink-based software solution using the infinity computer methodology for higher order differentiation. Appl. Math. Comput. **409**, 125606 (2021)
27. Fiaschi, L., Cococcioni, M.: Numerical asymptotic results in game theory using Sergeyev's infinity computing. Int. J. Unconv. Comput. **14**(1), 1–25 (2018)
28. Fiaschi, L., Cococcioni, M.: Non-archimedean game theory: a numerical approach. Appl. Math. Comput. **393**, Article Number 125356 (2021). https://doi.org/10.1016/j.amc.2020.125356
29. Gangle, R., Caterina, G., Tohmé, F.: A constructive sequence algebra for the calculus of indications. Soft. Comput. **24**(23), 17621–17629 (2020)
30. Gaudioso, M., Giallombardo, G., Mukhametzhanov, M.S.: Numerical infinitesimals in a variable metric method for convex nonsmooth optimization. Appl. Math. Comput. **318**, 312–320 (2018)
31. Gödel, K.: The Consistency of the Continuum-Hypothesis. Princeton University Press, Princeton (1940)
32. Gordon, P.: Numerical cognition without words: evidence from Amazonia. Science **306**(15), 496–499 (2004)
33. Hardy, G.H.: Orders of Infinity. Cambridge University Press, Cambridge (1910)
34. Heller, M., Woodin, W.H. (eds.): Infinity: New Research Frontiers. Cambridge University Press, Cambridge (2011)
35. Hilbert, D.: Mathematical problems: lecture delivered before the international congress of mathematicians at Paris in 1900. Bull. Am. Math. Soc. **8**, 437–479 (1902)
36. https://www.numericalinfinities.com
37. https://www.theinfinitycomputer.com
38. Iannone, P., Rizza, D., Thoma, A.: Investigating secondary school students' epistemologies through a class activity concerning infinity. In: Bergqvist, E., Österholm, M., Granberg, C., Sumpter, L., (eds.), Proceedings of the 42nd Conference of the International Group for the Psychology of Math. Education, vol. 3, pp. 131–138. PME, Umeå (2018)
39. Iavernaro, F., Mazzia, F., Mukhametzhanov, M.S., Sergeyev, Y.D.: Computation of higher order Lie derivatives on the infinity computer. J. Comput. Appl. Math. **383**, Article Number 113135 (2021)

40. Ingarozza, F., Adamo, M., Martino, M., Piscitelli, A.: A grossone-based numerical model for computations with infinity: a case study in an Italian high school. In: Sergeyev, Y.D., Kvasov, D.E., (eds.), Numerical Computations: Theory and Algorithms. NUMTA 2019, LNCS, vol. 11973, pp. 451–462. Springer (2020)
41. Iudin, D.I., Sergeyev, Y.D., Hayakawa, M.: Interpretation of percolation in terms of infinity computations. Appl. Math. Comput. **218**(16), 8099–8111 (2012)
42. Iudin, D.I., Sergeyev, Y.D., Hayakawa, M.: Infinity computations in cellular automaton forest-fire model. Commun. Nonlinear Sci. Numer. Simul. **20**(3), 861–870 (2015)
43. Kanamori, A.: The Higher Infinite: Large Cardinals in Set Theory from Their Beginnings, 2nd edn. Springer, Berlin (2003)
44. Knopp, K.: Theory and Application of Infinite Series. Dover Publications, New York (1990)
45. Kolmogorov, A.N.: Foundations of the Theory of Probability, 2nd English edn. Dover Publications, New York (2018)
46. Lai, L., Fiaschi, L., Cococcioni, M.: Solving mixed Pareto-lexicographic multi-objective optimization problems: The case of priority chains. Swarm Evol. Comput. **55**, 100687 (2020)
47. Leder, G.C.: Mathematics for all? The case for and against national testing. In: Cho, S., (ed.), The Proceedings of the 12th International Congress on Mathematical Education: Intellectual and Attitudinal Chalenges, pp. 189–207. Springer, New York (2015)
48. Leibniz, G.W., Child, J.M.: The Early Mathematical Manuscripts of Leibniz. Dover Publications, New York (2005)
49. Levi-Civita, T.: Sui numeri transfiniti. Rend. Acc. Lincei, Serie 5a **7**, 91–113 (1898)
50. Lolli, G.: Infinitesimals and infinites in the history of mathematics: A brief survey. Appl. Math. Comput. **218**(16), 7979–7988 (2012)
51. Lolli, G.: Metamathematical investigations on the theory of grossone. Appl. Math. Comput. **255**, 3–14 (2015)
52. Mancosu, P.: Abstraction and Infinity. Oxford University Press, Oxford (2016)
53. Margenstern, M.: Using grossone to count the number of elements of infinite sets and the connection with bijections. p-Adic Numb., Ultrametric Anal. Appl. **3**(3), 196–204 (2011)
54. Margenstern, M.: An application of grossone to the study of a family of tilings of the hyperbolic plane. Appl. Math. Comput. **218**(16), 8005–8018 (2012)
55. Montagna, F., Simi, G., Sorbi, A.: Taking the Pirahã seriously. Commun. Nonlinear Sci. Numer. Simul. **21**(1–3), 52–69 (2015)
56. Newton, I.: Method of Fluxions (1671)
57. Parikh, R.: Existence and feasibility in arithmetic. J. Symb. Log. **36**(3), 494–508 (1971)
58. Pepelyshev, A., Zhigljavsky, A.: Discrete uniform and binomial distributions with infinite support. Soft. Comput. **24**, 17517–17524 (2020)
59. Pica, P., Lemer, C., Izard, V., Dehaene, S.: Exact and approximate arithmetic in an Amazonian indigene group. Science **306**(15), 499–503 (2004)
60. Rizza, D.: Supertasks and numeral systems. In: Sergeyev, Y.D., Kvasov, D.E., Dell'Accio, F., Mukhametzhanov, M.S., (eds.), Proceedings of the 2nd International Conference "Numerical Computations: Theory and Algorithms", vol. 1776. AIP Publishing, New York (2016). https://doi.org/10.1063/1.4965369.090005
61. Rizza, D.: A study of mathematical determination through Bertrand's Paradox. Philos. Math. **26**(3), 375–395 (2018)

62. Rizza, D.: Numerical methods for infinite decision-making processes. Int. J. Unconv. Comput. **14**(2), 139–158 (2019)
63. Robinson, A.: Non-standard Analysis. Princeton University Press, Princeton (1996)
64. Sazonov, V.Y.: On feasible numbers. In: D. Leivant (ed.) Logic and Computational Complexity: LNCS, vol. 960, pp. 30–51. Springer (1995)
65. Sergeyev, Y.D.: Numerical infinities applied for studying Riemann series theorem and Ramanujan summation. In: AIP Conference Proceedings of ICNAAM 2017, vol. 1978, p. 020004. AIP Publishing, New York (2018). https://doi.org/10.1063/1.5043649
66. Sergeyev, Y.D.: Some paradoxes of infinity revisited. Mediterranean J. Math. (to appear)
67. Sergeyev, Y.D.: Arithmetic of Infinity. Edizioni Orizzonti Meridionali, CS 2003, 2nd edn. (2013)
68. Sergeyev, Y.D.: Blinking fractals and their quantitative analysis using infinite and infinitesimal numbers. Chaos, Solitons Fractals **33**(1), 50–75 (2007)
69. Sergeyev, Y.D.: Evaluating the exact infinitesimal values of area of Sierpinski's carpet and volume of Menger's sponge. Chaos, Solitons Fractals **42**(5), 3042–3046 (2009)
70. Sergeyev, Y.D.: Computer system for storing infinite, infinitesimal, and finite quantities and executing arithmetical operations with them. USA patent 7,860,914 (2010)
71. Sergeyev, Y.D.: Counting systems and the First Hilbert problem. Nonlinear Anal. Ser. A: Theory, Methods Appl. **72**(3–4), 1701–1708 (2010)
72. Sergeyev, Y.D.: Using blinking fractals for mathematical modelling of processes of growth in biological systems. Informatica **22**(4), 559–576 (2011)
73. Sergeyev, Y.D.: The Olympic medals ranks, lexicographic ordering, and numerical infinities. Math. Intell. **37**(2), 4–8 (2015)
74. Sergeyev, Y.D.: The exact (up to infinitesimals) infinite perimeter of the Koch snowflake and its finite area. Commun. Nonlinear Sci. Numer. Simul. **31**(1–3), 21–29 (2016)
75. Sergeyev, Y.D.: Numerical infinities and infinitesimals: methodology, applications, and repercussions on two Hilbert problems. EMS Surv. Math. Sci. **4**(2), 219–320 (2017)
76. Sergeyev, Y.D.: Independence of the grossone-based infinity methodology from non-standard analysis and comments upon logical fallacies in some texts asserting the opposite. Found. Sci. **24**(1), 153–170 (2019)
77. Sergeyev, Y.D., Garro, A.: Observability of Turing machines: a refinement of the theory of computation. Informatica **21**(3), 425–454 (2010)
78. Sergeyev, Y.D., Garro, A.: Single-tape and multi-tape Turing machines through the lens of the Grossone methodology. J. Supercomput. **65**(2), 645–663 (2013)
79. Sergeyev, Y.D., Kvasov, D.E., Mukhametzhanov, M.S.: On strong homogeneity of a class of global optimization algorithms working with infinite and infinitesimal scales. Commun. Nonlinear Sci. Numer. Simul. **59**, 319–330 (2018)
80. Sergeyev, Y.D., Mukhametzhanov, M.S., Mazzia, F., Iavernaro, F., Amodio, P.: Numerical methods for solving initial value problems on the infinity computer. Int. J. Unconv. Comput. **12**(1), 3–23 (2016)
81. Tohmé, F., Caterina, G., Gangle, R.: Computing truth values in the topos of infinite Peirce's α-existential graphs. Appl. Math. Comput. **385**, article number 125343 (2020)
82. Wallis, J.: Arithmetica infinitorum (1656)
83. Woodin, W.H.: The continuum hypothesis. Part I. Notices AMS **48**(6), 567–576 (2001)

84. Zhigljavsky, A.: Computing sums of conditionally convergent and divergent series using the concept of grossone. Appl. Math. Comput. **218**(16), 8064–8076 (2012)
85. Žilinskas, A.: On strong homogeneity of two global optimization algorithms based on statistical models of multimodal objective functions. Appl. Math. Comput. **218**(16), 8131–8136 (2012)

Nonlinear Optimization: A Brief Overview

Renato De Leone

Abstract In this chapter some of the most important results for unconstrained and constrained optimization problems are discussed. This chapter does not claim to cover all the aspects in nonlinear optimization that will require more than one complete book. We decided, instead, to concentrate our attention on few fundamental topics that are also at the basis of the new results in nonlinear optimization using *grossone* introduced in the successive chapters.

1 Introduction

The aim of this chapter is to introduce the reader to the most important results and algorithms in unconstrained and constrained optimization. However, it would be impossible to discuss the wide range of results in this area. We decided to concentrate the attention on few fundamental topics that are also at the basis of the new results obtained using ① and introduced in the successive chapters. The interested reader can refer to various classical and recent books to deepen his/her knowledge. We also omitted almost all proofs of the results, but complete references are always provided.

After introducing the concepts of convex set and functions, in Sect. 3 optimality conditions for unconstrained optimization are presented as well as the most important algorithmic techniques: gradient method, conjugate gradient method, Newton's and Quasi-Newton's methods. In the successive Sect. 4 we concentrate the attention on optimality conditions for constrained optimiza-

R. De Leone (✉)
School of Science and Technology, University of Camerino, Camerino (MC), Italy
e-mail: renato.deleone@unicam.it

Y. D. Sergeyev and R. De Leone (eds.), *Numerical Infinities and Infinitesimals in Optimization*, Emergence, Complexity and Computation 43,
https://doi.org/10.1007/978-3-030-93642-6_2

tion and the construction of the dual of nonlinear optimization problems. Finally, some important algorithms for constrained nonlinear optimization problems are presented.

Three important aspects in optimization are not presented here: global optimization, multi–objective optimization, and non–smooth optimization. In this chapter we only discuss first and second order optimality conditions for local optima, and all algorithms will compute a local minimum, or a stationary point (for the unconstrained case) or a Karush–Kuhn–Tucker point (for the constrained case). Global optimization is an important area, theoretically and practically, for which several different approaches have been suggested: space covering methods, trajectory methods, random sampling, random search, etc. A basic reference on various aspects of global optimization is [40], while practical applications are discussed in [51]. Finally, a comprehensive archive of online information can be found at the web page http://www.globaloptimization.org/.

Many optimization problems are multi–objective in nature, and different techniques have been proposed to deal with this important aspect of optimization. In general, due to the presence of conflicting objectives, there is no single solution that simultaneously optimizes all the objectives, and, hence, the aim is to determine non–dominated, Pareto optimal solutions [16, 23]. Two general approaches to multiple-objective optimization are present: combine the individual objective functions into a single function, or move all but one objective to the constraint set. More recently algorithms based on different meta-heuristics have also been proposed in literature [41].

Non–smooth optimization problems arise in many important practical applications. Here it is assumed that the function is continuous, but not differentiable. The methods for non–smooth optimization can be roughly divided into two main classes: subgradient methods and bundle methods. Both classes of methods are based on the assumption that, at each point, the objective function value and one subgradient can be computed. A classical book on this topic is [43]. A survey on different numerical methods for non–smooth optimization and the most recent developments presented in [5]. Finally, a recent compact survey on non–smooth optimization can be found in [28].

We briefly describe our notation now. All vectors are column vectors and will be indicated with lower case Latin letter (x, y, ...). Subscripts indicate components of a vector, while superscripts are used to identify different vectors. Matrices will be indicated with upper case roman letter (A, B, ...). For a $m \times n$ matrix A, A_{ij} is the element in the ith row, jth column, $A_{.j}$ is the j–th column of A, while $A_{i.}$ is its i–th row. The set of real numbers and the set of nonnegative real numbers will be denoted by $\mathbb{R}$ and $\mathbb{R}_+$ respectively. The space of the n–dimensional vectors with real components will be indicated by $\mathbb{R}^n$ and $\mathbb{R}^n_+$ is an abbreviation for the nonnegative orthant in $\mathbb{R}^n$. The symbol $\|x\|$ indicates the norm of the vector x. In particular, the Euclidean norm is denoted by $\|x\|_2$, and $\|x\|_2^2 = x^T x$. Superscript ^{T} indicates transpose. The scalar product of two vectors x and y in $\mathbb{R}^n$ will be denoted by $x^T y$. The space of the $m \times n$ matrices with real components will be indicated by $\mathbb{R}^{m\times n}$. The rank of a matrix A will be indicated by rank$A$. A square matrix $A \in \mathbb{R}^{n\times n}$ is positive semidefinite if $x^T Ax \geq 0$ for all $x \in \mathbb{R}^n$ and positive definite if $x^T Ax > 0$ for all $0 \neq x \in \mathbb{R}^n$. If $f : S \subseteq \mathbb{R}^n \to \mathbb{R}\cup\{\pm\infty\}$, dom f is the set of points x for which $f(x)$ is defined and $f(x) \in \mathbb{R}$. The gradient $\nabla f(x)$ of a continuously differentiable function $f : \mathbb{R}^n \to \mathbb{R}$ at a point $x \in \mathbb{R}^n$ is a column vector with components $[\nabla f(x)]_j = \frac{\partial f(x)}{\partial x_j}$. For a twice differentiable function $f : \mathbb{R}^n \to \mathbb{R}$, the Hessian $\nabla^2 f(x)$ belongs to $\mathbb{R}^{n\times n}$ and $\left[\nabla^2 f(x)\right]_{ij} = \frac{\partial^2 f(x)}{\partial x_i \partial x_j}$. If $F : \mathbb{R}^n \to \mathbb{R}^m$ is a continuously differentiable vector–valued function, then $\nabla F(x) \in \mathbb{R}^{m\times n}$ denotes the Jacobian matrix of F at $x \in \mathbb{R}^n$. Here and throughout the symbols := and =: denote definition of the term on the left and the right sides of each symbol, respectively.

2 Convex Sets and Functions

Definition 1 ([56]) A set $C \subseteq \mathbb{R}^n$ is a convex set if

$$x^1, x^2 \in C, \quad \lambda \in [0, 1] \qquad \Longrightarrow \qquad (1-\lambda)x^1 + \lambda x^2 \in C.$$

Therefore, for a convex set C the segment joining any two distinct points in C is all contained in C.

Definition 2 A point $x \in \mathbb{R}^n$ is a convex combination of $x^1, x^2, \dots, x^k \in \mathbb{R}^n$ if there exist scalars $\lambda_1, \lambda_2, \dots, \lambda_k$ such that

$$x = \lambda_1 x^1 + \lambda_2 x^2 + \ldots + \lambda_k x^k \text{ with } \lambda_1 + \lambda_2 + \ldots + \lambda_k = 1,\ \lambda_j \geq 0,\ j = 1, \dots, k.$$

The concept of convex hull of a set is extremely important in optimization.

Definition 3 ([46, Definition 3.1.16]) Given $S \subseteq \mathbb{R}^n$ the convex hull of S is defined as

$$\text{conv}\, S := \left\{ \begin{array}{c} x \in \mathbb{R}^n : x = \lambda_1 x^1 + \lambda_2 x^2 + \ldots + \lambda_k x^k \\ x^1, \dots, x^k \in S \\ \lambda_j \in [0, 1],\ j = 1, \dots, k \\ \lambda_1 + \lambda_2 + \ldots + \lambda_k = 1 \end{array} \right\}$$

Clearly, if S is a convex set, then $S = \text{conv}\, S$.

Hyperplanes

$$H := \left\{ x \in \mathbb{R}^n : p^T x = \mu \right\},$$

and halfspaces

$$S := \left\{ x \in \mathbb{R}^n : p^T x \leq \mu \right\}$$

are examples of convex sets (here $0 \neq p \in \mathbb{R}^n$, and $\mu \in \mathbb{R}$). The set of symmetric positive semidefinite matrices is a convex subsets of $\mathbb{R}^{n \times n}$ the set of square matrices of dimension n. Another interesting example of convex set is the norm cone

$$C := \left\{ \begin{bmatrix} x \\ t \end{bmatrix} \in \mathbb{R}^{n+1} : \|x\|_2 \leq t \right\} \subseteq \mathbb{R}^{n+1}.$$

Definition 4 ([46, Definition 4.1.1]) Let $f : \mathbb{R}^n \to \mathbb{R} \cup \{+\infty\}$, f is a convex function if $\text{dom}\, f$ is a convex set and

$$f\Big[(1-\lambda)x + \lambda y\Big] \leq (1-\lambda) f(x) + \lambda f(y) \quad \forall x, y \in \text{dom}\, f,\ \lambda \in [0, 1] \tag{1}$$

The following propositions provide necessary and sufficient conditions for convexity for differentiable and twice differentiable functions.

Proposition 1 ([8, 3.1.3], [46, Theorem 6.1.2]) *Let $f : \mathbb{R}^n \to \mathbb{R} \cup \{+\infty\}$ be a differentiable function. The function f is convex if and only if* $\text{dom}\, f$ *is convex and*

$$f(y) \geq f(x) + \nabla f(x)^T (y - x) \quad \forall x, y \in \text{dom}\, f. \tag{2}$$

Proposition 2 ([8, 3.1.4], [46, Theorem 6.3.1]) *Let $f : \mathbb{R}^n \to \mathbb{R} \cup \{+\infty\}$, be twice differentiable. The function f is convex if and only if* $\text{dom}\, f$ *is convex and $\nabla^2 f(x)$ is positive semidefinite* $\forall x \in \text{dom}\, f$.

Finally, let introduce two additional weaker classes of functions whose importance will be clear when necessary optimality conditions for unconstrained optimization problems will be introduced.

Definition 5 ([46, Definition 9.3.1]) Let $f : \mathbb{R}^n \to \mathbb{R} \cup \{+\infty\}$ be a differentiable function. The function f is pseudo–convex if

$$x, y \in \text{dom}\, f, \quad \nabla f(x)^T (y - x) \geq 0 \Rightarrow f(y) \geq f(x)$$

A convex function is also pseudo–convex.

Definition 6 ([8, 3.4.1], [46, *Definition 9.1.1*]) Let $f : \mathbb{R}^n \to \mathbb{R}\cup\{+\infty\}$. The function f is quasi–convex if

$$x, y \in \text{dom } f, \quad f\big((1-\lambda)x + \lambda y\big) \le \max\big\{f(x), f(y)\big\} \text{ for } \lambda \in [0, 1]$$

Clearly, a convex function is also quasi–convex. Moreover, a differentiable pseudo-convex function is also quasi–convex. The opposite is not true.

For simplicity, in the sequel, we will only consider functions whose domain is $\mathbb{R}^n$.

3 Unconstrained Optimization

In this section we will present optimality conditions and algorithms for the unconstrained optimization problems

$$\min_{x\in\mathcal{F}} f(x) \tag{3}$$

where $f : \mathbb{R}^n \to \mathbb{R}$ is the objective function and the feasible set $\mathcal{F} \subseteq \mathbb{R}^n$ is an open subset of $\mathbb{R}^n$ (note that $\mathcal{F}$ may coincide with $\mathbb{R}^n$).

First, let introduce the following definitions.

Definition 7 ([6, Definition 3.4.1]) Let $\bar{x} \in \mathcal{F}$. The point $\bar{x}$ is a global minimum of Problem (3) if

$$f(\bar{x}) \le f(x), \ \forall x \in \mathcal{F}.$$

The point $\bar{x}$ is a strict global minimum of Problem (3) if

$$f(\bar{x}) < f(x), \ \forall x \in \mathcal{F}, \ x \ne \bar{x}.$$

Definition 8 ([6, Definition 3.4.1]) Let $\bar{x} \in \mathcal{F}$. The point $\bar{x}$ is a local minimum of Problem (3) if $\exists\ \gamma > 0$ such that

$$f(\bar{x}) \le f(x) \quad \forall x \in \mathcal{F} \text{ and } \|x - \bar{x}\| \le \gamma.$$

The point $\bar{x}$ is a strict local minimum if $\exists\ \gamma > 0$ such that

$$f(\bar{x}) < f(x) \quad \forall\, x \in \mathcal{F} : \ 0 < \|x - \bar{x}\| \le \gamma.$$

The point $\bar{x}$ is a strong or isolated local minimum if $\exists\ \gamma > 0$ such that $\bar{x}$ is the only local minimum in $\{x \in \mathbb{R}^n : \|x - \bar{x}\| \le \gamma\} \cap \mathcal{F}$.

If $\bar{x}$ is a strong local minimum, then it is also a strict local minimum. On the other hand, a strict local minimum may not be necessarily a strong local minimum. For example, the function [7, p. 21]

$$f(x) = \begin{cases} x^2 \left(\sqrt{2} - \sin\left(\frac{4}{3}\pi - \sqrt{3}\ln(x^2)\right)\right) & \text{if } x \neq 0 \\ 0 & \text{otherwise} \end{cases}$$

has in $x = 0$ a strict local minimum that is not a strong (isolated) local minimum. In the remaining of the section, without loss of generality, we assume that $\mathcal{F} = \mathbb{R}^n$

3.1 Necessary and Sufficient Optimality Conditions

Consider the nonlinear optimization problem (3) where $f : \mathbb{R}^n \to \mathbb{R}$ is a continuously differentiable function. The directional derivatives [3, Chap. 8] of f at $\bar{x}$ along the direction d is given by

$$\lim_{t \to 0^+} \frac{f(\bar{x} + td) - f(\bar{x})}{t} = \nabla f(\bar{x})^T d.$$

A direction $d \in \mathbb{R}^n$ is a descent direction for $f(x)$ at $\bar{x}$ if

$$\exists\, \delta > 0 : f(\bar{x} + td) < f(\bar{x}) \quad \forall\, 0 < t \leq \delta \tag{4}$$

If f is continuously differentiable, then d is a descent direction at $\bar{x}$ if and only if $\nabla f(\bar{x})^T d < 0$.

A point x^* is a local minimum if at x^* there are no descent directions. The following theorems provide first and second order necessary optimality conditions for a point x^*.

Theorem 1 *Let $f : \mathbb{R}^n \to \mathbb{R}$ be a continuously differentiable function and let $x^* \in \mathbb{R}^n$. If x^* is a local minimum then*

$$\nabla f(x^*) = 0. \tag{5}$$

The above result can be easy proved by contradiction; in fact, if $\nabla f(x^*) \neq 0$, then $d = -\nabla f(x^*)$ is a descent direction. This observation is the basis of the most used technique for function minimization: the gradient method.

Theorem 2 ([35, Proposition 2.5]) *Let $f : \mathbb{R}^n \to \mathbb{R}$ be twice continuously differentiable and let $x^* \in \mathbb{R}^n$. If x^* is a local minimum then*

Algorithm 1: Generic Minimization Algorithm

1 **Choose** $x^0 \in \mathbb{R}^n$, **Set** $k = 0$;
2 **if** $x^k \in \Omega$ **then Stop**;
3 **Compute** a search direction $d^k \in \mathbb{R}^n$;
4 **Compute** a stepsize $\alpha_k > 0$ along d^k;
5 **Set** $x^{k+1} = x^k + \alpha_k d^k$, $k = k + 1$, **return** to 2;

$$\nabla f(x^*) = 0 \tag{6}$$

$$\nabla^2 f(x^*) \text{ is positive semidefinite.} \tag{7}$$

The above conditions cannot exactly be reversed to obtain sufficiency optimality conditions. In fact, positive definiteness of the Hessian matrix is required to obtain a strict local minimum.

Theorem 3 ([35, Proposition 2.6]) *Let $f : \mathbb{R}^n \to \mathbb{R}$ be twice continuously differentiable and let $x^* \in \mathbb{R}^n$. Suppose that:*

$$\nabla f(x^*) = 0 \tag{8}$$

$$\nabla^2 f(x^*) \text{ is positive definite.} \tag{9}$$

Then x^* is a strict local minimum.

Conditions for global optima are much more complex to obtain. However, if $f : \mathbb{R}^n \to \mathbb{R}$ is a differentiable pseudo–convex function, then a local minimum is also a global minimum [46, Theorem 9.3.7].

3.2 *Algoritms for Unconstrained Optimization*

A very general minimization algorithm is presented in Algorithm 1 where

$$\Omega := \left\{ x \in \mathbb{R}^n : \nabla f(x) = 0 \right\} \tag{10}$$

is the set of stationary points. The algorithm, starting from the initial point x^0, constructs either a finite sequence terminating in a stationary point or an infinite sequence which, under adequate conditions, converges to a stationary point, or, has at least an accumulation point that is also a stationary point.

Given the current point x^k and a search direction d^k, the new point x^{k+1} is obtained by moving along d^k with a stepsize α_k.

The following general theorem establishes conditions on the direction d^k and the stepsize α_k ensuring that, in the case an infinite sequence is obtained by Algorithm 1, the sequence has at least one accumulation point and each accumulation point is a stationary point. Before stating the theorem, it is necessary to introduce the concept of forcing function.

Definition 9 ([50, Definition 14.2.1]) A function $\sigma : \mathbb{R}_+ \to \mathbb{R}_+$ is a forcing function if

$$\left.\begin{array}{c} \forall \text{ sequence } \ \{t_k\} \subseteq \mathbb{R}_+ \\ \\ \lim_k \sigma(t_k) = 0 \end{array}\right\} \implies \lim_k t_k = 0.$$

Note that any non–decreasing function $\sigma : \mathbb{R}_+ \to \mathbb{R}_+$ such that $\sigma(0) = 0$ and $\sigma(t) > 0$ when $t > 0$ is a forcing function. The functions

$$\sigma(t) = t, \qquad \sigma(t) = ct^q, \quad with \quad c > 0, \quad q > 0$$

are examples of forcing functions.

Theorem 4 *Let $\{x^k\}$ be obtained by Algorithm 1 and assume that*

(i) the level set

$$\mathcal{L}_0 := \left\{x \in \mathbb{R}^n : f(x) \le f(x^0)\right\}$$

is compact,

(ii) $d^k \neq 0$ when $\nabla f(x^k) \neq 0$,

(iii) $f(x^{k+1}) \le f(x^k)$,

(iv) if $\nabla f(x^k) \neq 0$ for all k then

$$\lim_k \frac{\nabla f(x^k)^T d^k}{\|d^k\|} = 0,$$

(v) when $d^k \neq 0$

$$\frac{|\nabla f(x^k)^T d^k|}{\|d^k\|} \ge \sigma\left(\left\|\nabla f(x^k)\right\|\right)$$

for some forcing function σ.

Then, either the algorithm terminates after a finite number of iterations in a stationary point or an infinite sequence $\{x^k\}$ is generated that satisfies the following properties:

(a) the sequence $\{x^k\}$ remains in $\mathcal{L}_0$ and has at least one accumulation point;

(b) each accumulation point of $\{x^k\}$ belongs to $\mathcal{L}_0$;

(c) *the sequence of real numbers* $\left\{f(x^k)\right\}$ *converges;*

(d) $\lim_k \left\|\nabla f(x^k)\right\| = 0$;

(e) *each accumulation point* x^* *of the sequence* $\{x^k\}$ *is a stationary point.*

With reference to Theorem 4,

- Conditions *(iii)* and *(v)* can be guaranteed by choosing an opportune descent direction d^k and a line–search along this direction. In particular, using the Euclidean norm and choosing $\sigma(t) = ct$ (with $c > 0$), Condition *(v)* can be written as
$$\nabla f(x^k)^T d^k \leq -c \left\|d^k\right\|_2 \left\|\nabla f(x^k)\right\|_2,$$
that is, when x^k is not a stationary point, a direction d^k must be chosen such that
$$\frac{\nabla f(x^k)^T d^k}{\left\|d^k\right\|_2 \left\|\nabla f(x^k)\right\|_2} \leq -c.$$
Geometrically, the above condition requires that the cosine of the angle between the direction d^k and the direction $-\nabla f(x^k)$ (the "antigradient") must be greater than a constant independent of k. This implies that d^k and the gradient direction cannot be orthogonal as k goes to infinity.
If $\sigma(t) = ct^q$ (with $c > 0$), Condition *(v)* becomes, instead,
$$\frac{\nabla f(x^k)^T d^k}{\left\|d^k\right\|} \leq -c \left\|\nabla f(x^k)\right\|^q.$$
- Condition *(iv)* can be guaranteed using specific safeguard rules on the line–search along the direction d^k.

From the above considerations, it is clear the importance of inexact line–search procedures for determining the stepsize α_k.

A simple, but also very effective, line–search procedure is the Armijo Backtracking method [4, 7, p. 29]. Given x^k and a descent search direction d^k, the algorithm starts with a fixed large value of α and decreases it (that is, the new values of α is obtained by multiplying the current value of α by a constant $\delta < 1$) until the stopping criterion

$$f(x^k + \alpha d^k) \leq f(x^k) + \gamma\alpha\nabla f(x^k)^T d^k \tag{11}$$

is satisfied, where $\gamma \in (0, \frac{1}{2})$. In other words, the Armijo procedures chooses as α_k the maximum value of α in the set

$$S = \left\{\alpha : \alpha = \delta^l \alpha_{\text{init}},\ l = 0, 1, \ldots\right\}$$

for which Condition (11) holds.

The following proposition demonstrates that, if the search direction d^k is opportunely chosen, and the stepsize is obtained according to the Armijo rule, then conditions *(iii)* and *(iv)* in Theorem 4 are automatically satisfied.

Proposition 3 *Let $f : \mathbb{R}^n \to \mathbb{R}$ be continuously differentiable. Suppose that:*

(i) *$\mathcal{L}_0$ is a compact set;*
(ii) $\nabla f\left(x^k\right)^T d^k < 0.\ \forall k$;
(iii) *there exists a forcing function $\sigma : \mathbb{R}_+ \to \mathbb{R}_+$ such that*

$$\left\|d^k\right\| \geq \sigma\left(\frac{\left|\nabla f\left(x^k\right)^T d^k\right|}{\left\|d^k\right\|}\right).$$

Then, if α_k is chosen according to the Armijo procedure, and $x^{k+1} = x^k + \alpha_k d^k$

(a) $f\left(x^{k+1}\right) < f\left(x^k\right)$;

(b) $\lim_k \dfrac{\left|\nabla f\left(x^k\right)^T d^k\right|}{\left\|d^k\right\|} = 0.$

A different line–search stopping criterion was proposed by Goldstein [31] considering two lines

$$f(x^k) + \gamma_1 \alpha \nabla f(x^k)^T d^k = 0,$$

$$f(x^k) + \gamma_2 \alpha \nabla f(x^k)^T d^k = 0$$

where $0 < \gamma_1 < \gamma_2 < \frac{1}{2}$. The chosen stepsize $\alpha_k > 0$ must satisfy, instead of (11), the following conditions:

$$f(x^k + \alpha_k d^k) \leq f(x^k) + \gamma_1 \alpha_k \nabla f(x_k)^T d^k, \tag{12a}$$

$$f(x^k + \alpha_k d^k) \geq f(x^k) + \gamma_2 \alpha_k \nabla f(x_k)^T d^k. \tag{12b}$$

From a geometrical point of view, this corresponds to choose a value for α_k for which $f(x^k + \alpha_k d^k)$ is between the two straight lines with slopes $\gamma_1 \nabla f(x^k)^T d^k$ and $\gamma_2 \nabla f(x_k)^T d^k$, and passing through the point $(0, f(x_k))$.

Moreover, let γ_1 and γ_2 such that $0 < \gamma_1 < \gamma_2 < 1$. The step length α_k satisfy the Wolfe conditions [62] if the following conditions are satisfied:

$$f(x^k + \alpha_k d^k) \leq f(x^k) + \gamma_1 \alpha_k \nabla f(x^k)^T d^k, \tag{13a}$$

$$\nabla f(x^k + \alpha_k d^k)^T d^k \geq \gamma_2 \nabla f(x^k)^T d^k. \tag{13b}$$

It is possible to show that, when d^k is an opportune descent direction, then using both the Goldstein and the Wolfe line–search methods, similar results to those in Proposition 3 can be obtained.

Various other line–search procedures have been proposed in literature, including line–search procedures that do not use information on the gradient of the function [15, 44] and/or non–monotone line–search procedures [33, 34, 63] for which the monotonicity of the objective function for the generated sequence is not required.

3.3 The Gradient Method

In the gradient method (also known as the steepest descent method) the search direction is given by

$$d^k = -\nabla f(x^k). \tag{14}$$

For this choice of the search direction, both Condition (iii) and Condition (v) in Theorem 4 are trivially satisfied with $\sigma(t) = t$ and, therefore, the method is globally convergent when one of line–search procedures outlined in the previous subsection is applied. However, the gradient method, even when applied to the minimization of a strictly convex quadratic function

$$\min_x \frac{1}{2} x^T M x + q^T x$$

($M \in \mathbb{R}^{n\times n}$ positive definite, and $q \in \mathbb{R}^n$) and exact line–search is utilized, exhibits linear rate of convergence. The theorem below shows that the gradient method has at least linear rate of convergence.

Theorem 5 ([45, Sect. 7.6, p. 218]) *Let $M \in \mathbb{R}^{n\times n}$ be a symmetric positive definite matrix and let $0 < \lambda_1 \leq \lambda_2 \leq \ldots \leq \lambda_n$ be the eigenvalues of M, let $q \in \mathbb{R}^n$ and let $f(x) := \frac{1}{2}x^T M x + q^T x$. Let $\{x^k\}$ be obtained by the gradient method with exact line–search:*

$$x^{k+1} = x^k - \frac{(Mx^k + q)^T (Mx^k + q)}{(Mx^k + q)^T M (Mx^k + q)} \left(Mx^k + q\right).$$

Then[1]

[1] Here $\|x\|_M^2 := x^T M x$.

$$\left\| x^{k+1} - x^* \right\|_M \le \frac{\lambda_n - \lambda_1}{\lambda_n + \lambda_1} \left\| x^k - x^* \right\|_M \tag{15a}$$

and

$$\left\| x^{k+1} - x^* \right\|_2 \le \left(\frac{\lambda_n}{\lambda_1} \right)^{\frac{1}{2}} \frac{\lambda_n - \lambda_1}{\lambda_n + \lambda_1} \left\| x^k - x^* \right\|_2 \tag{15b}$$

where $x^* = -M^{-1}q$ *is the unique minimum point.*

Moreover, it is not difficult to construct an examples (see [8, Example 9.3.2, p. 469]) of quadratic problems where the inequalities (15) above are satisfied as equality, demonstrating that, in fact, the gradient method has linear rate of convergence.

3.4 The Newton's Method

As stated before the gradient method exhibits a slow rate of convergence. Here we consider a method that, instead, has quadratic rate of convergence at the expenses of a higher cost per iteration, requiring, in addition to calculate the gradient at each point, also the calculation of the Hessian function. Moreover, only local convergence can be established, that is convergence is guaranteed only if the initial point x^0 is chosen sufficiently close to the stationary point x^*.

Let $f : \mathbb{R}^n \to \mathbb{R}$ be twice continuously differentiable. From Taylor's series expansion we have that

$$f(x+s) = f(x) + \nabla f(x)^T s + \frac{1}{2} s^T \nabla^2 f(x) s + \beta(x,s)$$

where

$$\lim_{s \to 0} \frac{\beta(x,s)}{\|s\|^2} = 0.$$

In Newton's method, at each iteration a quadratic approximation of the function f around the current point x^k is constructed:

$$q_k(s) := f(x^k) + \nabla f(x^k)^T s + \frac{1}{2} s^T \nabla^2 f(x^k) s,$$

and the new point x^{k+1} is obtained as $x^{k+1} = x^k + s^k$ where

$$s^k \in \operatorname*{argmin}_s q_k(s).$$

The following theorem provides conditions for local convergence of the Newton's method for minimization problems.

Theorem 6 ([25, Theorem 3.1.1] [35, Proposizione 7.2]) *Let $f : \mathbb{R}^n \to \mathbb{R}$ be a twice continuously differentiable function, and assume that*

(a) there exists $x^* \in \mathbb{R}^n$: $\nabla f(x^*) = 0$,
(b) $\nabla^2 f(x^*)$ is non singular,
(c) $\nabla^2 f(x)$ is a Lipschitz–continuous function, i.e.,

$$\exists\, L > 0 : \forall x, y \in \mathbb{R}^n \qquad \left\|\nabla^2 f(x) - \nabla^2 f(y)\right\| \leq L \left\|x - y\right\|.$$

Then, there exists $B(x^*, \rho) := \{x \in \mathbb{R}^n : \|x - x^*\| \leq \rho\}$ with $\rho > 0$ such that, if $x^0 \in B(x^*, \rho)$,

(i) the Newton's iterate is well defined,
(ii) the sequence $\{x^k\}$ remains in $B(x^*, \rho)$,
(iii) the sequence $\{x^k\}$ converges to x^* with at least a quadratic rate of convergence

$$\left\|x^{k+1} - x^*\right\| \leq \alpha \left\|x^k - x^*\right\|^2, \text{ for some } \alpha > 0.$$

3.5 The Conjugate Gradient Method

The slow convergence of the gradient method and the heavy requirements of the fast converging Newton's method where, at each iteration, it is necessary to calculate the Hessian matrix, set the stage for new classes of methods that only require to compute the gradient of the function to minimize, and, at the same time, exhibit q–superlinear rate of convergence [50]. In this section the important class of conjugate gradient methods is introduced, while the next section is devoted to Quasi-Newton's methods.

Definition 10 Let $M \in \mathbb{R}^{n \times n}$ be a symmetric positive define matrix. The n nonzero vectors $d^0, d^1, \ldots, d^{n-1}$ in $\mathbb{R}^n$ are conjugate directions with respect to the matrix M if

$$d^{i^T} M d^j = 0 \qquad \forall\, i, j = 0, 1, \ldots, n-1, i \neq j. \tag{16}$$

Consider the quadratic function

$$f(x) = \frac{1}{2} x^T M x + q^T x \tag{17}$$

Algorithm 2: Generic Conjugate Direction Method for Quadratic Problems

1 **Choose** $x^0 \in \mathbb{R}^n$;
2 **Choose** $d^0, d^1, \ldots, d^{n-1}$ nonzero conjugate directions with respect to M;
3 **for** k=0 **to** n-1 **do**
4   **Set** $\alpha_k = \dfrac{-(Mx^k + q)^T d^k}{{d^k}^T M d^k}$;
5   **Set** $x^{k+1} = x^k + \alpha_k d^k$
6 **end**

where M is a symmetric positive matrix, $M \in \mathbb{R}^{n \times n}$ and $q \in \mathbb{R}^n$; the following algorithm utilizes n conjugate directions to determine the unique point minimizing $f(x)$.

One important result for this algorithm is that for quadratic strictly convex problems, when conjugate directions are utilized as search directions, then convergence is achieved in a finite number of iterations.

Theorem 7 *Consider the quadratic function* $f(x) = \frac{1}{2}x^T Mx + q^T x$ *where* M *is a symmetric positive matrix,* $M \in \mathbb{R}^{n \times n}$ *and* $q \in \mathbb{R}^n$*. Starting from any* $x^0 \in \mathbb{R}^n$ *apply Algorithm 2. Then* x^{k+1} *is the minimizer of* $f(x)$ *over the affine set*

$$S_k := \left\{ x \in \mathbb{R}^n : x = x^0 + \sum_{j=0}^{k} \lambda_j d^j, \ \lambda_j \in \mathbb{R} \right\}$$

for $k = 0, 1, \ldots, n-1$*. Moreover,* $x^n = -M^{-1}q$ *is the unique minimizers of* $f(x)$ *over* $\mathbb{R}^n$*.*

The above Algorithm 2 allows different choices for the set of conjugate directions. A interesting choice is to link these conjugate directions to the gradient of the function $f(x)$. Algorithm 3 explores this possibility.

Theorem 8 ([25, Theorem 4.1.1]) *Consider the quadratic function* $f(x) = \frac{1}{2}x^T Mx$
$+ q^T x$ *where* $M \in \mathbb{R}^{n \times n}$ *is a symmetric positive matrix, and* $q \in \mathbb{R}^n$*. The Conjugate Gradient Algorithm 3 terminates after* $m < n$ *iterations and for all* $i = 0, \ldots, m$

$$ {d^i}^T M d^j = 0, \qquad j = 0, \ldots, i-1, \tag{18a}$$

Algorithm 3: Conjugate Gradient Algorithm

1 **Choose** $x^0 \in \mathbb{R}^n$, **Set** $k = 0$;
2 **while** $x^k \notin \Omega$ **do**
3   **if** $k = 0$ **then**
4   **Set** $d^k = -\nabla f(x^k)$;
5   **else**
6   **Set** $\beta_{k-1} = \dfrac{\nabla f(x^k)^T M d^{k-1}}{d^{k-1^T} M d^{k-1}}$;
7   **Set** $d^k = -\nabla f(x^k) + \beta_{k-1} d^{k-1}$;
8   **end**
9   **Set** $\alpha_k = \dfrac{-\nabla f(x^k)^T d^k}{d^{k^T} M d^k}$;
10   **Set** $x^{k+1} = x^k + \alpha_k d^k$;
11   **Set** $k = k + 1$;
12 **end**

$$\nabla f(x^i)^T \nabla f(x^j) = 0, \qquad j = 0, \ldots, i-1, \tag{18b}$$

$$\nabla f(x^i)^T d^i = -\nabla f(x^i)^T \nabla f(x^i). \tag{18c}$$

The theorem above shows that the directions generated by Algorithm 3 are conjugate directions with respect to the matrix M, and, therefore, finite termination of algorithm follows from Theorem 7.

It is important to note that

$$\alpha_k = -\frac{\nabla f(x^k)^T d_1^k}{d^{k^T} M d^k} = \frac{\nabla f(x^k)^T \nabla f(x^k)}{d^{k^T} M d^k} > 0. \tag{19}$$

Moreover, from $x^k = x^{k-1} + \alpha_{k-1} d^{k-1}$ it follows that

$$\alpha_{k-1} M d^{k-1} = M x^k - M x^{k-1} = \nabla f(x^k) - \nabla f(x^{k-1})$$

and hence

$$\nabla f(x^k)^T M d^{k-1} = \frac{1}{\alpha_{k-1}} \nabla f(x^k)^T \left(\nabla f(x^k) - \nabla f(x^{k-1}) \right) = \frac{1}{\alpha_{k-1}} \nabla f(x^k)^T \nabla f(x^k).$$

Now, from (19), $\alpha_{k-1} d^{k-1^T} M d^{k-1} = \nabla f(x^{k-1})^T \nabla f(x^{k-1})$, and hence

$$\begin{aligned}\beta_{k-1} &= \frac{\nabla f(x^k)^T M d^{k-1}}{d^{k-1^T} M d^{k-1}} = \frac{1}{\alpha_{k-1}} \frac{\nabla f(x^k)^T \nabla f(x^k)}{d^{k-1^T} M d^{k-1}} \\ &= \frac{\nabla f(x^k)^T \nabla f(x^k)}{\nabla f(x^{k-1})^T \nabla f(x^{k-1})} =: \beta_{k-1}^{\text{FR}}\end{aligned} \tag{20}$$

This formula was first proposed by Fletcher and Reeves [27]. Alternative choices, all equivalent in the case of strictly convex quadratic problems, for β_{k-1} are

$$\beta_{k-1}^{\text{PRP}} = \frac{\nabla f(x^k)^T \left[\nabla f(x^k) - \nabla f(x^{k-1})\right]}{\nabla f(x^{k-1})^T \nabla f(x^{k-1})}. \tag{21a}$$

$$\beta_{k-1}^{\text{HS}} = -\frac{\nabla f(x^k)^T \left[\nabla f(x^k) - \nabla f(x^{k-1})\right]}{d^{k-1^T} \left[\nabla f(x^k) - \nabla f(x^{k-1})\right]} \tag{21b}$$

proposed, respectively, by Polak and Ribiére [52] and Polyak [53] and Hestenes and Stiefel [39].

For general non–quadratic functions, there is no guarantee that the conjugate gradient method will terminate in a finite number of steps. Moreover, it is extremely difficult to exactly solve the one–dimensional subproblem and inexact line–search methods such as Armijo or Goldstein or Wolfe methods must be utilized. Moreover, each n iterations or when a non–descent direction is generated, a reset step should be performed using as search direction the negative gradient direction. Computational results, however, show that the use of a restarting procedure is not convenient and is better to opportunely modify the formula for β_{k-1} and to choose specific line-search procedure to globalize the overall scheme.

The first global convergence result of the Fletcher-Reeves method with inexact line search was given by Al-Baali [2]. Under strong Wolfe conditions, he demonstrated that the method generates sufficient descent directions and, therefore, global convergence can be established.

For the Polak-Ribiére-Polyak method the convergence for general nonlinear function is uncertain. While global convergence can be proved in the case of strongly convex functions, there are examples of not strongly convex functions, for which the method may not converge, even with an exact line search. Instead, convergence can be proved when

$$\beta_{k-1}^{\text{PRP+}} = \min\{\beta_{k-1}^{\text{PRP}}, 0\}$$

is utilized [29]. Additional information on conjugate gradient method for general nonlinear functions can be found in [35, 38, 54].

3.6 Quasi-Newton's Methods

As already noted earlier, the Newton's method is locally convergent at quadratic rate and requires, at each iteration, to compute the Hessian function. On the other hand, the gradient method for which global convergence can be established, only requires to compute the gradient of the function; however, its convergence can be quite slow (at most linear, in some cases).

The aim of Quasi-Newton's methods is to define procedures that

- will not require to compute the Hessian function,
- exhibits q–superlinear rate of convergence.

Two classes of Quasi–Newton methods have been studied in literature:

- direct methods for which

$$\begin{cases} x^{k+1} = x^k - \left[B^k\right]^{-1} \nabla f(x^k), \\ B^{k+1} = B^k + \Delta B^k, \qquad\qquad B^{k+1} \approx \nabla^2 f(x^{k+1}); \end{cases}$$

- inverse methods where

$$\begin{cases} x^{k+1} = x^k - H^k \nabla f(x^k), \\ H^{k+1} = H^k + \Delta H^k, \qquad\qquad H^{k+1} \approx \left[\nabla^2 f(x^{k+1})\right]^{-1}. \end{cases}$$

In order to derive updating formulas for B^k and H^k, consider again the quadratic function

$$f(x) = \frac{1}{2} x^T M x + q^T x$$

where $M \in \mathbb{R}^{n \times n}$ is a symmetric positive matrix, and $q \in \mathbb{R}^n$. Let now $x, y \in \mathbb{R}^n$, then

$$\nabla f(y) = \nabla f(x) + M(y - x)$$

or equivalently

$$M^{-1} \left(\nabla f(y) - \nabla f(x)\right) = y - x.$$

Therefore, since B^{k+1} (resp. H^{k+1}) must be an approximation of M (resp. M^{-1}), it is reasonable to require that

Algorithm 4: Generic (Direct) Quasi–Newton Method

1 **Choose** $x^0 \in \mathbb{R}^n$, **Choose** $B^0 \in \mathbb{R}^{n\times n}$, **Set** $k = 0$;
2 **while** $x^k \notin \Omega$ **do**
3 **Set** $d^k = -\left(B^k\right)^{-1} \nabla f(x^k)$;
4 **Compute** α_k using a line–search procedure;
5 **Set** $x^{k+1} = x^k + \alpha_k d^k$;
6 **Compute** $B^{k+1} = B^k + \Delta B^k$ **Set** $k = k + 1$;
7 **end**

$$\nabla f(x^{k+1}) - \nabla f(x^k) = B^{k+1}\left(x^{k+1} - x^k\right) \tag{22a}$$

or

$$H^{k+1}\left(\nabla f(x^{k+1}) - \nabla f(x^k)\right) = x^{k+1} - x^k. \tag{22b}$$

After defining $\gamma^k := \nabla f(x^{k+1}) - \nabla f(x^k)$ and $\delta^k := x^{k+1} - x^k$ the above conditions become

$$\gamma^k = B^{k+1}\delta^k \tag{23a}$$

and

$$H^{k+1}\gamma^k = \delta^k. \tag{23b}$$

These conditions are known as "secant conditions". Furthermore, it is reasonable to require that B^{k+1} (resp. H^{k+1}) be symmetric and as close as possible to B^k (resp. H^k) imposing that B^{k+1} (resp. H^{k+1})) differs from B^k (resp. H^k by a matrix of rank 1 or 2 and a minimality condition in some norm, for example in Frobenius norm[2] [18]. The generic direct Quasi–Newton method is reported in Algorithm 4.

A similar algorithm can be easily constructed for a Quasi–Newton generic inverse method.

[2] For a matrix $A \in \mathbb{R}^{m\times n}$, the Frobenius norm $\|A\|_F$ of A is defined as [48]

$$\|A\|_F := \sqrt{\sum_{i=1}^{m}\sum_{j=1}^{n} A_{ij}^2} = \sqrt{\sum_{i=1}^{m} \|A_{i.}\|_2^2} = \sqrt{\sum_{j=1}^{n} \|A_{.j}\|_2^2}$$
$$= \sqrt{\text{trace}\left(A^T A\right)} = \sqrt{\text{trace}\left(A A^T\right)}.$$

In the sequel, unless strictly necessary we will drop the superscripts k and $k+1$; the current matrix will be indicated simply by B (resp. H) while the updated matrix will be indicated by $\bar{B}$ (resp. $\bar{H}$). A similar rule will be used for γ^k and δ^k.

The simplest updating direct formula for B is a rank-1 updating formula

$$\bar{B} = B + \rho u v^T \tag{24}$$

where $\rho \in \mathbb{R}$ and $u, v \in \mathbb{R}^n$, $u, v \neq 0$.

For the symmetric case (i.e., $u = v$), we have

$$\bar{B} = B + \rho u u^T \tag{25}$$

and, imposing the secant condition $\bar{B}\delta = \gamma$ either $\bar{B} = B$ (which happens if already $B\delta = \gamma$) or, under the hypothesis that $(\gamma - B\delta)^T\delta \neq 0$, the following formula, known as Symmetric Rank-1 updating formula (SR1), is obtained

$$\bar{B} = B + \frac{(\gamma - B\delta)(\gamma - B\delta)^T}{(\gamma - B\delta)^T\delta}. \tag{26}$$

This formula was first proposed by Davidon [14] and later rediscovered by Broyden [10].

Using $H = B^{-1}$ and the Sherman–Morrison–Woodbury [32, 37] formula, it is possible to derive the inverse Symmetric Rank–1 updating scheme:

$$\bar{H} = H + \frac{(\delta - H\gamma)(\delta - H\gamma)^T}{(\delta - H\gamma)^T\gamma}. \tag{27}$$

Note that in this case the secant condition completely determines the updating formula under the hypothesis that $(\gamma - B\delta)^T\delta \neq 0$ (resp. $(H\gamma - \delta)^T\gamma \neq 0$). However, it must be noticed that there is no guarantee that the search direction that is obtained is a descent direction.

The symmetric rank–1 updating formula (27) has a very interesting behaviour when applied to a quadratic function.

Theorem 9 ([17, Theorem 7.1]) *Let $A \in \mathbb{R}^{n\times n}$ be symmetric and nonsingular. Let $\{s^0, s^1, \ldots, s^m\}$ be m vectors spanning $\mathbb{R}^n$ and let $y^k = As^k$, $k = 0, \ldots, m$. Again, let H^0 be a symmetric $n \times n$ real matrix and for $k = 0, \ldots, m$ let*

$$H^{k+1} = H^k + \frac{(s^k - H^k y^k)(s^k - H^k y^k)^T}{(s^k - H^k y^k)^T y^k}$$

assuming that $\left(s^k - H^k y^k\right)^T y^k \neq 0$. *Then*

$$H^{m+1} = A^{-1}.$$

The theorem above shows that, given a nonsingular matrix A, the inverse A^{-1} can be calculated using the SR1 formula which only involves matrix–vector multiplications

For general nonlinear functions, under specific assumptions, including that

$$\left| \left(\gamma^k - B^k \delta^k\right)^T \delta^k \right| \geq c \left\| \gamma^k - B^k \delta^k \right\|_2 \left\| \delta^k \right\|_2$$

and that the sequence $\{x^k\}$ has at least one accumulation point x^*, then [13, Theorem 2]

$$\lim_k \left\| B^k - \nabla^2 f(x^*) \right\| = 0.$$

In the non–symmetric case (mostly utilized for solving system of nonlinear equations) the updating formula require to choose both vectors u and v and the secant condition does not define uniquely both vectors. Therefore, the new matrix $\bar{B}$ is required to be as close as possible to the current matrix B. The following lemma shows that a rank-1 updating formula is obtained when the closest matrix (in Frobenius norm) to the current matrix B is calculated among all matrices for which the secant condition is satisfied.

Theorem 10 ([17, Theorem 4.1], [18, Lemma 8.1.1]) *Let* $B \in \mathbb{R}^{n \times n}$ *and let* $\gamma, \delta \in \mathbb{R}^n$. *Then, the solution of the minimization problem*

$$\begin{array}{ll} \min\limits_A & \|A - B\|_F \\ *\text{subject to} & \gamma = A\delta \end{array} \tag{28}$$

is given by

$$\bar{B} = B + \frac{(\gamma - B\delta)\,\delta^T}{\delta^T \delta}. \tag{29}$$

Using again the Sherman–Morrison–Woodbury formula, it possible (under the hypothesis that $\gamma^T H \delta \neq 0$) to derive the inverse updating formula:

$$\bar{H} = H + \frac{(\delta - H\gamma)\,\delta^T H}{\delta^T H \gamma} \tag{30}$$

The updating formula (30) was first proposed by Broyden [9] in the context of solving systems of nonlinear equations. Locally q–superlinear convergence for this method was then proved by Broyden, Dennis and Morè in [12], see also [19, Theorem 3.3.3] and [19, Theorem 8.2.2].

The generic rank–2 updating formula is given by

$$\bar{B} = B + auu^T + bvv^T$$

where $a, b \in \mathbb{R}$ and $u, v \in \mathbb{R}^n$; similar rank–2 updating formulas can be defined for $\bar{H}$.

The use of a rank–2 updating formula allows to imposes important conditions, in addition to the secant property, on the updated matrix such as symmetry and positive definiteness. Theorem 11 show that the closest matrix to B for which symmetry and secant condition are satisfied can be obtained via a rank–2 modification. Here a weighted Frobenius norm is utilized, where W is the weight matrix:

$$\|A - B\|_{\mathrm{W,F}} := \|W(A - B)W\|_{\mathrm{F}}.$$

Theorem 11 ([17, Theorem 7.3]) *Let $B \in \mathbb{R}^{n\times n}$ be a symmetric matrix and let $c, \delta, \gamma \in \mathbb{R}^n$ with $c^T\delta > 0$. Let $W \in R^{n\times n}$ be a symmetric nonsingular matrix such that $Wc = W^{-1}\delta$. Then, the unique solution of*

$$\begin{array}{ll} \min\limits_{A} & \|A - B\|_{W,F} \\ *\text{subject to} & A\delta = \gamma \\ & A \text{ symmetric} \end{array} \tag{31}$$

is given by

$$\bar{B} = B + \frac{(\gamma - B\delta)\, c^T + c\,(\gamma - B\delta)^T}{c^T\delta} - \frac{(\gamma - B\delta)^T\,\delta}{\left(c^T\delta\right)^2} cc^T. \tag{32}$$

The above theorem leaves space for opportune choices of the matrix W and the vector c, which can be utilized to impose positive definiteness of the updating formula, i.e., conditions that ensure that the matrix $\bar{B}$ be positive definite, provided that the same property holds for B. Positive definiteness of the matrix $\bar{B}$ ensures that the search direction utilized in the Quasi–Newton method (see Algorithm 4) is a descent direction.

In view of the fact that $\bar{B}\delta = \gamma$, if $\bar{B}$ is positive definite then $0 < \delta^T\bar{B}\delta = \delta^T\gamma$. It is not difficult to show that this condition is also necessary for the positive definiteness of $\bar{B}$ [17, Theorem 7.5].

A simple, but extremely interesting, choice for the vector c is $c = \gamma$. This specific choice leads to the updating formula

$$\bar{B} = B + \frac{(\gamma - B\delta)\gamma^T + \gamma(\gamma - B\delta)^T}{\gamma^T\delta} - \frac{(\gamma - B\delta)^T\delta}{(\gamma^T\delta)^2}\gamma\gamma^T. \tag{33}$$

Again, using the Sherman–Morrison–Woodbury formula, it is possible to compute the corresponding inverse updating formula (here $H = B^{-1}$ and $\bar{H} = \bar{B}^{-1}$)

$$\bar{H} = H + \frac{\delta\delta^T}{\gamma^T\delta} - \frac{H\gamma\gamma^T H}{\gamma^T H\gamma}. \tag{34}$$

These formulas, known as direct and inverse DFP formulas, were firstly proposed by Davidon [14] and later rediscovered by Fletcher and Powell that also clarified and improved them [26].

Moreover, it is also possible to construct updating formulas directly for H similarly requiring that $\bar{H}$ be symmetric, that the secant condition be satisfied and $\bar{H}$ be as close as possible to H in a specific norm as shown in the theorem below, in a similar way as done for the direct updating formula.

Theorem 12 *Let $H \in \mathbb{R}^{n\times n}$ be a symmetric matrix and let $d, \delta, \gamma \in \mathbb{R}^n$ with $d^T\gamma > 0$. Let $W \in R^{n\times n}$ be a symmetric nonsingular matrix such that $Wd = W^{-1}\gamma$. Then, the unique solution of*

$$\begin{array}{ll} \min\limits_{A} & \|W(A - H)W\|_F \\ *\text{subject to} & A\gamma = \delta \\ & A \text{ symmetric} \end{array} \tag{35}$$

is given by

$$\bar{H} = H + \frac{(\delta - H\gamma)d^T + d(\delta - H\gamma)^T}{d^T\gamma} - \frac{(\delta - H\gamma)^T\gamma}{(d^T\gamma)^2}dd^T. \tag{36}$$

Similarly to what done in the direct case, it is possible to impose that the new matrix $\bar{H}$ be positive definite. Moreover, the choice $d = \delta$ brings to the following updating formula

$$\bar{H} = H + \frac{(\delta - H\gamma)\delta^T + \delta(\delta - H\gamma)^T}{\delta^T\gamma} - \frac{(\delta - H\gamma)^T\gamma}{(\delta^T\gamma)^2}\delta\delta^T. \tag{37}$$

Using the Sherman–Morrison–Woodbury formula the corresponding direct updating formula can be obtained

Algorithm 5: Quasi–Newton method (Broyden class updating scheme)

1 **Choose** $x^0 \in \mathbb{R}^n$, **Choose** $H^0 \in \mathbb{R}^{n\times n}$, **Set** $k = 0$;
2 **while** $x^k \notin \Omega$ **do**
3 **Set** $d^k = -H^k \nabla f(x^k)$;
4 **Compute** α_k using a line–search procedure;
5 **Set** $x^{k+1} = x^k + \alpha_k d^k$;
6 **Set** $\delta = x^{k+1} - x^k$, $\gamma = \nabla f(x^{k+1}) - \nabla f(x^k)$,
$v = \left(\gamma^T H^k \gamma\right)^{\frac{1}{2}} \left(\frac{\delta}{\gamma^T \delta} - \frac{H^k \gamma}{\gamma^T H^k \gamma}\right)$;
7 **Choose** $\phi_k \geq 0$;
8 **Compute** $H^{k+1} = H^k + \frac{\delta\delta^T}{\gamma^T\delta} - \frac{H^k\gamma\gamma^T H^k}{\gamma^T H^k \gamma} + \phi_k v v^T$;
9 **Set** $k = k + 1$;
10 **end**

$$\bar{B} = B + \frac{\gamma\gamma^T}{\delta^T\gamma} - \frac{B\delta\delta^T B}{\delta^T B\delta}. \tag{38}$$

The above formulas are known as (inverse and direct) BFGS updating formulas from Broyden [11], Fletcher [24], Goldfarb [30] and Shanno [57] that discovered them.

Starting from the DFP and BFGS updating formulas a whole class (known as Broyden class) of updating methods can be constructed:

$$\bar{H}^{\phi} = (1 - \phi)\bar{H}^{\text{DFP}} + \phi\bar{H}^{\text{BFGS}}$$

where $\bar{H}^{\text{DFP}}$ and $\bar{H}^{\text{BFGS}}$ are given by (34) and (37) respectively, and $\phi \in \mathbb{R}$.

Note that $\bar{H}^{\phi}$ can be easily rewritten as

$$\bar{H}^{\phi} = \bar{H}^{\text{DFP}} + \phi v v^T$$

where

$$v = \left(\gamma^T H\gamma\right)^{\frac{1}{2}} \left(\frac{\delta}{\gamma^T\delta} - \frac{H\gamma}{\gamma^T H\gamma}\right)$$

clearly showing the relationship between $\bar{H}^{\phi}$ and $\bar{H}^{\text{DFP}}$.

The generic Quasi–Newton algorithm using the Broyden updating scheme is reported in Algorithm 5 (here we return to use again the index k).

For quadratic strictly convex problems

$$\min f(x) := \frac{1}{2} x^T M x + q^T x$$

where $M \in \mathbb{R}^{n \times n}$ is a positive definite matrix and $q \in \mathbb{R}^n$, finite termination is guaranteed when $\phi_k \geq 0$ and α_k is obtained via exact line–search [17, Theorem 8.1], [59, Theorem 5.2.1]. Finite termination follows from the observation that for all $k = 0, \ldots, n$

$$\left(x^{k+1} - x^k\right)^T M \left(x^{i+1} - x^i\right) = 0, \quad \forall i < k$$

i.e., the directions $\delta^0 = x^1 - x^0, \delta^1 = x^2 - x^1, \ldots, \delta^k = x^{k+1} - x^k$ are M-conjugate and x^{k+1} minimizes $f(x)$ in the affine space the affine set

$$S_k := \left\{ x \in \mathbb{R}^n : x = x^0 + \sum_{j=0}^{k} \sigma_j \delta^j, \ \sigma_j \in \mathbb{R} \right\}$$

The following theorem, originally due to Powell, shows global convergence for rank–2 symmetric updating schemes for generic convex optimization problems.

Theorem 13 ([17, Theorem 8.2]) *Let $f : \mathbb{R}^n \to \mathbb{R}$ be twice differentiable and convex. Let $x^0 \in \mathbb{R}^n$ and assume the level set $\mathcal{L}_0$ be bounded. Let H^0 be symmetric and positive definite and let $\{x^k\}$ be generated by Algorithm 5 with either*

- $\phi_k = 0$ *(DFP updating formula) and exact line–search, or*
- $\phi_k = 1$ *(BFGS updating formula) and inexact line–search satisfying the Wolfe condition.*

Then, for any $\epsilon > 0$ there exists k such that $\left\|\nabla f(x^k)\right\| < \epsilon$.

Dixon [22] demonstrated that in case of exact line–search, the sequence $\{x^k\}$ is independent of $\phi_k \geq 0$. Furthermore, in [24], it is shown that, for stability reasons, it is better to choose $\phi_k \in [0, 1]$. However, from a computational point of view the choice $\phi_k = 1$ appears better than $\phi_k = 0$.

Finally, q–superlinear convergence of rank–2 updating methods is obtained in the case $\phi_k = 0$ or $\phi_k = 1$ [17, Theorem 8.9] if the stepsize α_k is chosen according to the Wolfe rule (in which case it is possible to show that $\alpha_k = 1$ for $k \geq k_0$) provided that $\sum_{k=1}^{+\infty} \left\|x^k - x^*\right\| < +\infty$.

We refer the interested reader to [35, 59] for additional insight on Quasi–Newton methods.

4 Constrained Optimization

In this section we concentrate our attention on the constrained optimization problem

$$\begin{array}{llll} \min\limits_{x} & f(x) & & \\ *\text{subject to} & c_i(x) \leq 0 & i = 1, \ldots, m & \\ & c_i(x) = 0 & i = m+1, \ldots, m+h & \end{array} \tag{39}$$

where $f : \mathbb{R}^n \to \mathbb{R}$ is the objective function and $c_i : \mathbb{R}^n \to \mathbb{R}, i = 1, \ldots, m+h$ are the constraints defining the feasible region X

$$X := \left\{x \in \mathbb{R}^n : c_i(x) \leq 0, i = 1, \ldots, m \text{ and } c_i(x) = 0, i = m+1, \ldots, m+h\right\}.$$

For simplicity, we will always assume that all the functions are at least twice continuously differentiable.

Definition 11 A feasible point $\bar{x}$ is a local minimum if $\exists\ \gamma > 0$ such that

$$f(\bar{x}) \leq f(x) \quad \forall x \in X \text{ and } \|x - \bar{x}\| \leq \gamma.$$

More generally, a point $\bar{x}$ is a local minimum if $\exists\ \mathcal{I}$ neighborhood of $\bar{x}$ such that

$$f(\bar{x}) \leq f(x) \quad \forall\, x \in X \cap \mathcal{I}.$$

A point $\bar{x}$ is a strict local minimum if $\exists\ \mathcal{I}$ neighborhood of $\bar{x}$ such that

$$f(\bar{x}) < f(x) \quad \forall\, x \in X \cap \mathcal{I}, x \neq \bar{x}.$$

A point $\bar{x}$ is a strong or isolated local minimum if $\exists\ \mathcal{I}$ neighborhood of $\bar{x}$ such that $\bar{x}$ is the only local minimum in $X \cap \mathcal{I}$.

For the above minimization problem, the Lagrangian function is defined as follows:

$$L(x, \lambda) := f(x) + \sum_{i=1}^{m+h} \lambda_i c_i(x) \tag{40}$$

where $\lambda \in \mathbb{R}^{m+h}$. First and second order necessary and sufficient optimality conditions will be expressed in terms of the above Lagrangian function.

Another important concept in constrained optimization is the concept of active constraints.

Definition 12 Let $\bar{x}$ be a feasible point for the constrained optimization problem (39), i.e., , $\bar{x} \in X$. The set of active constraints at $\bar{x}$ is

$$\mathcal{A}(\bar{x}) := \left\{i : 1 \le i \le m : c_i(\bar{x}) = 0\right\} \cup \left\{m+1, \ldots, m+h\right\}.$$

4.1 Necessary and Sufficient Optimality Conditions

In order to derive optimality conditions for the constrained optimization problem (39), two sets must be introduced: the tangent cone and the set of linearized feasible directions.

Definition 13 Let $\bar{x} \in X$. A direction $d \in \mathbb{R}^n$ is tangent to the set X at $\bar{x}$ if there exists a sequence $\left\{z^k\right\} \subseteq X$ with $\lim_k z^k = \bar{x}$ and a sequence $\{\theta_k\}$ of positive scalars with $\lim_k \theta_k = 0$ such that

$$\lim_k \frac{z^k - \bar{x}}{\theta_k} = d. \tag{41}$$

The set of all tangent vectors to X at $\bar{x}$ is called the tangent cone at $\bar{x}$ and is denoted by $\mathcal{T}_X(\bar{x})$.

Definition 14 Let $\bar{x} \in X$ and let $\mathcal{A}(\bar{x})$ be the set of indices of active constraints. The set of linearized feasible directions $\mathcal{F}(\bar{x})$ is the set

$$\mathcal{F}(\bar{x}) := \left\{ \begin{array}{ll} d \in \mathbb{R}^n : \nabla c_i(\bar{x})^T d \le 0, & i = 1, \ldots, m, i \in \mathcal{A}(\bar{x}), \\ \qquad\qquad \nabla c_i(\bar{x})^T d = 0, & i = m+1, \ldots, m+h \end{array} \right\}.$$

Clearly, if $d \in \mathcal{F}(\bar{x})$ also $\alpha d \in \mathcal{F}(\bar{x})$ for all nonnegative α and hence also the set $\mathcal{F}(\bar{x})$ is a cone.

These two sets can be viewed as local approximations of the feasible region X around $\bar{x}$.

Note that, while the tangent cone only depends on the geometry of the feasible region, the set of linearized feasible directions depends on the specific formulation of the constraints utilized to define the feasible region. Moreover, for all $\bar{x} \in X$, $\mathcal{T}_X(\bar{x}) \subseteq \mathcal{F}(\bar{x})$.

A fundamental question is under which conditions the two sets coincide, that is $\mathcal{T}_X(\bar{x}) = \mathcal{F}(\bar{x})$ or, to be more precise, under which conditions the dual cones of the two sets coincide.[3] Since the seminal work of Kuhn and Tucker [42], a number of different (inter–related) Constraint Qualification

---

[3] Give a cone K, the dual cone of K is the set

$$K^* := \left\{d \in \mathbb{R}^n : d^T x \ge 0, \forall\, x \in K\right\}$$

conditions have been proposed in literature both for the case of only inequality constraints and for the more general case of equality of inequality constraints (see [60] and references therein, including the two schemes showing the relationships between the different Constraint Qualification conditions, and [46]).

In [60] the various Constraint Qualification conditions are partitioned in 4 different levels with weakest (less stringent) conditions being at level 1 and strongest (more stingent) conditions at level 4. In the following, we state some of the most common Constraint Qualification conditions.

Definition 15 (*Guignard's Constraint Qualification, GCQ*) [36] Let $\bar{x} \in X$. The Guignard's Constraint Qualification conditions are satisfied at $\bar{x}$ if $\mathcal{F}(\bar{x}) = \text{cl}$
$(\text{conv}(\mathcal{T}_X(\bar{x})))$.[4]

The GCQ is considered the weakest possible Constraint Qualification.

Definition 16 (*Abadie's Constraint Qualification, ACQ*) [1] Let $\bar{x} \in X$. The Abadie's Constraint Qualification conditions are satisfied at $\bar{x}$ if $\mathcal{T}_X(\bar{x}) = \mathcal{F}(\bar{x})$.

The Abadie's Constraint Qualification requires that $\mathcal{T}_X(\bar{x})$ be a convex cone. Hence this condition is stronger than Guignard's Constraint Qualification.

Definition 17 (*Slater's Constraint Qualification, SCQ*) [58] Let $\bar{x} \in X$. The Slater's Constraint Qualification conditions are satisfied at $\bar{x}$ if

- $c_i(x)$ is pseudo–convex at $\bar{x}$ for all $1 \le i \le m$ and $i \in \mathcal{A}(\bar{x})$,
- $c_i(x)$ is both quasi–convex and quasi–concave[5] at $\bar{x}$ for all $m+1 \le i \le m+h$,
- the vectors $\left\{\nabla c_i(\bar{x}),\quad m+1 \le i \le m+h\right\}$ are linearly independent,
- there exists $\hat{x}$ such that $c_i\left(\hat{x}\right) < 0$, for all $1 \le i \le m$ and $i \in \mathcal{A}(\bar{x})$ and $c_i\left(\hat{x}\right) = 0$ for all $m+1 \le i \le m+h$.

Definition 18 (*Mangasarian–Fromovitz's Constraint Qualification, MFCQ*) [47, 3.4–3.6] Let $\bar{x} \in X$. The Mangasarian–Fromovitz's Constraint Qualification conditions are satisfied at $\bar{x}$ if

- the vectors $\left\{\nabla c_i(\bar{x}),\quad m+1 \le i \le m+h\right\}$ are linearly independent,
- there exists d such that $\nabla c_i(\bar{x})^T d < 0$, for all $1 \le i \le m$ and $i \in \mathcal{A}(\bar{x})$ and $\nabla c_i(\bar{x})^T d = 0$ for all $m+1 \le i \le m+h$.

---

[4] For a set S, cl (S) indicates its closure.

[5] A generic function g is quasi–concave if and only if $-g$ is quasi–convex.

Definition 19 (*Linear Independence Constraint Qualification, LICQ*) [60, 5.4.6] Let $\bar{x} \in X$. Linear Independence Constraint Qualification (LICQ) conditions are satisfied at $\bar{x}$ if the vectors

$$\left\{\nabla c_i\left(\bar{x}\right), 1 \leq i \leq m, \quad i \in \mathcal{A}\left(\bar{x}\right)\right\} \cup \left\{\nabla c_i\left(\bar{x}\right), m+1 \leq i \leq m+h\right\}$$

are linearly independent.

Both LICQ and SCQ imply MFCQ. Moreover, MFCQ implies ACQ. Therefore, if any of these Constraint Qualification conditions hold at a feasible point $\bar{x}$, then $\mathcal{T}_X\left(\bar{x}\right) = \mathcal{F}\left(\bar{x}\right)$.

The following lemma provides necessary conditions for a local optimal solution.

Lemma 1 *Let x^* be a local optimal solution of Problem* (39). *Then*

$$\nabla f(x^*)^T d \geq 0 \qquad \forall\, d \in \mathcal{T}_X\left(x^*\right) \tag{42}$$

Proof Suppose, by contradiction, that at x^* there is a direction $d \in \mathcal{T}_X\left(x^*\right)$ such that

$$\nabla f(x^*)^T d < 0.$$

Then, there exists a sequence $\left\{d^k\right\}$ converging to d and a sequence $\{\theta_k\}$ of positive scalars converging to zero with $z^k = x^* + \theta_k d^k \in X$ for all k. Therefore,

$$\begin{aligned} f(z^k) = f(x^* + \theta_k d^k) &= f(x^*) + \nabla f(x^*)^T \left(z^k - x^*\right) + o\left(\left\|z^k - x^*\right\|\right) \\ &= f(x^*) + \theta_k \nabla f(x^*)^T d^k + o(\theta_k). \end{aligned}$$

But

$$\lim_k \nabla f(x^*)^T d^k = \nabla f(x^*)^T d < 0,$$

and hence

$$f(z^k) - f(x^*) = \theta_k \nabla f(x^*)^T d^k + o(\theta_k) < 0$$

for k sufficiently large. Therefore, for each neighborhood of x^*, there exists a sufficiently large index k such that the point z^k belongs to such neighborhood and, hence, x^* is not a local minimum.

In case that Constraint Qualification conditions are satisfied, the necessary optimality conditions can be expressed in a more convenient way, in terms of the Karush–Kuhn–Tucker conditions. This results utilizes Motzkin's theorem of the alternative, see [46, Chap. 2].

Theorem 14 ([46, 7.3.7][49, Theorem 12.1]) *Let x^* be a local minimum for the constrained optimization Problem* (39) *and assume that at x^* some Constraint Qualification conditions are satisfied. Then there exists a Lagrange multiplier vector λ^* such that the following conditions, known as Karush–Kuhn–Tucker (or KKT) conditions, hold:*

$$\nabla_x L(x^*, \lambda^*) = 0, \tag{43a}$$

$$c_i(x^*) \leq 0 \ \forall i = 1, \ldots, m, \tag{43b}$$

$$c_i(x^*) = 0 \ \forall i = m+1, \ldots, m+h, \tag{43c}$$

$$\lambda_i^* \geq 0 \ \forall i = 1, \ldots, m, \tag{43d}$$

$$\lambda_i^* c_i(x^*) = 0 \ \forall i = 1, \ldots, m, \tag{43e}$$

where $L(.,.)$ is the Lagrangian function (40).

Note that Condition 43a can be rewritten as

$$\nabla f(x^*) + \sum_{i=1}^{m+h} \lambda_i^* \nabla c_i(x^*) = 0.$$

Conditions (43e) are called complementarity conditions and they impose that for all $i = 1, \ldots, m$ either $\lambda_i^* = 0$ or $c_i(x^*) = 0$ or both.

In order to derive second order necessary and sufficient conditions, we need to introduce first the concept of critical cone.

Definition 20 Let x^* be a local minimum for the constrained optimization Problem (39) and suppose that the pair (x^*, λ^*) satisfies the Karush–Kuhn–Tucker conditions (43) for some $\lambda^* \in \mathbb{R}^{m+h}$. The Critical Cone is defined as:

$$C(x^*, \lambda^*) := \Big\{ d \in \mathcal{F}(x^*) : \nabla c_i(x^*)^T d = 0, \text{ for all } i = 1, \ldots, m, i \in \mathcal{A}(\bar{x}) \text{ with } \lambda_i^* > 0 \Big\}. \tag{44}$$

Equivalently, $d \in C(x^*, \lambda^*)$ if only if

$$\nabla c_i(x^*)^T d = 0 \ i = 1, \ldots, m, i \in \mathcal{A}(\bar{x}) \text{ with } \lambda_i^* > 0 \tag{45a}$$

$$\nabla c_i(x^*)^T d \leq 0 \ i = 1, \ldots, m, i \in \mathcal{A}(\bar{x}) \text{ with } \lambda_i^* = 0 \tag{45b}$$

$$\nabla c_i(x^*)^T d = 0 \ i = m+1, \ldots, m+h \tag{45c}$$

Note that the Hessian of the Lagrangian function $L(x, \lambda)$ is given by

$$\nabla^2_{xx} L(x, \lambda) = \nabla^2 f(x) + \sum_{i=1}^{m+h} \lambda_i \nabla^2 c_i(x).$$

This Hessian is of fundamental importance in the second order necessary and sufficient conditions.

Theorem 15 ([25, Theorem 9.3.1] [49, Theorem 12.5]) *Let x^* be a local minimum for the constrained optimization Problem* (39) *and assume that at x^* some Constrained Qualification Conditions are satisfied. Then, there exists a Lagrange multiplier vector λ^* such that the Karush–Kuhn–Tucker conditions* (43) *are satisfied and*

$$d^T \nabla^2_{xx} L\left(x^*, \lambda^*\right) d \geq 0, \text{ for all } d \in C\left(x^*, \lambda^*\right). \tag{46}$$

Theorem 16 ([25, Theorem 9.3.2] [49, Theorem 12.6]) *Let x^* be a feasible point for the constrained optimization Problem* (39) *and assume that there exists a vector λ^* such that the pait (x^*, λ^*) satifies the Karush–Kuhn–Tucker conditions* (43). *Furthermore, suppose that*

$$d^T \nabla^2_{xx} L\left(x^*, \lambda^*\right) d > 0, \text{ for all } 0 \neq d \in C\left(x^*, \lambda^*\right). \tag{47}$$

Then, x^* is a strict local minimum of the constrained optimization Problem (39)

As for the unconstrained case, also here there is a significative difference between necessary and sufficient optimality conditions. In fact, for necessary optimality condition the Hessian of the Lagrangian must be positive semidefinite in the Critical Cone. Instead, for sufficient optimality condition, the Hessian of the Lagrangian must be positive definite in the same Critical Cone.

4.2 Duality in Constrained Optimization

The concept of duality is central in constrained optimization as well as in other fields of Operations Research (e.g., in Linear programming) and, in general, in mathematics. Let

$$f^* = \inf_{x \in X} f(x). \tag{48}$$

The dual function is given by

$$q(\lambda) := \inf_x L(x, \lambda) \tag{49}$$

where $L(x, \lambda)$ is the Lagrangian function (40). Note that this dual function is a concave function (i.e., $-q(\lambda)$ is a convex function) and its domain $\mathcal{D}$ is a convex set. The dual problem is defined as

$$\begin{array}{ll} \max_{\lambda} & q(\lambda) \\ \text{subject to } \lambda_i \geq 0, & i = 1, \ldots, m. \end{array} \tag{50}$$

Moreover, let

$$\begin{array}{l} q^* = \sup_{\lambda} q(\lambda) \\ \qquad \lambda_i \geq 0, i = 1, \ldots, m. \end{array} \tag{51}$$

It is not difficult to show that (Weak Duality), if $\bar{\lambda} \in \mathbb{R}^{m+h}$ with $\bar{\lambda}_i \geq 0$, $i = 1, \ldots, m$ (that is, $\bar{\lambda}$ is dual feasible), and $\bar{x} \in X$ (that is, $\bar{x}$ is primal feasible), then

$$\begin{aligned} q(\bar{\lambda}) = \inf_x L(x, \bar{\lambda}) \leq L(\bar{x}, \bar{\lambda}) &= f(\bar{x}) + \sum_{i=1}^{m+h} \bar{\lambda}_i c_i(\bar{x}) \\ &= f(\bar{x}) + \sum_{i=1}^{m} \bar{\lambda}_i c_i(\bar{x}) \leq f(\bar{x}) \end{aligned}$$

and hence

$$q^* \leq f^*. \tag{52}$$

A more convenient form for the dual problem can be derived for convex problems. The theorem below shows the relationship between points satisfying Karush–Kuhn–Tucker conditions and optimal solutions of the dual problem. Moreover, it shows that in convex optimization there is no duality gap.

Theorem 17 ([49, Theorem 12.12]) *Let x^* be an optimal solution of the convex optimization problem* (39) *where $f : \mathbb{R}^n \to \mathbb{R}$ and $c_i : \mathbb{R}^n \to \mathbb{R}$, $i = 1, \ldots, m$ are convex functions and $c_i : \mathbb{R}^n \to \mathbb{R}$, $i = m+1, \ldots, m+h$ are linear functions. Assume, further, that at x^* some Constraint Qualification conditions are satisfied. Then, there exists λ^* such that the pair (x^*, λ^*) satisfies the Karush–Kuhn–Tucker conditions* (43) *and λ^* solves the dual problem* (50), *i.e.,*

$$
\begin{aligned}
q(\lambda^*) = \max_{\lambda} \; & q(\lambda) \\
& \lambda_i \geq 0, i = 1, \ldots, m.
\end{aligned}
\tag{53}
$$

Moreover,

$$
f^* = f(x^*) = L(x^*, \lambda^*) = q(\lambda^*) = q^*
$$

In order to revert, at least partially, this condition, strict convexity is required.

Theorem 18 ([49, Theorem 12.13]) *Let x^* be an optimal solution of the convex optimization problem* (39) *where $f : \mathbb{R}^n \to \mathbb{R}$ and $c_i : \mathbb{R}^n \to \mathbb{R}, i = 1, \ldots, m$ are convex functions and $c_i : \mathbb{R}^n \to \mathbb{R},\; i = m+1, \ldots, m+h$ are linear functions. Assume, further, that at x^* some Constraint Qualification conditions are satisfied. Moreover, suppose that $\bar{\lambda}$ is a solution of the dual problem* (50) *and that the function $L(x, \bar{\lambda})$ is strict convex in x and*

$$
L(\bar{x}, \bar{\lambda}) = \inf_{x} L(x, \bar{\lambda}).
$$

Then $x^* = \bar{x}$ and $f(x^*) = L(x^*, \bar{\lambda})$.

The two theorems above are the basis for a different, more convenient, form for the dual problem, known as Wolfe dual [61]:

$$
\begin{aligned}
\max_{x,\lambda} \quad & L(x, \lambda) \\
\text{subject to} \quad & \nabla_x L(x, \lambda) = 0 \\
& \lambda_i \geq 0, i = 1, \ldots, m.
\end{aligned}
\tag{54}
$$

For this dual optimization problem it is possible to show that, if x^* is a local minimum at which Constraint Qualification conditions are satisfied, there exists $\lambda^* \in \mathbb{R}^{m+h}$ such that the pair (x^*, λ^*) satisfies the Karush–Kuhn–Tucker conditions (43) and, furthermore, solves the Wolfe dual problem (54).

4.3 Penalty and Augmented Lagrangian Methods

An important and well studied class of methods for solving nonlinear optimization problems is based on the idea of replacing the original constrained problem by a single or a sequence of unconstrained problems. For these problems the new objective function will contain terms penalizing the violation of the original constraints.

An important issue, both from a theoretical and a practical point of view, is to determine if the minimizer of the penalty function and the solution of the original optimization problem coincide. This property is called exactness of the penalty function [20].

The simplest approach is to use a quadratic term to penalize the violation of the constraints. For the constrained problem (39), the quadratic penalty function is given by

$$\psi(x, \mu) := f(x) + \mu\left[\sum_{i=1}^{m} \max\{0, c_i(x)\}^2 + \sum_{i=m+1}^{m+h} c_i(x)^2\right], \tag{55}$$

where $\mu > 0$ is the penalty parameter. If the original functions $f(x)$ and $c_i(x)$, $i = 1, \ldots, m+h$ are sufficiently smooth, the function $\psi(x, \mu)$ is differentiable (and continuously differentiable, if $m = 0$, i.e., there are only equality constraints) and, hence, standard algorithms for unconstrained minimization can be applied to calculate its minimizer.

However, when quadratic penalty terms are utilized, it is necessary to drive the penalty term to $+\infty$. A general scheme require to:

- choose a priori a sequence $\{\mu_k\} \to +\infty$,
- for each value of μ_k, calculate x^{μ_k}, a minimizer of $\psi(x, \mu_k)$.

The procedure terminates when the violation of the constraints at x^{μ_k} is sufficiently small.

For this simple scheme it is possible to show [25, Theorem 12.1.1] [49, Theorem 17.1] that

1. $\{\psi(x^{\mu_k}, \mu_k)\}$ is non–decreasing,
2. $\{f(x^{\mu_k})\}$ is non–increasing,
3. the constraints violation is non–increasing,
4. every accumulation point x^* of the sequence $\{x^{\mu_k}\}$ is a solution of Problem (39).

Another widely used penalty function is the 1–norm penalty function defined as

$$\phi(x, \mu) := f(x) + \mu\left[\sum_{i=1}^{m} \max\{0, c_i(x)\} + \sum_{i=m+1}^{m+h} |c_i(x)|\right]. \tag{56}$$

This penalty function is exact in the sense that there exists a finite $\mu^* > 0$ such that for all values of $\mu \geq \mu^*$, if x^* is a strict local solution of the nonlinear problem (39) at which first-order necessary optimality conditions are satisfied, then x^* is a local minimizer of $\phi(x, \mu)$ [49, Theorem 17.3]. However, the 1–norm penalty function is non–smooth and, therefore, specific algorithms must be utilized and convergence is in many cases quite slow.

Continuously differentiable exact penalty functions can also be constructed [20] by using an Augmented Lagrangian function that includes additional terms penalizing the violation of the Karush–Kuhn–Tucker conditions. Under specific assumptions, stationary point, local and global minimizers of the Augmented Lagrangian function exactly correspond to Karush–Kuhn–Tucker points, local and global solutions of the constrained problem [21].

4.4 Sequential Quadratic Programming

A very effective method for solving constrained optimization problems is based on sequentially solving quadratic subproblems (Sequential Quadratic Programming, SQP). The idea behind the SQP approach is to construct, at each iteration, a quadratic approximation of Problem (39) around the current point x^k, and then use the minimizer of this subproblem as the new iterate x^{k+1} [49, Chap. 18].

More specifically, at the current point x^k we construct a quadratic approximation of the problem by a quadratic approximation for the objective function using the Hessian of the Lagrangian function with respect to the x variables and linear approximation of the constraints:

$$\begin{array}{ll} \min\limits_{d} & f(x^k) + \nabla f(x^k)^T d + \frac{1}{2} d^T \nabla^2_{xx} L(x^k, \lambda^k) d \\ *\text{subject to} & \begin{array}{ll} \nabla c_i(x^k)^T d + c_i(x^k) \leq 0, & i = 1, \ldots, m \\ \nabla c_i(x^k)^T d + c_i(x^k) = 0, & i = m+1, \ldots, m+h. \end{array} \end{array} \tag{57}$$

Very fast and efficient algorithms exist for solving the above problem that produce the vector d^k, multipliers associated to the linearized constraints λ^{k+1}, and an estimate of the active constraints.[6] Then the new point is obtained as $x^{k+1} = x^k + d^k$.

Under specific assumptions, it is possible to show that if x^* is a local solution of Problem (39), if, at x^* and some λ^*, Karush–Kuhn–Tucker conditions are satisfied, and if $(x^k, \lambda^k$ is sufficiently close to (x^*, λ^*), then there is a local solution of the subproblem (57) whose active set is the same as the active set of the nonlinear optimization problem (39) at x^* [55].

[6] One of the most important technique for solving convex quadratic programming problems with equality and inequality constraints is based on Active Set strategy where, at each iteration, some of the inequality constraints, and all the equality constraints, are imposed as equalities (the "Working Set") and a simpler quadratic problem with only equality constraints is solved. Then the Working Set is update and a new iteration is performed. For further details refer to [25, 10.3] and [49, 16.5].

The correct identification of the active constraints when x^k is sufficiently close to x^* is at the basis of the proof of local convergence for the SQP method.

Acknowledgements The author wants to express his gratitude to prof. Nadaniela Egidi for reading a first version of the manuscript and providing many useful suggestions.

References

1. Abadie, J.: On the Kuhn-Tucker Theorem. Operations Research Center University of Calif Berkeley, Technical report (1966)
2. Al-Baali, M.: Descent property and global convergence of the Fletcher' Reeves method with inexact line search. IMA J. Numer. Anal. **5**(1), 121–124 (1985)
3. Apostol, T.M.: Calculus, 2nd edn. Wiley (1967)
4. Armijo, L.: Minimization of functions having Lipschitz continuous first partial derivatives. Pac. J. Math. **16**(1), 1–3 (1966)
5. Bagirov, A.M., Gaudioso, M., Karmitsa, N., Mäkelä, M.M., Taheri, S.: Numerical Nonsmooth Optimization: State of the Art Algorithms. Springer Nature (2020)
6. Bazaraa, M.S., Sherali, H.D., Shetty, C.M.: Nonlinear Programming: Theory and Algorithms, 3rd edn. Wiley, Newark, NJ (2006)
7. Bertsekas, D.P.: Nonlinear Programming. Athena Scientific (1999)
8. Boyd, S., Vandenberghe, L.: Convex Optimization. Cambridge University Press, New York, NY, USA (2004)
9. Broyden, C.G.: A class of methods for solving nonlinear simultaneous equations. Math. Comput. **19**, 577–593 (1965)
10. Broyden, C.G.: Quasi-Newton methods and their application to function minimization. Math. Comput. **21**, 368–381 (1967)
11. Broyden, C.G.: The convergence of single-rank quasi-Newton methods. Math. Comput. **24**, 365–382 (1970)
12. Broyden, C.G., Dennis, J.E., Moré, J.J.: On the local and superlinear convergence of quasi-Newton methods. IMA J. Appl. Math. **12**(3), 223–245 (1973)
13. Conn, A.R., Gould, N.I.M., Toint, P.L.: Convergence of Quasi-Newton matrices generated by the symmetric rank one update. Math. Program. **50**(2), 177–195 (1991)
14. Davidon, W.C.: Variable metric method for minimization. AEC Research and Development Report ANL-5990, Argonne National Laboratory (1959)
15. De Leone, R., Gaudioso, M., Grippo, L.: Stopping criteria for linesearch methods without derivatives. Math. Program. **30**(3), 285–300 (1984)
16. Deb, K., Deb, K.: Multi-objective optimization. In: Burke, E., Kendall, G. (eds.) Search Methodologies: Introductory Tutorials in Optimization and Decision Support Techniques, pp. 403–449. Springer, US, Boston, MA (2014)
17. Dennis, J.E., Moré, J.J.: Quasi-Newton methods, motivations and theory. SIAM Rev. **19**, 46–89 (1977)
18. Dennis, J.E., Schnabel, R.B.: Numerical Methods for Unconstrained Optimization and Nonlinear Equations. Prentice-Hall, Englewook Cliffs, New Jersey (1983)
19. Dennis, J.E., Schnabel, R.B.: Chapter I A view of unconstrained optimization. In: Optimization. Handbooks in Operations Research and Management Science, vol. 1, pp. 1–72. Elsevier (1989)

20. Di Pillo, G., Grippo, L.: Exact penalty functions in constrained optimization. SIAM J. Control Optim. **27**(6), 1333–1360 (1989)
21. Di Pillo, G., Liuzzi, G., Lucidi, S., Palagi, L.: An exact augmented Lagrangian function for nonlinear programming with two-sided constraints. Comput. Optim. Appl. **25**(1), 57–83 (2003)
22. Dixon, L.C.W.: Variable metric algorithms: necessary and sufficient conditions for identical behavior of nonquadratic functions. J. Optim. Theory Appl. **10**, 34–40 (1972)
23. Ehrgott, M.: Multicriteria Optimization, vol. 491. Springer Science & Business Media (2005)
24. Fletcher, R.: A new approach to variable metric algorithms. Comput. J. **13**, 317–322 (1970)
25. Fletcher, R.: Practical Methods of Optimization, 2nd edn. Wiley (1990)
26. Fletcher, R., Powell, M.J.D.: A rapidly convergent descent method for minimization. Comput. J. **6**(2), 163–168 (1963)
27. Fletcher, R., Reeves, C.M.: Function minimization by conjugate gradients. Comput. J. **7**(2), 149–154 (1964)
28. Gaudioso, M., Giallombardo, G., Miglionico, G.: Essentials of numerical nonsmooth optimization. 4OR **18**(1), 1–47 (2020)
29. Gilbert, J.C., Nocedal, J.: Global convergence properties of conjugate gradient methods for optimization. SIAM J. Optim. **2**(1), 21–42 (1992)
30. Goldfarb, D.: A family of variable metric updates derived by variational means. Math. Comput. **24**, 23–26 (1970)
31. Goldstein, A.A.: Constructive Real Analysis. Harper and Row, London (1967)
32. Golub, G.H., Van Loan, C.F.: Matrix Computations. Johns Hopkins University Press, Baltimore, Maryland (1983)
33. Grippo, L., Lampariello, F., Lucidi, S.: A nonmonotone line search technique for Newton's method. SIAM J. Numer. Anal. **23**(4), 707–716 (1986)
34. Grippo, L., Sciandrone, M.: Nonmonotone globalization techniques for the Barzilai-Borwein gradient method. Comput. Optim. Appl. **23**(2), 143–169 (2002)
35. Grippo, L., Sciandrone, M.: Methods of unconstrained optimization. (Metodi di ottimizzazione non vincolata.). Unitext 53. La Matematica per il 3+2. Springer, Italia (2011)
36. Guignard, M.: Generalized Kuhn-Tucker conditions for mathematical programming problems in a Banach space. SIAM J. Control **7**(2), 232–241 (1969)
37. Hager, W.W.: Updating the inverse of a matrix. SIAM Rev. **31**(2), 221–239 (1989)
38. Hager, W.W., Zhang, H.C.: A survey of nonlinear conjugate gradient methods. Pac. J. Optim. **2**(1), 35–58 (2006)
39. Hestenes, M.R., Stiefel, E.: Methods of conjugate gradients for solving linear systems. J. Res. Natl. Bur. Stand. **49**, 409–436 (1952)
40. Horst, R., Pardalos, P.M.: Handbook of Global Optimization, vol. 2. Springer Science & Business Media (2013)
41. Konak, A., Coit, D.W., Smith, A.E.: Multi-objective optimization using genetic algorithms: a tutorial. Reliab. Eng. Syst. Saf. **91**(9), 992–1007 (2006)
42. Kuhn, H.W., Tucker, A.W.: Nonlinear Programming, pp. 481–492. University of California Press, Berkeley, California (1951)
43. Lemarechal, C., Mifflin, R.: Nonsmooth optimization: proceedings of a IIASA workshop, March 28–April 8, 1977. Elsevier (2014)

44. Li, D.H., Fukushima, M.: A derivative-free line search and global convergence of Broyden-like method for nonlinear equations. Optim. Methods Softw. **13**(3), 181–201 (2000)
45. Luenberger, D.G.: Linear and Nonlinear Programming, 2nd edn. Addison–Wesley (1984)
46. Mangasarian, O.L.: Nonlinear Programming. McGraw-Hill, New York (1969)
47. Mangasarian, O.L., Fromovitz, S.: The Fritz John necessary optimality conditions in the presence of equality and inequality constraints. J. Math. Anal. Appl. **17**, 37–47 (1967)
48. Meyer, C.D.: Matrix Analysis and Applied Linear Algebra. Society for Industrial and Applied Mathematics, Philadelphia, PA, USA (2000)
49. Nocedal, J., Wright, S.: Numerical Optimization. Springer Science & Business Media (2006)
50. Ortega, J.M., Rheinboldt, W.C.: Iterative Solution of Nonlinear Equations in Several Variables. Academic Press (1970)
51. Pinter, J.: Continuous global optimization: applications. In: C. Floudas, E.P.M. Pardalos (eds.) Encyclopedia of Optimization, pp. 482–486. Springer (2008)
52. Polak, E., Ribière, G.: Note sur la convergence de méthodes de directions conjuguées. ESAIM: Math. Model. Numer. Anal.-Modélisation Mathématique et Anal. Numérique **3**(R1), 35–43 (1969)
53. Polyak, B.T.: The conjugate gradient method in extremal problems. USSR Comput. Math. Math. Phys. **9**(4), 94–112 (1969)
54. Pytlak, R.: Conjugate Gradient Algorithms in Nonconvex Optimization, vol. 89. Springer Science & Business Media (2008)
55. Robinson, S.M.: Perturbed Kuhn-Tucker points and rates of convergence for a class of nonlinear-programming algorithms. Math. Program. **7**(1), 1–16 (1974)
56. Rockafellar, R.T.: Convex Analysis. Princeton Mathematical Series, Princeton University Press, Princeton, N. J. (1970)
57. Shanno, D.F.: Conditioning of Quasi-Newton methods for function minimization. Math. Comput. **24**, 647–656 (1970)
58. Slater, M.: Lagrange multipliers revisited. Technical report, Cowles Foundation Discussion Paper No. 80, Cowles Foundation for Research in Economics, Yale University (1950)
59. Sun, W., Yuan, Y.X.: Optimization Theory and Methods: Nonlinear Programming, vol. 1. Springer Science & Business Media (2006)
60. Wang, Z., Fang, S.C., Xing, W.: On constraint qualifications: motivation, design and inter-relations. J. Ind. Manag. Optim. **9**, 983–1001 (2013)
61. Wolfe, P.: A duality theorem for non-linear programming. Q. Appl. Math. **19**, 239–244 (1961)
62. Wolfe, P.: Convergence conditions for ascent methods. SIAM Rev. **11**(2), 226–235 (1969)
63. Zhang, H.C., Hager, W.W.: A nonmonotone line search technique and its application to unconstrained optimization. SIAM J. Optim. **14**(4), 1043–1056 (2004)

New Computational Tools in Optimization

The Role of *grossone* in Nonlinear Programming and Exact Penalty Methods

Renato De Leone

Abstract Exact penalty methods form an important class of methods for solving constrained optimization problems. Using penalty functions, the original constrained optimization problem can be transformed in an "equivalent" unconstrained problem. In this chapter we show how *grossone* can be utilized in constructing exact differentiable penalty functions for the case of only equality constraints, the general case of equality and inequality constraints, and quadratic problems. These new penalty functions allow to recover the solution of the unconstrained problem from the finite term (in its *grossone* expansion) of the optimal solution of the unconstrained problem. Moreover, Lagrangian duals associated to the constraints are also automatically obtained from the infinitesimal terms. Finally a new algorithmic scheme is presented.

1 Introduction

Penalty methods represent an important class of methods for solving constrained optimization problems. These functions, in their simplest form, are composed of two parts: the original objective function and a penalty function, usually multiplied by a positive scalar called penalty parameter, which penalize the violation of the constraints. Using penalty functions, the original constrained optimization problem can be transformed in an "equivalent" unconstrained problem. Two major issues must be taken into account when constructing such penalty functions: exactness and differentiability. Roughly speaking, exactness requires that the global or local solution of the uncon-

---

R. De Leone (✉)
School of Science and Technology, University of Camerino, Camerino, MC, Italy
e-mail: renato.deleone@unicam.it

Y. D. Sergeyev and R. De Leone (eds.), *Numerical Infinities and Infinitesimals in Optimization*, Emergence, Complexity and Computation 43,
https://doi.org/10.1007/978-3-030-93642-6_3

strained problem or a stationary point of it, must correspond to a global or local minimum of the constrained problem or to a point satisfying Karush–Kuhn–Tucker conditions. Using the Euclidean norm, it is quite simple to construct continuously differentiable, or at least differentiable penalty functions, as long as the functions in the original problem are sufficiently smooth. However, in these cases the penalty parameter must be driven to $+\infty$, thus generating numerical instability. Using the 1–norm, instead, it is possible to construct penalty functions that are exact, that is they do not require that the penalty parameter goes to $+\infty$. The difficulty in this case arises from the non–differentiability of the function.

Recently, Sergeyev introduced a new approach to infinite and infinitesimals. The proposed numeral system is based on ①, the number of elements of $\mathbb{N}$, the set of natural numbers. We refer the reader to Chap. 1 for insights on the arithmetic of infinity and the properties of ①. Here we want to stress that ① is not a symbol and is not used to make symbolic calculation. In fact, the new numeral ① is a natural number, and it has both cardinal and ordinal properties, exactly as the "standard", finite natural numbers. Moreover, the new proposed approach is far apart from non–Standard Analysis, as clearly shown in [30]. A comprehensive description of the grossone–based methodology can also be found in [29].

In this chapter we discuss the use of ① in constructing exact differentiable penalty functions for the case of only equality constraints, the general cases of equality and inequality constraints, and quadratic problems. Using this novel penalty function, it is possible to recover the solution of the unconstrained problem from the finite term (in its ① expansion) of the optimal solution of the unconstrained problem. Moreover, Lagrangian duals are also automatically and at no additional cost obtained just considering the ①$^{-1}$ grossdigits in their expansion in term ①.

While this chapter only concentrates the attention on the use of ① to define novel exact penalty functions for constrained optimization problems, it must be noted that the use of ① has been beneficial in many other areas in optimization. Already in [8], the authors demonstrated how the classical simplex method for linear programming can be modified, using ① to overcome the difficulties due to degenerate steps. Along this line of research, more recently in [4], the authors proposed the Infinitely-Big-M method, a re–visitation of the Big–M method for the Infinity Computer. Various different optimization problems have been successfully tackled using this new methodology: multiobjective optimization problems [3, 5, 6, 21], the use of negative curvature directions in large-scale unconstrained optimization [11, 12], variable metric methods in nonsmooth optimization [16]. Recently, this computational methodology has also been also utilized in the field of Machine Learning allowing to construct new spherical separations for classification problems [2], and novel sparse Support Vector Machines [10]. Furthermore, the use of ① has given rise to a variety of applications in several fields of pure and applied mathematics, providing new and alternative approaches. Here we only mention numerical differentiation [26], ODE [1, 20, 31], hyperbolic geometry [23], infinite series and the Riemann zeta function [25, 27], biology [28], and cellular automata [7].

We briefly describe our notation now. All vectors are column vectors and will be indicated with lower case Latin letter ($x, y, \ldots$). Subscripts indicate components of a vector, while superscripts are used to identify different vectors. Matrices will be indicated with upper case roman letter ($A, B, \ldots$). The set of real numbers and the set of nonnegative real numbers will be denoted by $\mathbb{R}$ and $\mathbb{R}_+$ respectively. The space of the n–dimensional vectors with real components will be indicated by $\mathbb{R}^n$. Superscript ^{T} indicates transpose. The scalar product of two vectors x and y in $\mathbb{R}^n$ will be denoted by $x^T y$. The norm of a vector x will be indicated by $\|x\|$. The space of the $m \times n$ matrices with real components will be indicated by $\mathbb{R}^{m\times n}$. Let $f : S \subseteq \mathbb{R}^n \to \mathbb{R}$, the gradient $\nabla f(x)$ of $f : \mathbb{R}^n \to \mathbb{R}$ at a point $x \in \mathbb{R}^n$ is a column vector with $[\nabla f(x)]_j = \frac{\partial f(x)}{\partial x_j}$.

For what is necessary in this chapter, in this new positional numeral system with base ① a value C is expressed as

$$C = C^{(1)}① + C^{(0)} + C^{(-1)}①^{-1}C^{(-2)}①^{-2} + \cdots$$

Here and throughout the symbols := and =: denote definition of the term on the left and the right sides of each symbol, respectively.

2 Exact Penalty Methods

In this section we will utilize the novel approach to infinite and infinitesimal numbers proposed by Sergeyev[1] to construct exact differentiable penalty functions for nonlinear optimization problems.

Consider the constrained optimization problem

[1] We refer the reader to Chap. 1 for an in–depth description of this new applied approach to infinite and infinitesimal quantities and the arithmetics of infinity.

$$\begin{array}{llll} \min\limits_{x} & f(x) & & \\ \text{subject to} & c_i(x) \leq 0 & i = 1, \ldots, m & \\ & c_i(x) = 0 & i = m+1, \ldots, m+h & \end{array} \quad (1)$$

where $f : \mathbb{R}^n \rightarrow \mathbb{R}$ is the objective function and $c_i : \mathbb{R}^n \rightarrow \mathbb{R}$, $i = 1, \ldots, m+h$ are the constraints defining the feasible region X

$$X := \left\{x \in \mathbb{R}^n : c_i(x) \leq 0, \ i = 1, \ldots, m \text{ and } c_i(x) = 0, \ i = m+1, \ldots, m+h\right\}.$$

For simplicity, we will assume that all the functions are at least twice continuously differentiable.

Let $\bar{x}$ be a feasible point for the above problem. The set of active constraints at $\bar{x}$ is defined as

$$\mathcal{A}(\bar{x}) := \Big\{i : 1 \leq i \leq m, c_i(\bar{x}) = 0\Big\} \cup \Big\{m+1, \ldots, m+h\Big\}.$$

In nonlinear optimization a key role is played by the Constraint Qualification conditions that ensure that the tangent cone and the cone of linearized feasible directions coincide and allow to express necessary and sufficient optimality conditions in terms of the well known Karush–Kuhn–Tucker conditions.[2] For reader's easiness we recall here the fundamental Linear Independence Constraint Qualification that will be heavily utilized in this form or in a modified form in this chapter.

Definition 1 Linear independence constraint qualification (LICQ) condition is said to hold true at $\bar{x} \in X$ if the gradients of the active constraints at $\bar{x}$ are linearly independent.

Note that weaker Constraint Qualification conditions can be imposed [32]. In [34] various Constraint Qualification conditions are stated and categorized into four levels by their relative strengths from weakest (less stringent) to strongest (more stringent, but easier to check).

The Lagrangian function associated to Problem (1) is given by

$$L(x, \lambda) := f(x) + \sum_{i=1}^{m+h} \lambda_i c_i(x). \quad (2)$$

Necessary and sufficient optimality conditions can be written in terms of the Lagrangian function. If $x^* \in X$ is a local minimum of Problem (1) at which LICQ condition holds true, then, there exist a vector λ^* such that the pair (x^*, λ^*) is a Karush–Kuhn–Tucker point.

[2] For further details we refer the reader to Chap. 2 and references therein.

Different algorithms have been proposed in literature for finding a local minimum of Problem (1). Among the most effective methods, penalty methods play a crucial role, obtaining the solution of the constrained problem by solving a single or a sequence of unconstrained optimization problems.

Let $\delta : \mathbb{R}^n \to \mathbb{R}_+$ be a function, when possible continuously differentiable, such that

$$\delta(x) \begin{cases} = 0 \text{ if } x \in X \\ > 0 \text{ otherwise.} \end{cases}$$

Then the constrained optimization problem (1) can be replaced by the following unconstrained problem

$$\min_x \psi(x, \mu). \tag{3}$$

where

$$\psi(x, \mu) := f(x) + \mu\delta(x), \tag{4}$$

and μ is a positive real number. Different choices for the function $\delta(x)$ conduct to different penalty methods. A convenient, highly utilized, choice is

$$\delta(x) = \frac{1}{2}\sum_{i=1}^{m} \max\{c_i(x), 0\}^2 + \frac{1}{2}\sum_{i=m+1}^{m+h} (c_i(x))^2 \tag{5}$$

which is differentiable but not twice differentiable. Therefore, this choice for $\delta(x)$ does not allow to utilize some of the most effective algorithms available for unconstrained optimization.

One key issue in exterior penalty methods is exactness. Roughly speaking, the penalty function is exact if, for some finite value of the parameter μ, a local (global) minimum of it corresponds to a local (global) minimum point of the constrained problem. As noted in [14] this characterization is satisfactory only when both the constrained problem and the penalty function are convex, while in the non–convex case a more precise definition of exactness is necessary [13, 14].

Unfortunately, for the penalty function (5) this exactness property does not hold, not even in the case of only equality constraints, that is when $m = 0$. In [15, p. 279] a simple 1–dimensional counter-example is reported, showing that the solution of the constraint problem is only obtained when $\mu \to +\infty$. Exact penalty functions can be constructed using 1-norm instead of the 2-norm utilized in (5) to penalize the violation of the constraints. However, in these cases the resulting function is nondifferentiable, and ad-hoc methods must be utilized for which convergence is often quite slow. Furthermore, exact differentiable penalty functions can be constructed [13] by introducing

---

Algorithm 1: Generic Sequential Minimization Algorithm

---

1 **Choose** $x^0 \in \mathbb{R}^n$. **Let** $\{\mu_k\}$ be a monotonically increasing sequence of positive real values, **Set** $k = 0$, $x^{\mu_0} = x^0$;
2 **while** $\delta(x^{\mu_k}) > \epsilon$ **do**
3 | **Set** $k = k + 1$;
4 | **Compute** x^{μ_k} an optimal solution of the unconstrained differentiable problem

$$\min_x \psi(x, \mu_k).$$

5 **end**

---

in the objective function additional terms related to first order optimality conditions, thus making the objective function more complicate to manage.

Sequential penalty methods require to solve a sequence of minimization problems with increasing values of the parameter μ as shown in Algorithm 1.

In the algorithm above, the point x^{μ_k} obtained at iteration k can be used as starting point for the minimization problem at iteration $k + 1$. Note that Step 4 cannot be, in general, completed in a finite number of steps and, hence, the algorithm must be modified requiring to calculate, at each iteration, only an approximation of the optimal solution.

In [8] a novel exact differentiable penalty method is introduced using the numeral grossone. In the next three sections, the equality constraints, the general equality and inequality constraints and quadratics case will be discussed. Finally, a simple new non–monotone algorithmic scheme is also proposed for the solution of penalty functions based on ①.

3 Equality Constraints Case

Consider the constrained optimization problem with only equality constraints (that is $m = 0$):

$$\begin{array}{lll} \min\limits_x & f(x) & \\ \text{subject to} & c_i(x) = 0 & i = 1, \ldots, h. \end{array} \tag{6}$$

Let

$$\delta(x) = \frac{1}{2} \sum_{i=1}^{h} (c_i(x))^2$$

and let $\psi(x, \mu)$ be given by (4). In this case, it is possible to show [15, Theorem 12.1.1] that for the sequence constructed by Algorithm 1

1. $\{\psi(x^{\mu_k}, \mu_k)\}$ is monotonically non–decreasing,
2. $\{\delta(x^{\mu_k})\}$ is monotonically non–increasing,
3. $\{f(x^{\mu_k})\}$ is monotonically non–decreasing.

Moreover, $\lim_k c_i(x^{\mu_k}) = 0$, $i = 1, \ldots, h$ and each accumulation point x^* of the sequence $\{x^{\mu_k}\}_k$ solves Problem (6).

For the equality constrained case, the penalty function proposed in [8] is defined as follows:

$$\min_x \quad \psi(x) := f(x) + \frac{①}{2} \sum_{i=1}^{h} (c_i(x))^2 . \tag{7}$$

Under the hypothesis that any generic point x, the function value $f(x)$, and the generic constraint $c_i(x)$ as well as the gradient $\nabla f(x)$ and $\nabla c_i(x)$ have a finite part (i.e., grossdigits corresponding to grosspower 0) and only infinitesimal terms, it is possible to show that [8, Theorem 3.3] if x^* is a stationary point for (7) then $\left(x^{*(0)}, \lambda^*\right)$ is a KKT point for (6), where $x^{*(0)}$ is the finite term in the representation of x^*, and for $i = 1, \ldots, h$, $\lambda_i^* = c_i^{(-1)}(x^{*(0)})$ where $c_i^{(-1)}(x^*)$ is the grossdigit corresponding to the grosspower -1 in the representation of $c_i(x^*)$.

The above result strongly relies on the fact that at x^* the LICQ condition is satisfied. The proof is based on the observation that grossdigits "do not mix", that is, they are kept well separated in the computations. Therefore, setting to 0 a grossnumber is equivalent to set to zero all grossdigits in its representation.

The fundamental aspect of this result is that, by solving the unconstrained optimization problem (where the objective function is twice continuously differentiable), an optimal solution of the constrained problem is obtained. Moreover, the multipliers associated to the equality constraints are automatically recovered at no additional cost, just from the representation of $c_i(x^*)$ in terms of powers of ①.

In [9] two simple examples are discussed, showing the importance of constraint qualification conditions. Here, we propose a novel example, similar in spirit to the first example discussed in [9]. The problem is originally studied in [15, pp. 279–280] to show the effectiveness and weakness of sequential penalty method.

Consider the problem

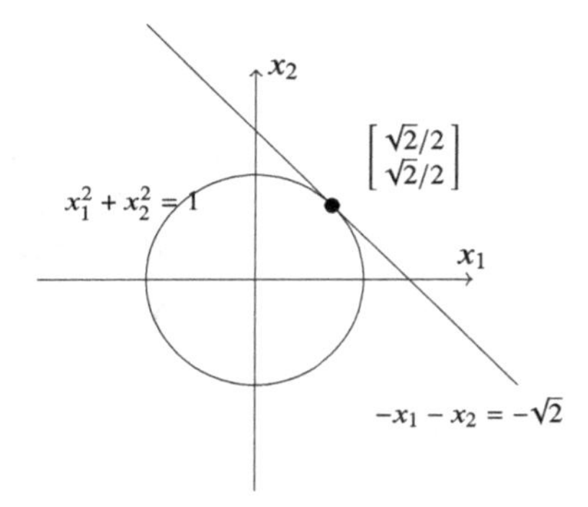

Fig. 1 Feasible region and optimal solution for Problem (8)

$$\begin{array}{ll} \min\limits_{x} & -x_1 - x_2 \\ \text{subject to} & x_1^2 + x_2^2 - 1 = 0. \end{array} \tag{8}$$

The feasible region and the optimal solution for this problem is shown in Fig. 1.

The Lagrangian function is given by

$$L(x, \lambda) = -x_1 - x_2 + \lambda \left(x_1^2 + x_2^2 - 1\right)$$

and the optimal solution is $x^* = \begin{bmatrix} \frac{1}{\sqrt{2}} \\ \frac{1}{\sqrt{2}} \end{bmatrix}$. Moreover, it is not difficult to show that the pair $\left(x^*, \lambda^* = \frac{1}{\sqrt{2}}\right)$ satisfies the KKT conditions.

In Fig. 2 the contour plots for the function $\psi(x, \mu)$ for different values of μ are shown.

The penalty function we construct is:

$$\psi\,(x, ①) := -x_1 - x_2 + \frac{①}{2}\left(x_1^2 + x_2^2 - 1\right)^2. \tag{9}$$

The First–Order Optimality Conditions $\nabla \psi\,(x, ①) = 0$ are:

$$\begin{cases} -1 + 2① x_1 \left(x_1^2 + x_2^2 - 1\right) = 0, \\ -1 + 2① x_2 \left(x_1^2 + x_2^2 - 1\right) = 0. \end{cases} \tag{10}$$

By symmetry,

$$x_1^* = x_2^* = R = R^{(0)} + ①^{-1} R^{(-1)} + ①^{-2} R^{(-2)} + \ldots$$

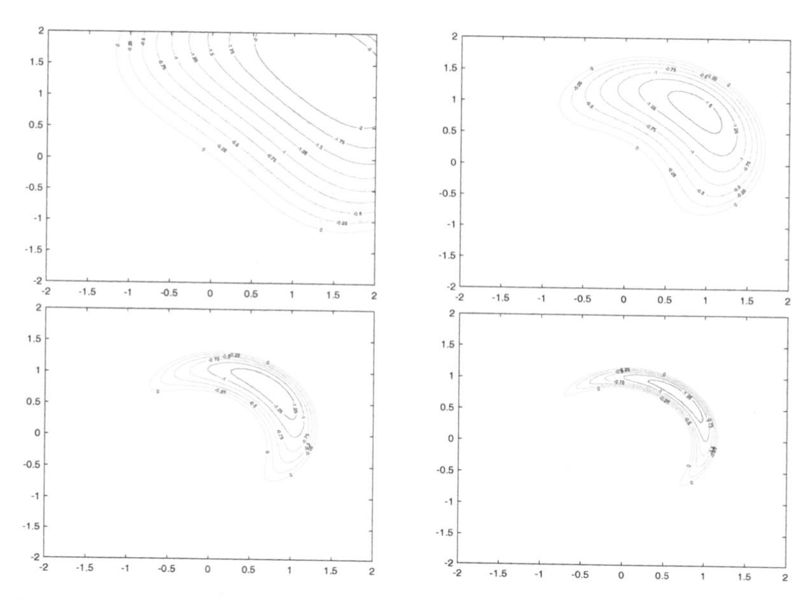

Fig. 2 Contour plots of $\psi(x, \mu)$ for $\mu = 0.1, 1, 5, 20$

and from (10) we have that

$$-1 + 4①R\left[R^2 - \frac{1}{2}\right] = 0. \tag{11}$$

From (11), by equating the term of order ① to 0, we obtain:

$$4R^{(0)}\left[\left(R^{(0)}\right)^2 - \frac{1}{2}\right] = 0$$

from which either $R^{(0)} = \frac{1}{\sqrt{2}}$ or $R^{(0)} = -\frac{1}{\sqrt{2}}$ or $R^{(0)} = 0$. The first choice for $R^{(0)}$ corresponds to a minimum of the unconstrained problem (and also for the constrained problem (8)), while the second corresponds to a maximum. Finally, the third choice of $R^{(0)}$ corresponds to a spurious solution whose presence is due to the fact that, for this point $\hat{x} = \begin{bmatrix} 0 \\ 0 \end{bmatrix}$, LICQ are not satisfied. In fact $\nabla h(\hat{x}) = \begin{bmatrix} 0 \\ 0 \end{bmatrix}$. Fixing now $R^{(0)} = \frac{1}{\sqrt{2}}$, and equating to 0 the finite terms in (11), i.e., the term corresponding to $①^0$, we obtain:

$$-1 + 4\frac{1}{\sqrt{2}}2\frac{1}{\sqrt{2}}R^{(-1)} = 0,$$

from which $R^{(-1)} = \frac{1}{4}$ and hence

$$x_1^* = x_2^* = R = \frac{1}{\sqrt{2}} + \frac{1}{4}①^{-1} + \cdots$$

where all remaining terms are infinitesimal of higher order.

Moreover,

$$\begin{aligned}\left(x_1^*\right)^2 + \left(x_2^*\right)^2 - 1 = 2R^2 - 1 = 2\left[\frac{1}{2} + \frac{2}{4\sqrt{2}}①^{-1} + \frac{1}{16}①^{-2} + \cdots\right] - 1 \\ = \frac{1}{\sqrt{2}}①^{-1} + \cdots\end{aligned}$$

where again we neglect infinitesimal of higher order than $①^{-1}$.

As expected from the theory, the $①^{-1}$ terms in the representation of the (unique) constraint provides automatically, and at no additional costs, the value of the Lagrangian multiplier.

Here we want to stress the importance of Constraint Qualification conditions that, when not satisfied, could bring to spurious results, as shown by this example.

This situation is even better clarified in the second example from [9]:

$$\begin{array}{ll}\min\limits_{x} & x_1 + x_2 \\ \text{subject to} & \left(x_1^2 + x_2^2 - 2\right)^2 = 0.\end{array} \tag{12}$$

Here the objective function and the feasible region are the same as in the first example in [9] and the optimal solution is $x^* = \begin{bmatrix} -1 \\ -1 \end{bmatrix}$ with Lagrangian multiplier $\lambda^* = \frac{1}{2}$. However, in this case

$$\nabla c_1(x) = \begin{bmatrix} 2x_1\left(x_1^2 + x_2^2 - 2\right) \\ \\ 2x_1\left(x_1^2 + x_2^2 - 2\right) \end{bmatrix}$$

which, calculated at the optimal solution, gives:

$$\nabla c_1\left(\begin{bmatrix} -1 \\ -1 \end{bmatrix}\right) = \begin{bmatrix} 0 \\ 0 \end{bmatrix}.$$

Therefore, the LICQ condition is not satisfied and the Karush–Kuhn–Tucker conditions

$$\begin{cases} 1 - 4\lambda x_1 \left(x_1^2 + x_2^2 - 2\right) = 0 \\ 1 - 4\lambda x_2 \left(x_1^2 + x_2^2 - 2\right) = 0 \\ \left(x_1^2 + x_2^2 - 2\right)^2 = 0 \end{cases} \tag{13}$$

have no solution. In this case (see [9]) the optimal solution $x^* = \begin{bmatrix} c-1 \\ -1 \end{bmatrix}$ of Problem (12) cannot be recovered from the exact penalty function

$$\psi(x, ①) = x_1 + x_2 + \frac{①}{2}\left(x_1^2 + x_2^2 - 2\right)^4$$

using first order optimality conditions.

4 Equality and Inequality Constraints Case

Consider now the more general nonlinear optimization problem (1) that includes equality and inequality constraints. In this case, the exact penalty function proposed in [8] is given by

$$\psi(x, ①) = f(x) + \frac{①}{2}\sum_{i=1}^{m} \max\{0, c_i(x)\}^2 + \frac{①}{2}\sum_{i=m+1}^{m+h} c_i(x)^2. \tag{14}$$

In order to derive the correspondence between stationary points of (14) and KKT points for Problem (1) a Modified LICQ condition is introduced in [8].

Definition 2 Let $\bar{x} \in \mathbb{R}^n$. The Modified LICQ (MLICQ) condition is said to hold true at $\bar{x}$ if the vectors

$$\left\{\nabla c_i(\bar{x}), i : 1 \le i \le m \text{ and } c_i(\bar{x}) \ge 0\right\} \cup \left\{\nabla c_i(\bar{x}), i = m+1, \ldots, m+h\right\}$$

are linearly independent.

If the above conditions are satisfied, and x^* is a stationary point for the unconstrained problem

$$\min_x \psi(x, ①)$$

then, it possible, again, to show [8, Theorem 3.4] that the pair $\left(x^{*(0)}, \lambda^*\right)$ is a KKT point for Problem (1), where $x^{*(0)}$ is the finite term in the representation of x^*, and for $i = 1, \ldots, m+h$, $\lambda_i^* = c_i^{(-1)}(x^{*(0)})$ where $c_i^{(-1)}(x^*)$ is the grossdigit corresponding to the grosspower -1 in the representation of $c_i(x^*)$.

5 Quadratic Case

In this section we apply the new exact penalty function to the quadratic problem

$$\begin{array}{ll} \min\limits_x & \frac{1}{2}x^T Mx + q^T x \\ \text{subject to} & Ax = b \\ & x \geq 0 \end{array} \tag{15}$$

where $M \in \mathbb{R}^{n\times n}$ is positive definite, $A \in \mathbb{R}^{m\times n}$ with $\text{rank}(A) = m$, $q \in \mathbb{R}^n$, and $b \in \mathbb{R}^m$. We assume that the feasible region is not empty.

For this linearly constrained problem, Constraint Qualification conditions are always satisfied and the Karush–Kuhn–Tucker conditions [22] are:

$$\begin{array}{ll} Mx + q - A^T u - v & = 0, \\ Ax - b & = 0, \\ x & \geq 0, \\ v & \geq 0, \\ x^T v & = 0. \end{array} \tag{16}$$

For this quadratic problem, the new exact penalty function using ① is given by:

$$\psi(x, ①) := \frac{1}{2}x^T Mx + q^T x + \frac{①}{2}\|Ax - b\|_2^2 + \frac{①}{2}\|\max\{0, -x\}\|_2^2 \tag{17}$$

and the corresponding unconstrained problem is:

$$\min_x \psi(x, ①). \tag{18}$$

The first order optimality conditions can be written as follows

$$0 = \nabla\psi(x, ①) = Mx + q + ①A^T(Ax - b) - ①\max\{0, -x\}. \tag{19}$$

Lemma 1 in [9] shows that the function

$$\delta(x) = \frac{1}{2}\|Ax - b\|_2^2 + \frac{1}{2}\|\max\{0, -x\}\|_2^2$$

is convex, and $\nabla\delta(x) = 0$ if and only if $Ax = b$ and $x \geq 0$.

Based on this lemma it is possible to show that, if x^* is a stationary point for the unconstrained problem (18), then $\left(x^{*(0)}, \pi^*, \mu^*\right)$ is a Karush–Kuhn–Tucker points for Problem (15), where again $x^{*(0)}$ is the finite term in the representation of x^*, and, taking into account the linearity of the original constraints,

- $\pi^* = Ax^{*(-1)} + b^{(-1)}$, where $x^{*(-1)}$ (resp. $b^{(-1)}$) is the grossdigit corresponding to the grossterm ①$^{-1}$ in the representation of x^* (resp. b), and
- $\mu_j^* = \max\{0, -x^{*(-1)}\}$.

Once again, the proof is based on the fact that during the computation grossdigits do not mix. Then, the results follow from (19)

- by setting first to 0 the terms with grosspower ①, in this way the feasibility of $x^{*(0)}$ is provided (second and third Karush–Kuhn–Tucker conditions in (16)),
- equating to zero the finite terms (that is the terms with grosspower ①^0) thus obtaining the first, fourth and last Karush–Kuhn–Tucker conditions in (16).

6 A General Scheme for the New Exact Penalty Function

In the previous sections, we have shown how the solution of the constrained minimization problem (1) can be obtained by solving an unconstrained minimization problem that uses ①. Then, any standard optimization algorithm can be utilized to obtain a stationary point for these problems from which the solution of the constrained problems as well the multipliers can be easily obtained.

In this section a simple novel general algorithmic scheme is proposed to solve the unconstrained minimization problem arising from penalizing the constraints using ①, as constructed in the previous sections. The algorithms belongs to the class of non–monotone descend algorithms [17–19, 33, 35], and does not require that the new calculated points be necessary better than the current one. This property is necessary since for a descent algorithm, when applied to determine the minimum of $\psi(x, ①)$, once a feasible point is obtained, then all remaining points generated by the algorithm should remain feasible. This, in general, could be quite complicate to ensure in practice.

For simplicity, consider the generic minimization problem:

$$\min_x f(x)$$

where f is continuously differentiable and

$$f(x) = ①f^{(1)}(x) + f^{(0)}(x) + ①^{-1} f^{(-1)}(x) + \ldots \tag{20}$$

$$\nabla f(x) = ①\nabla f^{(1)}(x) + \nabla f^{(0)}(x) + ①^{-1}\nabla f^{(-1)}(x) + \ldots \tag{21}$$

The proposed algorithm utilizes (as customary in non–monotone algorithms) two sequences $\{t_k\}_k$ and $\{l_k\}_k$ of positive integers such that

$$t_0 = 0, \quad t_{k+1} \leq \max\{t_k + 1, T\},$$

$$l_0 = 0, \quad l_{k+1} \leq \max\{l_k + 1, L\},$$

where T and L are two fixed positive integers.

At the generic iteration k, having x^k, it is necessary first to verify if the stopping criterion

$$\nabla f^{(1)}(x^k) = 0 \text{ and } \nabla f^{(0)}(x^k) = 0$$

is satisfied. Otherwise, the next iterate x^{k+1} is calculated in the following way:

- if $\nabla f^{(1)}(x^k) \neq 0$, choose x^{k+1} such that

$$f^{(1)}(x^{k+1}) \leq f^{(1)}(x^k) + \sigma\left(\left\|\nabla f^{(1)}(x^k)\right\|\right),$$

and

$$f^{(0)}(x^{k+1}) \leq \max_{0 \leq j \leq l_k} f^{(0)}(x^{k-j}) + \sigma\left(\left\|\nabla f^{(0)}(x^k)\right\|\right);$$

- If $\nabla f^{(1)}(x^k) = 0$, choose x^{k+1} such that

$$f^{(0)}(x^{k+1}) \leq f^{(0)}(x^k) + \sigma\left(\left\|\nabla f^{(0)}(x^k)\right\|\right),$$

$$f^{(1)}(x^{k+1}) \leq \max_{0 \leq j \leq t_k} f^{(1)}(x^{k-j})$$

where $\sigma(.)$ is a forcing function [24].

In other words, when $\nabla f^{(1)}(x^k) \neq 0$ the infinite term cannot grow more than a quantity that depends on the norm of $\nabla f^{(1)}(x^k)$. For the finite term, instead, the new value $f^{(0)}(x^{k+1})$ cannot be bigger than the worst value of $f^{(0)}(x^{k-j})$ at l_k previous steps plus a quantity that depends on the norm of $\nabla f^{(0)}(x^k)$.

Instead, when $\nabla f^{(1)}(x^k) = 0$, the finite term $f^{(0)}(x^{k+1})$ cannot grow bigger than a quantity that depends on the norm of $\nabla f^{(0)}(x^k)$, and the infinite term $f^{(1)}(x^{k+1})$ must be better than the worst of the previous t_k values $f^{(1)}(x^{k-j})$.

In order to demonstrate convergence of the above scheme, we need to consider two different cases.

Case 1: there exists $\bar{k}$ such that $\nabla f^{(1)}(x^k) = 0$, for all $k \geq \bar{k}$. Then

$$f^{(1)}(x^{k+1}) \leq \max_{0 \leq j \leq t_k} f^{(1)}(x^{k-j}), \qquad k \geq \bar{k}.$$

Therefore, in this case:

$$\max_{0 \leq i \leq T} f^{(1)}(x^{\bar{k}+Tj+i}) \leq \max_{0 \leq i \leq T} f^{(1)}(x^{\bar{k}+T(j-1)+i})$$

and the sequence $\left\{ \max_{0 \leq i \leq T} f^{(1)}(x^{\bar{k}+Tj+i}) \right\}_j$ is monotonically decreasing. Moreover,

$$f^{(0)}(x^{k+1}) \leq f^{(0)}(x^k) + \sigma\left(\left\| \nabla f^{(0)}(x^k) \right\|\right), \qquad k \geq \bar{k}.$$

Assuming that the level sets for $f^{(1)}(x^0)$ and $f^{(0)}(x^0)$ are compact sets, the sequence $\left\{ \max_{0 \leq i \leq T} f^{(1)}(x^{\bar{k}+Tj+i}) \right\}_j$ has at least one accumulation point x^* and any accumulation point (according to the second condition) satisfies $\nabla f^{(0)}(x^*) = 0$ in addition to $\nabla f^{(1)}(x^*) = 0$.

Case 2: there exists a subsequence j_k such that $\nabla f^{(1)}(x^{j_k}) \neq 0$.
In this case we have that

$$f^{(1)}(x^{j_k+1}) \leq f^{(1)}(x^{j_k}) + \sigma\left(\left\| \nabla f^{(1)}(x^{j_k}) \right\|\right).$$

Again, taking into account that it is possible that for some index i bigger than j_k, $\nabla f^{(1)}(x^i) = 0$

$$\max_{0 \leq i \leq M} f^{(1)}(x^{j_k+Tj+i}) \leq \max_{0 \leq i \leq M} f^{(1)}(x^{j_k+T(j-1)+i}) + \sigma\left(\left\| \nabla f^{(1)}(x^{j_k}) \right\|\right)$$

and hence, assuming again that the level sets for $f^{(1)}(x^0)$ and $f^{(0)}(x^0)$ are compact sets, we have that $\nabla f^{(1)}(x^{j_k})$ goes to 0.
Moreover,

$$\max_{0\le i\le L} f^{(0)}(x^{j_k+Lj+i}) \le \max_{0\le i\le L} f^{(0)}(x^{j_k+L(j-1)+i}) + \sigma\left(\left\|\nabla f^{(0)}(x^{j_k})\right\|\right)$$

and hence also $\nabla f^{(0)}(x^{j_k})$ goes to 0.

7 Conclusions

Penalty methods are an important and widely studied class of algorithms in nonlinear optimization. Using penalty functions the solution of the original constrained optimization problem could be obtained by solving an unconstrained problem. The main issues with this class of methods are exactness and differentiability. In this chapter we presented some recent development on penalty functions based on the use of ①. The solution of the proposed unconstrained problem provides not only the solution of the original problem but also the Lagrangian dual variables associated to the constraints at no additional cost, from the expansion of the constraints in terms of ①. Some simple examples are also reported, showing the effectiveness of the method. Finally a general non–monotone scheme is presented for the minimization of functions that include ① grossterms.

References

1. Amodio, P., Iavernaro, F., Mazzia, F., Mukhametzhanov, M.S., Sergeyev, Y.D.: A generalized Taylor method of order three for the solution of initial value problems in standard and infinity floating-point arithmetic. Math. Comput. Simul. **141**, 24–39 (2017)
2. Astorino, A., Fuduli, A.: Spherical separation with infinitely far center. Soft. Comput. **24**(23), 17751–17759 (2020)
3. Cococcioni, M., Cudazzo, A., Pappalardo, M., Sergeyev, Y.D.: Solving the lexicographic multi-objective mixed-integer linear programming problem using branch-and-bound and grossone methodology. Commun. Nonlinear Sci. Numer. Simul. **84**, 105177 (2020)
4. Cococcioni, M., Fiaschi, L.: The Big-M method with the numerical infinite M. Optim. Lett. **15**, 2455–2468 (2021)
5. Cococcioni, M., Pappalardo, M., Sergeyev, Y.D.: Towards lexicographic multi-objective linear programming using grossone methodology. In: Y.D. Sergeyev, D.E. Kvasov, F. Dell'Accio, M.S. Mukhametzhanov (eds.) Proceedings of the 2nd International Conference. Numerical Computations: Theory and Algorithms, vol. 1776, p. 090040. AIP Publishing, New York (2016)
6. Cococcioni, M., Pappalardo, M., Sergeyev, Y.D.: Lexicographic multi-objective linear programming using grossone methodology: theory and algorithm. Appl. Math. Comput. **318**, 298–311 (2018)

7. D'Alotto, L.: Cellular automata using infinite computations. Appl. Math. Comput. **218**(16), 8077–8082 (2012)
8. De Cosmis, S., De Leone, R.: The use of grossone in mathematical programming and operations research. Appl. Math. Comput. **218**(16), 8029–8038 (2012)
9. De Leone, R.: Nonlinear programming and grossone: quadratic programming and the role of constraint qualifications. Appl. Math. Comput. **318**, 290–297 (2018)
10. De Leone, R., Egidi, N., Fatone, L.: The use of grossone in elastic net regularization and sparse support vector machines. Soft. Comput. **24**(23), 17669–17677 (2020)
11. De Leone, R., Fasano, G., Roma, M., Sergeyev, Y.D.: How grossone can be helpful to iteratively compute negative curvature directions. In: Lecture Notes in Computer Science (including subseries Lecture Notes in Artificial Intelligence and Lecture Notes in Bioinformatics) **11353 LNCS**, pp. 180–183 (2019)
12. De Leone, R., Fasano, G., Sergeyev, Y.D.: Planar methods and grossone for the conjugate gradient breakdown in nonlinear programming. Comput. Optim. Appl. **71**(1), 73–93 (2018)
13. Di Pillo, G., Grippo, L.: An exact penalty method with global convergence properties for nonlinear programming problems. Math. Program. **36**, 1–18 (1986)
14. Di Pillo, G., Grippo, L.: Exact penalty functions in constrained optimization. SIAM J. Control Optim. **27**(6), 1333–1360 (1989)
15. Fletcher, R.: Practical Methods of Optimization, 2nd edn. Wiley (1990)
16. Gaudioso, M., Giallombardo, G., Mukhametzhanov, M.S.: Numerical infinitesimals in a variable metric method for convex nonsmooth optimization. Appl. Math. Comput. **318**, 312–320 (2018)
17. Grippo, L., Lampariello, F., Lucidi, S.: A truncated Newton method with nonmonotone line search for unconstrained optimization. J. Optim. Theory Appl. **60**(3), 401–419 (1989)
18. Grippo, L., Sciandrone, M.: Nonmonotone globalization techniques for the Barzilai-Borwein gradient method. Comput. Optim. Appl. **23**(2), 143–169 (2002)
19. Grippo, L., Sciandrone, M.: Nonmonotone derivative-free methods for nonlinear equations. Comput. Optim. Appl. **37**(3), 297–328 (2007)
20. Iavernaro, F., Mazzia, F., Mukhametzhanov, M.S., Sergeyev, Y.D.: Computation of higher order Lie derivatives on the Infinity Computer. J. Comput. Appl. Math. **383**(113135) (2021)
21. Lai, L., Fiaschi, L., Cococcioni, M.: Solving mixed Pareto-lexicographic multi-objective optimization problems: the case of priority chains. Swarm Evol. Comput. **55** (2020)
22. Mangasarian, O.L.: Nonlinear programming. McGraw-Hill Series in Systems Science. McGraw-Hill, New York (1969)
23. Margenstern, M.: An application of grossone to the study of a family of tilings of the hyperbolic plane. Appl. Math. Comput. **218**(16), 8005–8018 (2012)
24. Ortega, J.M., Rheinboldt, W.C.: Iterative solution of nonlinear equations in several variables. Academic Press (1970)
25. Sergeyev, Y.D.: Numerical point of view on Calculus for functions assuming finite, infinite, and infinitesimal values over finite, infinite, and infinitesimal domains. Nonlinear Anal. Ser. A: Theory, Methods Appl. **71(12)**, e1688–e1707 (2009)
26. Sergeyev, Y.D.: Higher order numerical differentiation on the infinity computer. Optim. Lett. **5**(4), 575–585 (2011)
27. Sergeyev, Y.D.: On accuracy of mathematical languages used to deal with the Riemann zeta function and the Dirichlet eta function. p-Adic Numbers, Ultrametric Anal. Appl. **3(2)**, 129–148 (2011)

28. Sergeyev, Y.D.: Using blinking fractals for mathematical modelling of processes of growth in biological systems. Informatica **22**(4), 559–576 (2011)
29. Sergeyev, Y.D.: Numerical infinities and infinitesimals: methodology, applications, and repercussions on two Hilbert problems. EMS Surv. Math. Sci. **4**(2), 219–320 (2017)
30. Sergeyev, Y.D.: Independence of the grossone-based infinity methodology from non-standard analysis and comments upon logical fallacies in some texts asserting the opposite. Found. Sci. **24**(1), 153–170 (2019)
31. Sergeyev, Y.D., Mukhametzhanov, M.S., Mazzia, F., Iavernaro, F., Amodio, P.: Numerical methods for solving initial value problems on the infinity computer. Int. J. Unconv. Comput. **12**(1), 3–23 (2016)
32. Solodov, M.V.: Constraint qualifications. In: Wiley Encyclopedia of Operations Research and Management Science. Wiley Online Library (2010)
33. Sun, W., Han, J., Sun, J.: Global convergence of nonmonotone descent methods for unconstrained optimization problems. J. Comput. Appl. Math. **146**(1), 89–98 (2002)
34. Wang, Z., Fang, S.C., Xing, W.: On constraint qualifications: motivation, design and inter-relations. J. Ind. Manag. Optim. **9**, 983–1001 (2013)
35. Zhang, H.C., Hager, W.W.: A nonmonotone line search technique and its application to unconstrained optimization. SIAM J. Optim. **14**(4), 1043–1056 (2004)

Krylov-Subspace Methods for Quadratic Hypersurfaces: A Grossone–based Perspective

Giovanni Fasano

Abstract We study the role of the recently introduced *infinite* number grossone, to deal with two renowned Krylov-subspace methods for symmetric (possibly indefinite) linear systems. We preliminarily explore the relationship between the Conjugate Gradient (CG) method and the Lanczos process, along with their specific role of yielding tridiagonal matrices which retain large information on the original linear system matrix. Then, we show that on one hand there is not immediate evidence of an advantage from embedding grossone within the Lanczos process. On the other hand, coupling the CG with grossone shows clear theoretical improvements. Furthermore, reformulating the CG iteration through a grossone-based framework allows to encompass also a certain number of Krylov-subspace methods relying on conjugacy among vectors. The last generalization remarkably justifies the use of a grossone-based reformulation of the CG to solve also indefinite linear systems. Finally, pairing the CG with the algebra of grossone easily provides relevant geometric properties of quadratic hypersurfaces.

1 Introduction

We consider the iterative solution of indefinite linear systems by Krylov-subspace methods. After a preliminary analysis, where a couple of renowned methods are briefly detailed and compared, we directly focus on those algorithms based on the generation of conjugate vectors, and we disregard those methods which rely on generating Lanczos vectors.

G. Fasano (✉)
Department of Management, University Ca' Foscari of Venice, Venice, Italy
e-mail: fasano@unive.it

Y. D. Sergeyev and R. De Leone (eds.), *Numerical Infinities and Infinitesimals in Optimization*, Emergence, Complexity and Computation 43,
https://doi.org/10.1007/978-3-030-93642-6_4

More specifically, we analyze the behaviour of the Conjugate Gradient (CG) method in case of degeneracy, since it yields relevant implications when solving symmetric linear systems within Nonconvex Optimization problems. In this regard, the current literature on Krylov-subspace methods (see e.g. [27]) reports plenty of applications in nonlinear programming, where the CG is used and it can possibly prematurely halt on the solution of indefinite linear systems (e.g. Newton's equation for nonconvex problems).

We recall that the CG iteratively computes the sequence $\{x_k\}$, where x_k approximates at step k the solution of the symmetric linear system $Ax = b$, being $A \in \mathbf{R}^{n \times n}$. The stopping rule of the CG is based on a Ritz-Galerkin condition, i.e. the norm of the current residual $r_k = b - Ax_k$ is checked, in order to evaluate the quality of the current approximate solution x_k. Unexpectedly, the unfortunate choice of the initial iterate x_1 may cause a premature undesired stop of the CG on specific indefinite linear systems. As well known, the last drawback may have a direct dramatic impact on optimization frameworks: a so called gradient-related direction cannot be computed and possibly inefficient arrangements need to be considered. When a premature stop of the CG occurs it corresponds to an unexpected numerical failure: namely a division by a small quantity is involved. This situation is usually addressed in the literature as a *pivot breakdown*, and corresponds to the fact that the steplength along the current search direction selected by the CG tends to be unbounded. As a consequence, the CG stops beforehand and the current iterate x_k may be far from a solution of the linear system (i.e. the quantity $\|r_k\|$ might be significantly nonzero).

This paper specifically addresses the pivot breakdown of the CG, from a perspective suggested by the recent introduction of the numeral *grossone* [37]. We urge to remark that a comprehensive description of the grossone-based methodology can be found in [42], and it should be stressed that it is not formally related to *non-standard analysis* (see [43]).
Our perspective is definitely unusual for the CG, since the literature of the last decades has mainly focused on its performance and stability, rather than on the way to recover its iteration in the indefinite case. Nevertheless, we are convinced that a proper investigation of the ultimate reasons of CG degeneracy might pursue a couple of essential tasks:

- to recover the degeneracy and *provide gradient-related directions* within optimization frameworks;
- to generate *negative curvature directions*, that allow convergence of optimization methods to solutions satisfying second order necessary optimality conditions.

As regards the organization of the paper, in Sect. 2 we detail similarities and dissimilarities of two well known Krylov-subspace methods: namely the CG and the Lanczos process. In Sect. 3 we describe one of the main conclusions in the current paper, i.e. the use of grossone with the CG can help overcoming problems of degeneracy in the indefinite case. Section 4 contains details on the second relevant contribution of this paper, namely the use of grossone for the iterative computation of negative curvature directions in large scale (unconstrained) optimization frameworks, where the objective function is twice continuously differentiable. Finally, a section of conclusions will complete the paper.

As regards the symbols adopted in the paper, we use $\mathbf{R}^p$ to represent the set of the real p-vectors, while for the sake of simplicity $\|x\|$ is used to indicate the Euclidean norm of the vector x, in place of $\|x\|_2$. Given the n-real vectors x and y, with $x^T y$ we indicate their standard inner product in $\mathbf{R}^n$. The symbol $f \in C^\ell(\mathcal{A})$ indicates that the function f is ℓ times continuously differentiable on the set $\mathcal{A}$. The symbol $B \succ 0$ (respectively $B \succeq 0$) indicates that the square matrix B is positive definite (respectively semidefinite). Finally, $\lambda_M(A)$ (respectively $\lambda_m(A)$) represents the largest (respectively smallest) eigenvalue of the square matrix A.

2 The CG Method and the Lanczos Process for Matrix Tridiagonalization

Let us consider the solution of the *symmetric* linear system

$$Ax = b, \qquad A \in \mathbf{R}^{n \times n}, \tag{1}$$

where the matrix A is possibly *indefinite* and *nonsingular*. As long as A in (1) is *positive definite*, the CG method [26] iteratively provides a tridiagonalization of it (see also [22]). A general description of the CG method for solving (1) is reported in Table 1 [25], where $r_{k+1} = b - Ax_{k+1}$ and the sequences $\{r_i\}$ and $\{p_i\}$ are such that after $k+1$ iterations:

$$\begin{aligned}
r_i^T r_j &= 0, \quad i \neq j \leq k+1, \qquad &&\text{(orthogonality among } \{r_i\}\text{)},\\
r_i^T p_j &= 0, \quad j < i \leq k+1, &&\\
p_i^T A p_j &= 0, \quad i \neq j \leq k+1, &&\text{(conjugacy among } \{p_i\}\text{)}.
\end{aligned}$$

Assume that after m steps the CG stops and $r_{m+1} = 0$ (i.e. a solution to the linear system (1) is found), then setting

Table 1 The CG method for solving (1) when $A \succ 0$

The Conjugate Gradient (CG) method
Step 1: $k = 1\ ,\ x_1 \in \mathbf{R}^n\ ,\ r_1 = b - Ax_1\ ,\ p_1 = r_1.$
Step k: If $r_k = 0$ then STOP, else
$x_{k+1} = x_k + \theta_k p_k, \qquad \theta_k = \frac{r_k^T p_k}{p_k^T A p_k},$
$r_{k+1} = r_k - \theta_k A p_k,$
$p_{k+1} = r_{k+1} + \beta_k p_k, \qquad \beta_k = -\frac{r_{k+1}^T A p_k}{p_k^T A p_k} = \frac{\Vert r_{k+1}\Vert^2}{\Vert r_k\Vert^2},$
$k \leftarrow k+1$ repeat **Step** k.
End if

$$R_m = \left(\frac{r_1}{\Vert r_1\Vert} \cdots \frac{r_m}{\Vert r_m\Vert} \right) \in \mathbf{R}^{n\times m}, \tag{2}$$

$$P_m = \left(\frac{p_1}{\Vert r_1\Vert} \cdots \frac{p_m}{\Vert r_m\Vert} \right) \in \mathbf{R}^{n\times m}, \tag{3}$$

along with

$$L_m = \begin{pmatrix} 1 & & & 0 \\ -\sqrt{\beta_1} & \ddots & & \\ & \ddots & \ddots & \\ 0 & & -\sqrt{\beta_{m-1}} & 1 \end{pmatrix} \in \mathbf{R}^{m\times m}, \tag{4}$$

and

$$D_m = \begin{pmatrix} \frac{1}{\theta_1} & & 0 \\ & \ddots & \\ & & \ddots & \\ 0 & & \frac{1}{\theta_m} \end{pmatrix} \in \mathbf{R}^{m\times m}, \tag{5}$$

after an easy computation we have from (2)–(5)

$$P_m L_m^T = R_m, \tag{6}$$

$$A P_m = R_m L_m D_m, \tag{7}$$

$$A P_m L_m^T = R_m L_m D_m L_m^T \implies A R_m = R_m T_m^{CG}, \tag{8}$$

being $T_m^{CG} = L_m D_m L_m^T \in \mathbf{R}^{m\times m}$ the symmetric tridiagonal matrix

$$T_m^{CG} = \begin{pmatrix} \frac{1}{\theta_1} & -\frac{\sqrt{\beta_1}}{\theta_1} & & & 0 \\ -\frac{\sqrt{\beta_1}}{\theta_1} & \left(\frac{1}{\theta_2}+\frac{\beta_1}{\theta_1}\right) & \ddots & & \\ & \ddots & \ddots & \ddots & \\ & & \ddots & \left(\frac{1}{\theta_{m-1}}+\frac{\beta_{m-2}}{\theta_{m-2}}\right) & -\frac{\sqrt{\beta_{m-1}}}{\theta_{m-1}} \\ 0 & & & -\frac{\sqrt{\beta_{m-1}}}{\theta_{m-1}} & \left(\frac{1}{\theta_m}+\frac{\beta_{m-1}}{\theta_{m-1}}\right) \end{pmatrix} \in \mathbf{R}^{m\times m}. \tag{9}$$

Remark 1 Relation (8) underlies a three-term recurrence among the residuals $\{r_i\}$, being

$$A\frac{r_i}{\|r_i\|} \in span\{r_{i-1}, r_i, r_{i+1}\}, \qquad i \in \{1, \ldots, m\}. \tag{10}$$

2.1 Basics on the Lanczos Process

Similarly to the previous section, let us now consider the Lanczos process which is reported in Table 2. Unlike the CG method, it was initially conceived to iteratively solve a symmetric eigenvalue problem in the indefinite case [29], so that after m steps it allows to reduce (8) into relation

$$AQ_m = Q_m T_m^L, \tag{11}$$

where A is the matrix in (1), $Q_m = (q_1 \cdots q_m)$ and T_m^L is again a tridiagonal matrix, such that (Sturm sequence of tridiagonal matrices)

$$\lambda_m(A) \le \lambda_m\left(T_m^L\right) \le \lambda_m\left(T_{m-1}^L\right) \le \cdots \le \lambda_M\left(T_{m-1}^L\right) \le \lambda_M\left(T_m^L\right) \le \lambda_M(A).$$

Moreover, coupling the Lanczos process with a suitable factorization of the matrix T_m^L, the iterative solution of (1) can be pursued.

Indeed, given a *symmetric indefinite* matrix A, after $m \ge 1$ steps the Lanczos process similarly to (2) generates the directions (the *Lanczos vectors*) $q_1, \ldots, q_m$ satisfying the orthogonality properties

$$q_i^T q_j = 0, \qquad i \ne j \le m.$$

In particular at step $k \le m$ of the iterative procedure, the basis $\{q_1, \ldots, q_k\}$ for the Krylov subspace $\mathcal{K}_k(q_1, A) \doteq span\{q_1, Aq_1, \ldots, A^{k-1}q_1\}$ is generated. Then, as for the CG, the Lanczos process provides a basis of $\mathcal{K}_k(q_1, A)$ which is used to solve the problem

Table 2 The Lanczos process for the tridiagonalization of (1), when A is possibly indefinite

The Lanczos process

Step 1: $k = 0\,,\ v_0 = b \in \mathbf{R}^n\,,\ q_0 = 0\,,\ \delta_0 = \|b\|$.
Step k: If $\delta_k = 0$ then STOP, else

$$q_{k+1} = \frac{v_k}{\delta_k},$$
$$k \leftarrow k+1,$$
$$\alpha_k = q_k^T A q_k,$$
$$v_k = (A - \alpha_k I)q_k - \delta_{k-1}q_{k-1},$$
$$\delta_k = \|v_k\|,$$

repeat **Step** k.
End if

$$\min_{x \in \mathcal{K}_k(q_1, A)} \|Ax - b\|.$$

Hence, since

$$\dim\left[\mathcal{K}_1(q_1, A)\right] < \dim\left[\mathcal{K}_2(q_1, A)\right] < \cdots ,$$

in at most n iterations of the CG or the Lanczos process a sufficient information is available to compute the solution of (1).

Similarly to the CG (see (8)), in case at step m of the Lanczos process we have $q_{m+1} = 0$ (i.e. $\mathcal{K}_m(q_1, A) \equiv \mathcal{K}_{m+1}(q_1, A)$), then relation (11) holds, where $T_m^L \in \mathbf{R}^{m \times m}$ is the *tridiagonal matrix*

$$T_m^L = \begin{pmatrix} \alpha_1 & \delta_1 & & 0 \\ \delta_1 & \ddots & \ddots & \\ & \ddots & \ddots & \delta_{m-1} \\ 0 & & \delta_{m-1} & \alpha_m \end{pmatrix},$$

and a conclusion similar to (10) holds, replacing R_m by Q_m and T_m^{CG} by T_m^L. Moreover, at step $k \geq 1$ of the Lanczos process we also have

$$T_k^L = Q_k^T A Q_k,$$

so that in case the Lanczos process performs n steps, the square matrix Q_n turns to be orthogonal and its columns span $\mathbf{R}^n$.

On the other hand, since A is nonsingular the problem (1) is equivalent to compute the stationary point of the quadratic functional

$$q(x) = \frac{1}{2}x^T Ax - b^T x, \qquad (12)$$

and the Lanczos method can be a natural candidate for its solution, too. Indeed, if the Lanczos process stops at step m (i.e. $\delta_m = 0$), then replacing $x = Q_m z$, with $z \in \mathbf{R}^m$, into (12) and recalling that $q_1 = b/\|b\|$, we obtain:

$$\nabla q(z) = Q_m^T A Q_m z - Q_m^T b = T_m^L z - \|b\| e_1.$$

Hence, if the solution z^* of the *tridiagonal* system

$$T_m^L z - \|b\| e_1 = 0, \qquad z \in \mathbf{R}^m, \qquad (13)$$

is available, the point $x^* = Q_m z^*$ is both a solution of the original system (1) and a stationary point of (12) over the Krylov subspace $\mathcal{K}_m(b, A) = span\{q_1, \ldots, q_m\}$.

2.2 How the CG and the Lanczos Process Compare: A Path to Degeneracy

We urge to give some considerations about the comparison between the CG and the Lanczos process, in the light of possibly introducing the issue of degeneracy for both these algorithms:

- the Lanczos process properly *does not solve* the linear system (1); it rather *reformulates* (1) into the tridiagonal one (13). This means that some further calculations are necessary (i.e. a factorization for the matrix T_m^L) in order to give the explicit solution of (13) and then backtracking to a solution of (1). The CG (similarly for the CG-based methods in [14–16, 18] – see the next sections) does not require the last two-step solution scheme, inasmuch as at step k it at once decomposes the matrix T_k^{CG} and computes x_{k+1} as

$$x_{k+1} \in \operatorname*{argmin}_{x \in \mathcal{K}_k(b,A)} \{\|Ax - b\|\};$$

- since the solution z^* of (13) yields the solution $x^* = Q_m z^*$ of (1), for the Lanczos process we apparently need to *store the matrix* Q_m, in order to calculate x^*. However, in case we are just interested about computing the solution x^*, the storage of Q_m can be avoided (see e.g. [27], the algorithm SYMMLQ [35] and the algorithm SYMMBK [3]), by means of a suitable recursion. On the other hand, in case the Lanczos process were also asked to provide information on negative curvature directions associated with $q(x)$

in (12), at x^*, then the storage of the full rank matrix Q_m seems mandatory (see also [30]) or an additional computational effort is required (see the more recent paper [6]). Both the CG and the CG-based schemes reported in this paper avoid the last additional effort. Hence, our great interest for specifically pairing grossone with conjugacy.

We also recall that the tridiagonal matrices T_m^{CG} and T_m^L are obtained in a similar fashion, by the CG and the Lanczos process, respectively. However, in general neither in case $A \succ 0$ nor in the indefinite case they coincide, as extensively motivated in the paper [17]. Furthermore, the CG explicitly performs the Cholesky-like factorization $T_m^{CG} = L_m D_m L_m^T$ of T_m^{CG} in (8), in order to solve the linear system (1). The last matrix decomposition always exists when $A \succ 0$; conversely, if A is indefinite this decomposition exists if and only if no *pivot breakdown* occurs, i.e. none of the diagonal entries of D_m is near zero (which causes a premature stop of the CG).

On the contrary, if the Lanczos process is applied it cannot stop beforehand also when A is indefinite, because it does rely on any matrix factorization of T_m^L, meaning that no pivot breakdown can occur (see also [35]). Therefore, the application of the Lanczos process is well-posed in the indefinite case, too. In the next sections we show that the last conclusion motivates the use of grossone to handle pivot breakdown for the CG. Conversely, no immediate application of grossone algebra for the Lanczos process seems advisable, inasmuch as no breakdown opportunity can take place.

3 Coupling the CG with Grossone: A Marriage of Interest

Here we motivate the importance of pairing the CG with grossone, in case the system matrix A in (1) is indefinite. We first give a geometric viewpoint of the CG degeneracy (see the next section), then we detail how to recover the last degeneracy using grossone: this yields a general framework, that is used to describe the issue of degeneracy also for several CG-based methods, as detailed in [12].

3.1 The Geometry Behind CG Degeneracy

When the CG is applied to solve (1), with A indefinite, by Sect. 2.2 a possible degenerate or nearly degenerate situation may occur, namely $p_k^T A p_k \approx 0$, with $p_k \neq 0$. This implies a couple of results we report here, that will be suitably reinterpreted in the next sections from an alternative standpoint, using *grossone*.

Observe that when A is positive definite, at any Step k of the CG we have $0 < \lambda_m(A)\|p_k\|^2 \leq p_k^T A p_k$, so that $p_k^T A p_k$ is suitably bounded from below. Conversely, in case A is indefinite (nonsingular), a similar bound does not hold and possibly we might have $p_k^T A p_k = 0$, for a nonzero vector p_k. Furthermore, in order to better analyze the (near) degenerate case, when A is indefinite nonsingular and at Step k we have $|p_k^T A p_k| \geq \varepsilon_k \|p_k\|^2$, $\varepsilon_k > 0$, with $\|p_k\|$, $\|p_{k+1}\| < +\infty$, then (see also [15]) the angle $\alpha_{k,k+1}$ between the vectors p_k and p_{k+1} satisfies

$$\frac{\pi}{2} - \arccos\left(\frac{\varepsilon_k}{|\lambda_M(A)|}\right) \leq |\alpha_{k,k+1}| \leq \frac{\pi}{2} + \arccos\left(\frac{\varepsilon_k}{|\lambda_m(A)|}\right). \tag{14}$$

The two side inequality (14) suggests that p_k and p_{k+1} may not become parallel as long as the constant value ε_k is sufficiently bounded away from zero. Conversely when p_k and p_{k+1} tend to be parallel, it implies from (14) that ε_k is approaching zero. As special cases, we report in Figs. 1 and 2 the geometry of the directions when $A \succ 0$ (Fig. 1) and A is indefinite (Fig. 2), respectively. In Fig. 1, when the eccentricity of the ellipse increases, then a (near) degeneracy may occur, but since $A \succ 0$ no degeneracy can be observed, i.e. p_k and p_{k+1} cannot become parallel. On the contrary, in Fig. 2 we have A indefinite, so that at Step k of the CG we can experience a degeneracy, with $p_k^T A p_k = 0$ and a premature CG halt. Equivalently, the point x_{k+1} approaches a point at infinity and the norm of $\|p_{k+1}\|$ becomes unbounded; moreover (see [12]), p_k and p_{k+1} tend to become parallel.

3.2 A New Perspective for CG Degeneracy Using Grossone

This section details how the recently defined extension of real numbers based on *grossone* (see e.g. [1, 7, 21, 28, 32, 36–41, 45], along with the related applications in optimization frameworks [2, 4, 5, 8–12, 23]), can be suitably used to model the CG degeneracy. In particular, we show that:

- adopting grossone algebra within the CG allows to recover the CG degeneracy in the indefinite case;
- coupling the CG with grossone provides results which exactly match the analysis carried on for the CG-based methods in [14, 31];

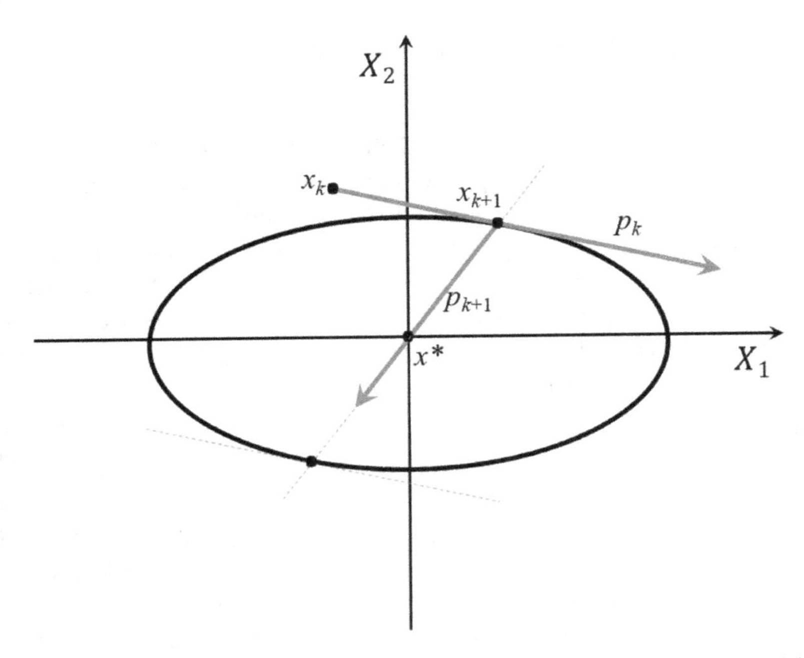

Fig. 1 The geometry behind the conjugate directions p_k and p_{k+1}: when $A \succ 0$ in (1) then without loss of generality in (14) we have $\varepsilon_k \geq \lambda_m(A) > 0$

- our approach confirms the geometry behind CG degeneracy in the indefinite case, as underlined by polarity for quadratic hypersurfaces (see also Fig. 2 and [18]).

On this purpose, let us consider the computation of the steplength θ_k at Step k of Table 1. Then, we set

$$p_k^T A p_k = s①, \tag{15}$$

where

- $s = O(①^{-1})$ if the Step k is a non-degenerate CG step (i.e. if $p_k^T A p_k \neq 0$),
- $s = O(①^{-2})$ if the Step k is a degenerate CG step (i.e. if $p_k^T A p_k = 0$).

In the last setting, following the standard Landau-Lifsitz notation we indicate with the symbol $O(①^{-2})$ *a term containing powers of* ① *at most equal to* -2. Observe that in the last case, standard results for grossone imply that the finite part of $p_k^T A p_k$ equals zero (or equivalently $p_k^T A p_k$ is infinitesimal). To large extent, the grossone-based expression on the righthand side of (15) can be further generalized; nevertheless, the setting (15) both seems simple enough and adequate to prove that the axioms and the basic algebra of grossone are well-suited to detail the behaviour of the CG, in the degenerate case.

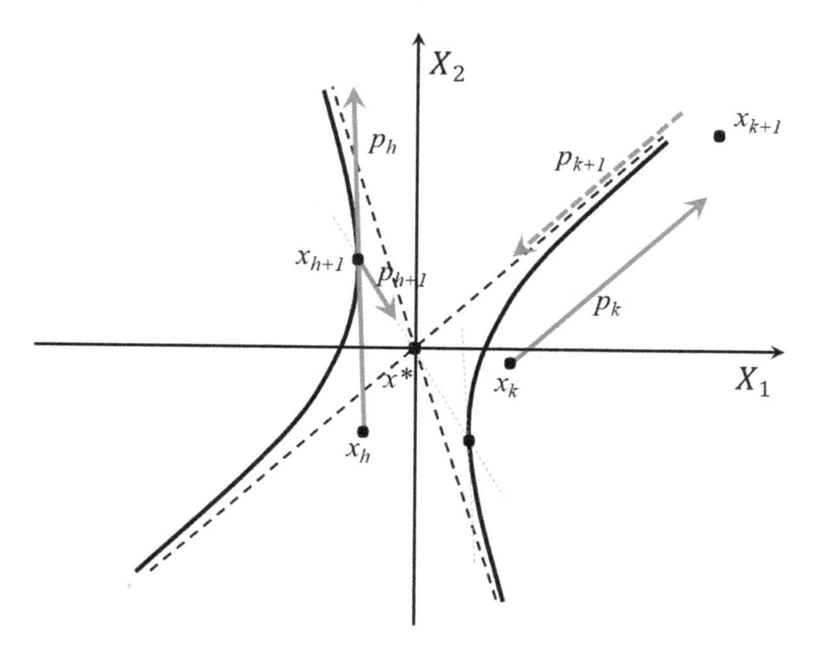

Fig. 2 The geometry of conjugate directions with A indefinite nonsingular in (1): since $|p_h^T A p_h|$ is sufficiently bounded away from zero then p_h and p_{h+1} are conjugate and *do not* tend to become parallel. Conversely, when $p_k^T A p_k \approx 0$ then p_k and p_{k+1} tend to become parallel

In particular, a remarkable aspect of our approach is that using grossone to cope with CG degeneracy does not require to alter the scheme in Table 1, which is therefore almost faithfully applied '*as is*'. This represents an undoubted advantage with respect to the CG-based methods (namely *planar methods*) in [14–16, 25, 31], that indeed need to suitably rearrange the CG iteration in order to dodge degeneracy. The consequence of introducing grossone in Table 1 is analyzed in the next Sect. 3.3, where in case of CG degeneracy at Step k, the expressions of the coefficients and vectors at Step k explicitly depend on ① and its powers.

3.3 Grossone for the Degenerate Step k of the CG

From Table 1, the position (15) and the properties of the CG, when A is indefinite and *the Step k is degenerate* we have

$$r_{k+1} = r_k - \theta_k A p_k = r_k - \frac{\|r_k\|^2}{s①} A p_k, \tag{16}$$

so that after a few arrangements (see [12])

$$p_{k+1} = r_{k+1} + \beta_k p_k = -\beta_{k-1} p_{k-1} - \frac{\|r_k\|^2}{s①} A p_k + \frac{\|r_k\|^2 \|A p_k\|^2}{s^2 ①^2} p_k. \tag{17}$$

We highlight that the CG degeneracy implies $s①$ to be infinitesimal, so that $\| p_{k+1} \|$ tends to be unbounded. The last result matches the geometric perspective reported in Sect. 3.1. Then, from (17) and the orthogonality/conjugacy conditions among vectors generated by the CG we can also infer

$$r_i^T r_j = 0, \qquad p_i^T A p_j = 0, \qquad \forall i \neq j,$$

along with

$$p_{k+1}^T A p_{k+1} = \frac{\|r_k\|^4}{s^2 ①^2} (A p_k)^T A (A p_k) - \frac{\|r_k\|^4 \|A p_k\|^4}{s^3 ①^3} + O(①), \tag{18}$$

(we recall that $O(①)$ in (18) sums up powers of ① equal to $+1$ and 0), and

$$r_{k+1}^T p_{k+1} = -\|r_k\|^2 + \frac{\|r_k\|^4 \|A p_k\|^2}{s^2 ①^2}. \tag{19}$$

From (16), (18)–(19) and recalling that when $p_k^T A p_k$ is infinitesimal so does $s①$, we obtain after some computation

$$r_{k+2} = r_k - \frac{\|r_k\|^2}{\|A p_k\|^2} A (A p_k) - \beta_{k-1} \frac{s①}{\|A p_k\|^2} A p_{k-1} + O(①^{-1}). \tag{20}$$

A noteworthy consequence of (16)–(17) and (20) is that in practice

- $r_1, \ldots, r_k$ are *independent* of ①,
- r_{k+1} and p_{k+1} heavily *depend* on ①,
- r_{k+2} is *independent* of negative powers of $s①$.

Thus, the geometric drawback detailed in Sect. 3.1, i.e. the CG degeneracy in the indefinite case, can be bypassed by exploiting the simple grossone algebra and neglecting the (infinitesimal) term with $s①$ in (20). This leaves the steps in the CG scheme of Table 1 fully unchanged.

Similarly, as regards the computation of the search direction p_{k+2}, by (16), (17) and (20) we have after some arrangements

$$p_{k+2} = r_k - \frac{\|r_k\|^2}{\|Ap_k\|^2} A(Ap_k) + \frac{\|r_{k+2}\|^2}{\|r_k\|^2} p_k - \beta_{k-1} \frac{s①}{\|Ap_k\|^2} Ap_{k-1} + O(①^{-1}). \tag{21}$$

Thus, similarly to r_{k+2}, in case of CG degeneracy also p_{k+2} is independent of negative powers of ① (i.e. equivalently $\|p_{k+2}\| < +\infty$, so that grossone algebra is able to bypass the degeneracy, recovering the CG iteration). After some computations it is also not difficult to verify that the vectors r_{k+1}, p_{k+1}, r_{k+2}, p_{k+2} in (16), (17), (20) and (21) satisfy the standard CG properties

$$\begin{cases} r_{k+2}^T r_i = 0, & i = 1, \dots, k+1, \\ p_{k+2}^T A p_i = 0, & i = 1, \dots, k+1. \end{cases} \tag{22}$$

An additional remarkable comment from (21) is that, neglecting the terms which contain powers of $s①$ larger or equal to 1 (i.e. neglecting infinitesimals in (21)), the vector p_{k+2} coincides with the one obtained in Algorithm CG_Plan of [14] (a similar result holds considering the algorithm by Luenberger in [31], too). Therefore, the use of grossone to cope with a CG degeneracy at Step k does not simply recover the theory and the results in [14, 31], but *it also retrieves the same scaling of the generated search directions*, which is a so relevant issue for large scale problems.
Also note that the expression of p_{k+1} in (17) explicitly includes negative powers of $s①$, showing that to large extent it can be assimilated to a vector with an unbounded norm, in accordance with Fig. 2, where x_{k+1} is a point at infinity.

Now, let us compute the iterate x_{k+2}, to verify to which extent using grossone may recover the CG iteration in case of degeneracy at Step k. By Table 1 and [12], and using

$$x_{k+2} = x_k + \theta_k p_k + \theta_{k+1} p_{k+1}$$

we obtain the final expression

$$x_{k+2} = x_k + \frac{\|r_k\|^2}{\|Ap_k\|^2} Ap_k - \frac{\|r_k\|^2}{\|Ap_k\|^4} (Ap_k)^T A(Ap_k) p_k + O(①^{-1}), \tag{23}$$

which perfectly matches the expression of x_{k+2} computed in [14, 31], as long as $O(①^{-1})$ is neglected. Therefore, in case of CG degeneracy at Step k, using grossone does not simply recover the residuals and search directions as in (22), but it also recovers the iterate x_{k+2}, since it is *independent of grossone*. Table 3 gives a formal description of CG$_①$, i.e. the CG method where ① is

Table 3 The $CG_{①}$ algorithm for solving the symmetric indefinite linear system $Ax = b$. In a practical implementation of Step k of $CG_{①}$, the test $p_k^T A p_k \neq 0$ may be replaced by the inequality $|p_k^T A p_k| \geq \varepsilon_k \|p_k\|^2$, with $\varepsilon_k > 0$ small

$CG_{①}$: the CG method coupled with grossone

Data: Set $k = 1$, $x_1 \in \mathbf{R}^n$, $r_1 = b - Ax_1$, $s = O(①^{-2})$.
If $\|r_1\| = 0$, then STOP. Else, set $p_1 = r_1$.

Step k: If $\|p_k\|$ is finite (bounded) and $p_k^T A p_k \neq 0$ then compute
$\alpha_k = r_k^T p_k / p_k^T A p_k$, $x_{k+1} = x_k + \alpha_k p_k$, $r_{k+1} = r_k - \alpha_k A p_k$.
If $\|r_{k+1}\| = 0$, then STOP.
Elseif $\|p_k\|$ is finite (bounded) set $p_k^T A p_k = s①$ and compute
$r_{k+1} = r_k - \|r_k\|^2/(s①) A p_k$.
Else compute $\alpha_k = r_k^T p_k / p_k^T A p_k$,
$x_{k+1} = x_k + \alpha_k p_k$, $r_{k+1} = r_k - \alpha_k A p_k$.
If the finite part of r_{k+1} satisfies $\|r_{k+1}\| = 0$, then STOP.
Endif
Set $\beta_k = -r_{k+1}^T A p_k / p_k^T A p_k = \|r_{k+1}\|^2 / \|r_k\|^2$, and
$p_{k+1} = r_{k+1} + \beta_k p_k$, $k = k + 1$.
Go to **Step** k.

introduced in case of degeneracy, while Proposition 1 summarizes the results in the current section.

Proposition 1 *Let be given the indefinite linear system $Ax = b$, with $A \in \mathbf{R}^{n \times n}$ and n large. Suppose the $CG_{①}$ method in Table 3 is applied for its solution, using the position* (15). *In case at Step $k \leq n$ the quantity $|p_k^T A p_k|$ is bounded away from zero, then the vectors generated by $CG_{①}$ exactly preserve the same properties of the corresponding vectors generated by the CG method in Table 1. Conversely, in case at Step k we have $p_k^T A p_k \approx 0$, then the vectors x_{k+2} in* (23) *and p_{k+2} in* (21) *computed by the $CG_{①}$ differ by infinitesimals from the corresponding vectors computed by the CG_Plan in [14] (or by the algorithm in [31]).*

4 Large Scale (unconstrained) Optimization Problems: The Need of Negative Curvatures

The solution of indefinite linear systems like (1) is almost ubiquitous in both constrained and unconstrained optimization frameworks. E.g. the iterative solution of the following problem

$$\min_{x \in \mathbb{R}^n} f(x), \tag{24}$$

where f is twice continuously differentiable and n is large, requires in a unified framework *(i)* to solve the associated Newton's equation (first order methods)

$$\nabla^2 f(x_j)\, s = -\nabla f(x_j), \tag{25}$$

and *(ii)* to identify those minima among the stationary points (second order methods – see also [6]). The task *(ii)* is often accomplished by selecting promising *negative curvature directions* for the function f at the current iterate x_j. In particular, for the sake of clarity here we restrict our attention to the Truncated Newton methods, that represent an efficient class of iterative methods to solve (24). Among them, the second order methods often rely on the theory in the seminal papers [33, 34], in order to assess algorithms generating negative curvature directions and converging to solutions where the Hessian matrix is positive semidefinite.

4.1 A Theoretical Path to the Assessment of Negative Curvature Directions

On the guidelines of the previous section, and with reference to [34], a sequence $\{d_j\}$ of effective negative curvature directions can be generated in accordance with the following assumption.

Assumption 1 Let us consider the optimization problem (24), with $f \in C^2(\mathbf{R}^n)$; the nonascent directions in the sequence $\{d_j\}$ are bounded and satisfy (see also [30])

(a) $\nabla f(x_j)^T d_j \leq 0, \;\; d_j^T \nabla^2 f(x_j) d_j \leq 0,$

(b) if $\lim_{j \to \infty} d_j^T \nabla^2 f(x_j) d_j = 0$ then $\lim_{j \to \infty} \min\left\{0,\ \lambda_m\left[\nabla^2 f(x_j)\right]\right\} = 0.$ □

In practice, (a) in Assumption 1 claims that at x_j the nonascent vector d_j cannot be a positive curvature direction for f. Conversely, condition (b) prevents the asymptotic convergence of the iterative algorithm to a region of concavity for the objective function. Evidently, on convex problems eventually the solutions of (24) both fulfill Newton's equation and satisfy second order stationarity conditions, without requiring the computation of negative curvature directions.

We remark that even in case the current iterate x_j is far from a stationary point, the use of the negative curvature direction d_j may considerably enhance efficiency in Truncated Newton methods. The latter fact was clearly evidenced in [19, 24, 30], and follows by considering at x_j the quadratic expansion along the vector d

$$q_j(d) = f(x_j) + \nabla f(x_j)^T d + \frac{1}{2} d^T \nabla^2 f(x_j) d,$$

which implies for the directional derivative of $q_j(d)$

$$\nabla q_j(d)^T d = \nabla f(x_j)^T d + d^T \nabla^2 f(x_j) d.$$

Thus, $\nabla q_j(d)^T d$ may strongly decrease both when d is of descent for f at x_j and is a negative curvature direction for f at x_j.

On large scale problems, we highlight that computing effective negative curvature directions for f at x_j, fulfilling Assumption 1, is a challenging issue, but it may become an easier task as long as *proper factorizations* of $\nabla^2 f(x_j)$ are available. Indeed, suppose at x_j the nonsingular matrices $M_j \in \mathbf{R}^{n \times k}$ and $C_j, Q_j, B_j \in \mathbf{R}^{k \times k}$ are available such that

$$M_j^T \nabla^2 f(x_j) M_j = C_j, \qquad C_j = Q_j B_j Q_j^T. \tag{26}$$

Then, for $y \in \mathbf{R}^k$ and assuming that $w \in \mathbf{R}^k$ is an eigenvector of B_j associated with the negative eigenvalue $\lambda < 0$, relations (26) yield

$$\begin{aligned}(M_j y)^T \nabla^2 f(x_j)(M_j y) = y^T \left[M_j^T \nabla^2 f(x_j) M_j \right] y \; &= \; y^T C_j y \\ = (Q_j^T y)^T B_j (Q_j^T y) \; = \; w^T B_j w \; &= \; \lambda \|w\|^2 \; < \; 0,\end{aligned}$$

so that $d_j = M_j y$ represents a negative curvature direction for f at x_j. Furthermore, if λ is the *smallest* negative eigenvalue of B_j, then $M_j y$ also represents an eigenvector of $\nabla^2 f(x_j)$ associated to it.

The most renowned Krylov-subspace methods for symmetric indefinite linear systems (i.e. SYMMLQ, SYMMBK, CG, Planar-CG methods [15–17]) are all able to provide the factorizations (26) (i.e. they fulfill item *(a)* in Assumption 1) when applied to Newton's equation at x_j. Nevertheless, fulfilling also *(b)* and the *boundedness* of the sequence of negative curvature directions is definitely a less trivial task (see e.g. the counterexample in Sect. 4 of [34]).

In this regard, we highlight that, under mild assumptions, by the use of grossone (namely $CG_{①}$ in Table 3) we can easily yield an implicit Hessian matrix factorization as in (26), fulfilling both *(a)* and *(b)*, as well as the boundedness of the negative curvature directions $\{d_j\}$ in Assumption 1. To accomplish this last task we need the next result (the proof follows from Lemma 4.3 in [34] and Theorem 3.2 in [20]).

Lemma 1 *Let problem* (24) *be given with* $f \in C^2(\mathbf{R}^n)$*, and consider any iterative method for solving* (24)*, which generates the sequence* $\{x_j\}$*. Let the level set* $\mathcal{L}_0 = \{x \in \mathbf{R}^n \ : \ f(x) \le f(x_0)\}$ *be compact, being any limit point* $\bar{x}$ *of* $\{x_j\}$ *a stationary point for* (24)*, with* $\left|\lambda[\nabla^2 f(\bar{x})]\right| > \bar{\lambda} > 0$*. Suppose* n *iterations of a Newton-Krylov method are performed to solve Newton's equation* (25) *at iterate* x_j*, so that the decompositions*

$$R_j^T \nabla^2 f(x_j) R_j = T_j, \qquad T_j = L_j B_j L_j^T \tag{27}$$

are available. Moreover, suppose $R_j \in \mathbf{R}^{n \times n}$ *is orthogonal,* $T_j \in \mathbf{R}^{n \times n}$ *has the same eigenvalues of* $\nabla^2 f(x_j)$*, with at least one negative eigenvalue, and* $L_j, B_j \in \mathbf{R}^{n \times n}$ *are nonsingular. Let* $\bar{z}$ *be the unit eigenvector corresponding to the smallest eigenvalue of* B_j*, and let* $\bar{y} \in \mathbf{R}^n$ *be the (bounded) solution of the linear system* $L_j^T y = \bar{z}$*. Then, the vector* $d_j = R_j \bar{y}$ *is bounded and satisfies Assumption 1.*

However, three main insidious drawbacks arise from Lemma 1:

- both computing the eigenvector $\bar{z}$ of B_j and solving the linear system $L_j^T y = \bar{z}$ might not be considered *easy tasks*;
- the vector $\bar{y}$ should be provably *bounded* (equivalently $|\det(L_j)|$ should be bounded away from zero, for any $j \ge 1$);
- at iterate x_j the Newton-Krylov method adopted to solve (25) possibly does not perform exactly n iterations.

4.2 CG$_{①}$ for the Computation of Negative Curvature Directions

Though the third issue raised in the end of the previous section remains of great theoretical interest, in practice when the sequence $\{x_j\}$ is approaching a stationary point, then we typically observe that Newton-Krylov methods tend to perform a large number of iterations to solve (25). On the contrary, the first two issues in the end of the previous section may be definitely more challenging, since they can be tackled only by a few Krylov-subspace methods, due to the structure and the complexity of the generated matrices L_j and B_j in Lemma 1. In our approach (see also [11] for more complete justifications), we can provably show that the use of CG$_{①}$ can fruitfully fulfill all the hypotheses of Lemma 1. In particular, in the current section we detail how to couple the Krylov-subspace method in [14] with $CG_{①}$, in the light of complying with the hypotheses of Lemma 1. Broadly speaking, observe that the Krylov-subspace method in [14] is basically a CG-based algorithm which performs exactly the CG iterations, as long as no degeneracy is experienced. On the contrary, in case at Step k the breakdown condition $p_k^T A p_k \approx 0$ occurs, a so called *planar step* is carried on to equivalently replace the Steps k and $k+1$ of the CG. Hence the taxonomy of *planar method.*

Assume without loss of generality that the Krylov-subspace method in [14] has performed n steps. Moreover, for the sake of simplicity hereafter in this section we drop the dependency of matrices on the iterate subscript j. After some computation the following matrices are generated by the method in [14] (see also [13])

$$L = \begin{pmatrix} L_{11} & 0 & 0 \\ L_{21} & L_{22} & 0 \\ 0 & L_{32} & L_{33} \end{pmatrix}, \qquad B = \begin{pmatrix} B_{11} & 0 & 0 \\ 0 & B_{22} & 0 \\ 0 & 0 & B_{33} \end{pmatrix},$$

where

$$L_{11} = \begin{pmatrix} 1 & & \\ -\sqrt{\beta_1} & \ddots & \\ & \ddots & 1 \end{pmatrix}, \; L_{21} = \begin{pmatrix} 0 \cdots & -\sqrt{\beta_{k-1}} \\ 0 \cdots & 0 \end{pmatrix}, \; L_{22} = \begin{pmatrix} 1 & 0 \\ 0 & 1 \end{pmatrix}, \tag{28}$$

$$L_{32} = \begin{pmatrix} -\sqrt{\beta_k \beta_{k+1}} & 0 \\ \vdots & \vdots \\ 0 & 0 \end{pmatrix}, \; L_{33} = \begin{pmatrix} 1 & & & \\ -\sqrt{\beta_{k+2}} & \ddots & & \\ & \ddots & 1 & \\ & & -\sqrt{\beta_{n-1}} & 1 \end{pmatrix}, \tag{29}$$

and

$$B_{11} = \begin{pmatrix} \frac{1}{\alpha_1} & & 0 \\ & \ddots & \\ 0 & & \frac{1}{\alpha_{k-1}} \end{pmatrix}, \; B_{22} = \begin{pmatrix} 0 & \sqrt{\beta_k} \\ \sqrt{\beta_k} & e_{k+1} \end{pmatrix}, \; B_{33} = \begin{pmatrix} \frac{1}{\alpha_{k+2}} & & 0 \\ & \ddots & \\ 0 & & \frac{1}{\alpha_n} \end{pmatrix}, \tag{30}$$

such that

$$AR = RT, \qquad T = LBL^T, \tag{31}$$

being the matrix $R \in \mathbf{R}^{n \times n}$ orthogonal with

$$R = \left(\frac{r_1}{\|r_1\|} \cdots \frac{r_n}{\|r_n\|} \right), \tag{32}$$

and $r_{k+1} = Ap_k$, while $T \in \mathbf{R}^{n \times n}$ is tridiagonal. Moreover, the quantities $\{\alpha_i\}$, $\{\beta_i\}$, e_{k+1} in (28)–(30) are suitable scalars (being in particular $e_{k+1} = (Ap_k)^T A(Ap_k)/\|Ap_k\|^2$). We also recall that $\beta_i > 0$, for any $i \geq 1$.

Furthermore, to simplify our analysis, the above matrices L and B are obtained applying the method in [14], assuming that it performed all CG steps, with the exception of only one planar iteration (namely the k-th iteration), corresponding to have indeed $p_k^T Ap_k \approx 0$. Then, our approach ultimately consists to introduce the numeral grossone, to exploit a suitable matrix factorization in place of (31), such that Lemma 1 is fulfilled. To this purpose, let us consider again the algorithm $CG_{①}$ in Table 3 (see also [12]), and assume that at Steps k and $k+1$ it generated the coefficients α_k and α_{k+1}. Thus, we have[1]

$$\begin{cases} \dfrac{1}{\alpha_k} = \dfrac{s①}{\|r_k\|^2} \\[2ex] \dfrac{1}{\alpha_{k+1}} = -\dfrac{\|Ap_k\|^2}{s①}. \end{cases} \tag{33}$$

Moreover, using the equivalence in Table 4 between the quantities computed by the algorithm in [14] and $CG_{①}$, we can compute the matrices

$$\hat{L} = \begin{pmatrix} L_{11} & 0 & 0 \\ L_{21} & V_k C_k^{-1} & 0 \\ 0 & \hat{L}_{32} & L_{33} \end{pmatrix}, \qquad \hat{D} = \begin{pmatrix} B_{11} & 0 & 0 \\ 0 & \hat{B}_{22} & 0 \\ 0 & 0 & B_{33} \end{pmatrix}, \tag{34}$$

where L_{11}, L_{21}, are defined in (28), L_{33} in (29), B_{11}, B_{33} in (30), and

$$\hat{L}_{32} = \begin{pmatrix} \left(-\sqrt{\beta_k \beta_{k+1}} \quad 0\right) \cdot V_k C_k^{-1} \\ \vdots \\ 0 \end{pmatrix}, \qquad \hat{B}_{22} = \begin{pmatrix} \dfrac{1}{\alpha_k s①} & 0 \\ 0 & \dfrac{s①}{\alpha_{k+1}} \end{pmatrix},$$

[1] More correctly, we urge to remark that the expressions (33) are obtained neglecting in the quantity α_{k+1} the infinitesimal terms, i.e. those terms containing negative powers of $s①$, that are indeed negligibly small due to the degenerate Step k in $CG_{①}$.

Table 4 Correspondence between quantities/vectors computed by the algorithm in [14] (*left*) and the algorithm CG$_{①}$ in [12] (*right*)

Algorithm in [14]		CG$_{①}$ in [12]
$r_i, \quad i = 1, \ldots, k$		$r_i, \quad i = 1, \ldots, k$
r_{k+1}		Ap_k
$r_i, \quad i \geq k+2$		$r_i, \quad i \geq k+2$ (neglecting the terms with $s①$)
$p_i, \quad i = 1, \ldots, k$		$p_i, \quad i = 1, \ldots, k$
p_{k+1}		$\frac{Ap_k}{\|Ap_k\|}$
$p_i, \quad i \geq k+2$		$p_i, \quad i \geq k+2$ (neglecting the terms with $s①$)
$\alpha_i, \quad i = 1, \ldots, k$		$\alpha_i, \quad i = 1, \ldots, k$
$\alpha_i, \quad i \geq k+2$		$\alpha_i, \quad i \geq k+2$ (neglecting the terms with $s①$)
$\beta_i, \quad i = 1, \ldots, k-1$		$\beta_i, \quad i = 1, \ldots, k-1$
β_k		$\frac{\|Ap_k\|^2}{\|r_k\|^2}$
$\beta_i, \quad i \geq k+1$		$\beta_i, \quad i \geq k+1$ (neglecting the terms with $s①$)

with

$$V_k C_k^{-1} = \begin{pmatrix} \frac{\|r_k\|\sqrt{\beta_k \lambda_k}}{\sqrt{\beta_k + \lambda_k^2}} & \frac{\sqrt{-\beta_k \lambda_{k+1}}}{\|Ap_k\|\sqrt{\beta_k + \lambda_{k+1}^2}} \\ \frac{\|r_k\|\lambda_k\sqrt{\lambda_k}}{\sqrt{\beta_k + \lambda_k^2}} & \frac{\lambda_{k+1}\sqrt{-\lambda_{k+1}}}{\|Ap_k\|\sqrt{\beta_k + \lambda_{k+1}^2}} \end{pmatrix}$$

and $\lambda_k,\ \lambda_{k+1}$ are the eigenvalues of B_{22} in (30). Thus, in Lemma 1 we have for matrix T_j the novel expression (see also (31))

$$T_j = LBL^T = \hat{L}\hat{D}\hat{L}^T.$$

We are now ready to compute at iterate x_j the negative curvature direction d_j which complies with Assumption 1, exploiting the decomposition $T_j =$

$\hat{L}\hat{D}\hat{L}^T$ from Lemma 1. The next proposition, whose proof can be found in [11], summarizes the last result.

Proposition 2 *Suppose n iterations of $CG_{①}$ algorithm are performed to solve Newton's equation* (25)*, at iterate x_j, so that the decompositions*

$$R^T \nabla^2 f(x_j) R = T, \qquad T = \hat{L}\hat{D}\hat{L}^T$$

exist, where R is defined in (32)*, and $\hat{L}$ along with $\hat{D}$ are defined in* (34)*. Let $\hat{z}$ be the unit eigenvector corresponding to the (negative) smallest eigenvalue of $\hat{D}$, and let $\hat{y}$ be the solution of the linear system $\hat{L}^T y = \hat{z}$. Then, the vector $d_j = R\hat{y}$ is bounded and satisfies Assumption 1. In addition, the computation of d_j requires the storage of at most two n-real vectors.*

Observe that the computation of the negative curvature direction d_j requires at most the additional storage of a couple of vectors, with respect to the mere computation of a solution for Newton's equation at x_j. This confirms the competitiveness with respect to the storage required in [20]. Thus, the approach in this paper does not only prove to be applicable to large-scale problems, but it also simplifies the theory in [20]. We remark that the theory in [20] is, to our knowledge, the only proposal in the literature of *iterative computation* of negative curvature directions for large-scale problems, such that

- it does not rely on any re-computation of quantities (as in [24]),
- it does not require any full matrix factorization,
- it does not need any matrix storage.

5 Conclusions

We propose an unconventional approach for a twofold purpose, within large-scale nonconvex optimization frameworks. On one hand we consider the efficient solution of symmetric linear systems. On the other hand, our proposal is also able to generate negative curvature directions for the objective function, allowing convergence towards stationary points satisfying second order necessary optimality conditions. Our idea exploits the simplicity of the algebra associated with the numeral grossone [37], which was recently introduced in the literature.

The theory in this paper also guarantees that the iterative computation of negative curvatures does not need any matrix storage, while preserving convergence. In addition, the proposed approach is independent under multiplication of the function by a positive scaling constant or adding a shifting constant. This is an important property that is specially exploited in global optimization frameworks (see e.g. [44, 45]), where *strongly homogeneous* algorithms are definitely appealing.

Acknowledgements The author is thankful to the Editors of the present volume for their great efforts and constant commitment. The author is also grateful for the support he received by both the National Research Council–Marine Technology Research Institute (CNR-INM), and the National Research Group GNCS (*Gruppo Nazionale per il Calcolo Scientifico*) within INδAM, Istituto Nazionale di Alta Matematica, Italy.

References

1. Antoniotti, L., Caldarola, F., Maiolo, M.: Infinite numerical computing applied to Hilbert's, Peano's, and Moore's curves. Mediterr. J. Math. **17**(3) (2020)
2. Astorino, A., Fuduli, A.: Spherical separation with infinitely far center. Soft Comput. **24**, 17751–17759 (2020)
3. Chandra, R.: Conjugate gradient methods for partial differential equations. Ph.D. thesis, Yale University, New Haven (1978)
4. Cococcioni, M., Fiaschi, L.: The Big-M method with the numerical infinite M. Optim. Lett. **15**(7) (2021)
5. Cococcioni, M., Pappalardo, M., Sergeyev, Y.D.: Lexicographic multi-objective linear programming using grossone methodology: theory and algorithm. Appl. Math. Comput. **318**, 298–311 (2018)
6. Curtis, F., Robinson, D.: Exploiting negative curvature in deterministic and stochastic optimization. Math. Program. **176**, 69–94 (1919)
7. D'Alotto, L.: Infinite games on finite graphs using grossone. Soft Comput. **55**, 143–158 (2020)
8. De Cosmis, S., De Leone, R.: The use of grossone in mathematical programming and operations research. Appl. Math. Comput. **218**(16), 8029–8038 (2012)
9. De Leone, R.: Nonlinear programming and grossone: quadratic programming and the role of constraint qualifications. Appl. Math. Comput. **318**, 290–297 (2018)
10. De Leone, R., Egidi, N., Fatone, L.: The use of grossone in elastic net regularization and sparse support vector machines. Soft Comput. **24**, 17669–17677 (2020)
11. De Leone, R., Fasano, G., Roma, M., Sergeyev, Y.D.: Iterative grossone-based computation of negative curvature directions in large-scale optimization. J. Optim. Theory Appl. **186**(2), 554–589 (2020)
12. De Leone, R., Fasano, G., Sergeyev, Y.D.: Planar methods and grossone for the conjugate gradient breakdown in nonlinear programming. Comput. Optim. Appl. **71**(1), 73–93 (2018)
13. Fasano, G.: Planar-CG methods and matrix tridiagonalization in large scale unconstrained optimization. In: Di Pillo, G., Murli, A. (eds.) In: High Performance Algorithms and Software for Nonlinear Optimization. Kluwer Academic Publishers, New York (2003)
14. Fasano, G.: Conjugate Gradient (CG)-type method for the solution of Newton's equation within optimization frameworks. Optim. Methods Softw. **19**(3–4), 267–290 (2004)

15. Fasano, G.: Planar-Conjugate gradient algorithm for large scale unconstrained optimization, part 1: theory. J. Optim. Theory Appl. **125**(3), 523–541 (2005)
16. Fasano, G.: Planar-Conjugate gradient algorithm for large scale unconstrained optimization, part 2: application. J. Optim. Theory Appl. **125**(3), 543–558 (2005)
17. Fasano, G.: Lanczos conjugate-gradient method and pseudoinverse computation on indefinite and singular systems. J. Optim. Theory Appl. **132**(2), 267–285 (2007)
18. Fasano, G.: A framework of conjugate direction methods for symmetric linear systems in optimization. J. Optim. Theory Appl. **164**(3), 883–914 (2015)
19. Fasano, G., Lucidi, S.: A nonmonotone truncated Newton-Krylov method exploiting negative curvature directions, for large scale unconstrained optimization. Optim. Lett. **3**(4), 521–535 (2009)
20. Fasano, G., Roma, M.: Iterative computation of negative curvature directions in large scale optimization. Comput. Optim. Appl. **38**(1), 81–104 (2007)
21. Fiaschi, L., Cococcioni, M.: Numerical asymptotic results in game theory using Sergeyev's Infinity Computing. Int. J. Unconv. Comput. **14**(1) (2018)
22. Fletcher, R.: Conjugate gradient methods for indefinite systems. In: Watson G.A. (ed.), Proceedings of the Dundee Biennal Conferences on Numerical Analysis. Springer, Berlin Heidelberg New York (1975)
23. Gaudioso, M., Giallombardo, G., Mukhametzhanov, M.S.: Numerical infinitesimals in a variable metric method for convex nonsmooth optimization. Appl. Math. Comput. **318**, 312–320 (2018)
24. Gould, N., Lucidi, S., Roma, M., Toint, P.: Exploiting negative curvature directions in linesearch methods for unconstrained optimization. Optim. Methods Softw. **14**, 75–98 (2000)
25. Hestenes, M.: Conjugate Direction Methods in Optimization. Springer, New York, Heidelberg, Berlin (1980)
26. Hestenes, M., Stiefel, E.: Methods of conjugate gradients for solving linear systems. J. Res. Nat. Bur. Stand. **49**, 409–436 (1952)
27. Higham, N.: Accuracy and Stability of Numerical Algorithms. SIAM, Philadelphia (1996)
28. Iavernaro, F., Mazzia, F., Mukhametzhanov, M.S., Sergeyev, Y.D.: Computation of higher order Lie derivatives on the Infinity Computer. J. Comput. Appl. Math. **383** (2021)
29. Lanczos, C.: An iterative method for the solution of the eigenvalue problem of linear differential and integral operators. J. Res. Nat. Bureau Stand. **45**(4), Research Paper 2133 (1950)
30. Lucidi, S., Rochetich, F., Roma, M.: Curvilinear stabilization techniques for Truncated Newton methods in large scale unconstrained optimization. SIAM J. Optim. **8**(4), 916–939 (1999)
31. Luenberger, D.G.: Hyperbolic Pairs in the method of conjugate gradients. SIAM J. Appl. Math. **17**, 1263–1267 (1996)
32. Mazzia, F., Sergeyev, Y.D., Iavernaro, F., Amodio, P., Mukhametzhanov, M.S.: Numerical methods for solving ODEs on the Infinity Computer. In: Sergeyev, Y.D., Kvasov, D.E., Dell'Accio, F., Mukhametzhanov, M.S. (eds.) Proceedings of the 2nd International Conferences "Numerical Computations: Theory and Algorithms", vol. 1776, p. 090033. AIP Publishing, New York (2016)
33. McCormick, G.: A modification of Armijo's step-size rule for negative curvature. Math. Program. **13**(1), 111–115 (1977)
34. Moré, J., Sorensen, D.: On the use of directions of negative curvature in a modified Newton method. Math. Program. **16**, 1–20 (1979)

35. Paige, C., Saunders, M.: Solution of sparse indefinite systems of linear equations. SIAM J. Numer. Anal. **12**, 617–629 (1975)
36. Pepelyshev, A., Zhigljavsky, A.: Discrete uniform and binomial distributions with infinite support. Soft Comput. **24**, 17517–17524 (2020)
37. Sergeyev, Y.D.: Arithmetic of Infinity. Edizioni Orizzonti Meridionali, CS, 2nd ed. (2013)
38. Sergeyev, Y.D.: Lagrange Lecture: methodology of numerical computations with infinities and infinitesimals. Rendiconti del Seminario Matematico dell'Università e del Politecnico di Torino **68**(2), 95–113 (2010)
39. Sergeyev, Y.D.: Higher order numerical differentiation on the Infinity Computer. Optim. Lett. **5**(4), 575–585 (2011)
40. Sergeyev, Y.D.: Computations with grossone-based infinities. In: Calude, C.S., Dinneen, M.J. (eds.), Unconventional Computation and Natural Computation: Proceedings of the 14th International Conference UCNC 2015, LNCS, vol. 9252 , pp. 89–106. Springer, New York (2015)
41. Sergeyev, Y.D.: Un semplice modo per trattare le grandezze infinite ed infinitesime. Matematica nella Società e nella Cultura: Rivista della Unione Matematica Italiana **8**(1), 111–147 (2015)
42. Sergeyev, Y.D.: Numerical infinities and infinitesimals: methodology, applications, and repercussions on two Hilbert problems. EMS Surv. Math. Sci. **4**(2), 219–320 (2017)
43. Sergeyev, Y.D.: Independence of the grossone-based infinity methodology from non-standard analysis and comments upon logical fallacies in some texts asserting the opposite. Found. Sci. **24**(1) (2019)
44. Sergeyev, Y.D., Kvasov, D.E., Mukhametzhanov, M.S.: On strong homogeneity of a class of global optimization algorithms working with infinite and infinitesimal scales. Commun. Nonlinear Sci. Numer. Simul. **59**, 319–330 (2018)
45. Žilinskas, A.: On strong homogeneity of two global optimization algorithms based on statistical models of multimodal objective functions. Appl. Math. Comput. **218**(16), 8131–8136 (2012)

Multi-objective Lexicographic Mixed-Integer Linear Programming: An Infinity Computer Approach

Marco Cococcioni, Alessandro Cudazzo, Massimo Pappalardo, and Yaroslav D. Sergeyev

Abstract In this chapter we show how a lexicographic multi-objective linear programming problem (LMOLP) can be transformed into an equivalent, single-objective one, by using the Grossone Methodology. Then we provide a simplex-like algorithm, called GrossSimplex, able to solve the original LMOLP problem using a single run of the algorithm (its theoretical correctness is also provided). In the second part, we tackle a Mixed-Integer Lexicographic Multi-Objective Linear Programming problem (LMOMILP) and we solve it in an exact way, by using a Grossone-version of the Branch-and-Bound scheme (called GrossBB). After proving the theoretical correctness

M. Cococcioni (✉)
Dipartimento di Ingegneria dell'Informazione, (University of Pisa), Largo Lucio Lazzarino, 1, 56122 Pisa, Italy
e-mail: marco.cococcioni@unipi.it

A. Cudazzo · M. Pappalardo
Dipartimento di Informatica, (University of Pisa), Largo B. Pontecorvo, 3, 56127 Pisa, Italy
e-mail: a.cudazzo1@studenti.unipi.it

M. Pappalardo
e-mail: massimo.pappalardo@unipi.it

Y. D. Sergeyev
Dipartimento di Ingegneria Informatica, Modellistica, Elettronica e Sistemistica, (University of Calabria), via P. Bucci, Cubo 41-C, 87036 Rende, Italy
e-mail: yaro@dimes.unical.it

Lobachevsky State University, Nizhny Novgorod, Russia

Y. D. Sergeyev and R. De Leone (eds.), *Numerical Infinities and Infinitesimals in Optimization*, Emergence, Complexity and Computation 43,
https://doi.org/10.1007/978-3-030-93642-6_5

of the associated pruning rules and terminating conditions, we show a few experimental results, run on an Infinity Computer simulator.

1 Introduction

In this contribution we survey recent achievements in the field of lexicographic linear programming by providing a coherent mathematical framework for the main results obtained in [2, 6].

Lexicographic multi-objective optimization problem consists of finding the solution that optimizes the first (most important) objective and, only if there are multiple equally-optimal solutions, find the one that optimizes the second most important objectives, and so on.

Lexicographic Multi-Objective Linear Programming (LMOLP) consists of the solution of an optimization problem having multiple linear objective function, sorted in order of priority (*lexicographic property*), over a region defined by linear constraints. If the search region is closed, bounded and not empty, then at least one solution exists. Sometimes the solution is also unique, despite the fact the problem is multi-objective. This is due to the lexicographic property. Indeed, while a Pareto Multi-Objective LP problem in general admits multiple Pareto optimal solutions (non-dominated solutions), the lexicographic attribute tends to reduce the number of solutions to a single optimal one.

In this book chapter we present a technique, based on Grossone [26], to transform an LMOLP problem into a single objective one (this technique is known as *scalarization*). Then we present a generalization of the simplex algorithm, that we have called GrossSimplex algorithm, which is able to solve the Grossone-based and scalarized single-objective LP problem. After showing an illustrative LMOLP problem, solved by the GrossSimplex algorithm, we move to introduce the LMOMILP problem. Then we show how to solve it by using a Grossone-based extension of the well-known Branch-and-Bound method: we called this algorithm *GrossBB*. Finally we present some test problems having a known solution, and then we demonstrate that the GrossBB algorithm, coupled with the GrossSimplex, is able to correctly solve all the considered problems. Before continuing, it is important to remind that the Grossone Methodology is independent from non-standard analysis, as discussed in [27].

Concerning related works, it must be acknowledged that the first successful attempt to use Grossone Methodology within LP problems is due to [7]. The same author together with his co-authors has proposed other applications of

Grossone to optimization problems, in [8–10]. Concerning global optimization, Grossone has been used in [28], while regarding convex non-smooth optimization it has been used in [13]. As regard as evolutionary optimization, four works have already been presented [15–18]. Grossone Methodology has also been applied to Game Theory [4, 11, 12], infinite decision processes science [21–23], ordinary differential equations [1, 24, 29], among other fields.

The chapter is organized as follows. In next Section we introduce the LMOLP problem, while in Sect. 3 we show its transformation into a single objective problem using the Grossone Methodology. Within the same Section, we prove that the original formulation and the Grossone-based one are equivalent. Then in Sect. 4 we introduce the GrossSimplex algorithm (an extension to the simplex algorithm able to work with Grossone-based numbers) and we show an example of its execution. In Sect. 5 we introduce the LMOMILP problem, and its reformulation using Grossone, again proving their equivalence. In Sect. 6 we provide a Grossone-based extension to a classical Branch-and-Bound algorithm. Section 7 is devoted to the experimental results: it shows how the GrossBB algorithm (coupled with the GrossSimplex one) is able to solve three LMOMILP problems. Finally, Sect. 8 provides a few conclusions.

2 Lexicographic Multi-objective Linear Programming

Given the LMOLP problem:

$$\begin{aligned} \text{LexMax} \quad & \mathbf{c}^{1T}\mathbf{x},\ \mathbf{c}^{2T}\mathbf{x},\ \dots,\ \mathbf{c}^{rT}\mathbf{x} \\ \text{s.t.} \quad & \left\{\mathbf{x}\in\mathbb{R}^n : \mathbf{A}\mathbf{x}=\mathbf{b},\ \mathbf{x}\geqslant 0\right\} \end{aligned} \qquad \text{P1}$$

where $\mathbf{x}$ and $\mathbf{c}^i$, $i = 1, \dots, r$, are column vectors $\in\mathbb{R}^n$, $\mathbf{A}$ is a full-rank matrix $\in\mathbb{R}^{m\times n}$, $\mathbf{b}$ is a column vector $\in\mathbb{R}^m$. LexMax in P1 denotes *lexicographic maximum* and means that the first objective is much more important than the second, which is, on its turn, much more important than the third one, and so on. Sometimes in the literature this is denoted as $\mathbf{c}^{1T}\mathbf{x} \gg \mathbf{c}^{2T}\mathbf{x} \gg \dots \gg \mathbf{c}^{rT}\mathbf{x}$.

As in any LP problem, the domain of P1 is a polyhedron:

$$\mathcal{S} \equiv \left\{x\in\mathbb{R}^n : \mathbf{A}\mathbf{x}=\mathbf{b},\ \mathbf{x}\geqslant 0\right\}. \qquad (1)$$

Notice that the formulation of P1 makes no use of Gross-numbers or Gross-arrays involving ①, namely, it involves finite numbers only. Hereinafter we assume that $\mathcal{S}$ is bounded and non-empty. In the literature (see, e.g., [14, 20, 30, 31]), there exists two approaches for solving the problem P1: the *preemptive scheme* and the *nonpreemptive scheme*. They are described in the following two subsections.

2.1 The Preemptive Scheme

The preemptive scheme introduced in [30] is an iterative method that attacks P1 by solving a series of single-objective LP problems. It starts by considering the first objective function alone, i.e., by solving the following problem:

$$\begin{aligned} \text{Max} \quad & \mathbf{c}^{1T}\mathbf{x} \\ \text{s.t.} \quad & \left\{\mathbf{x} \in \mathbb{R}^n : \mathbf{A}\mathbf{x} = \mathbf{b},\ \mathbf{x} \geqslant 0\right\}. \end{aligned} \qquad \text{P2.1}$$

Since P2.1 is a canonical LP problem, it can be solved algorithmically using any standard method (e.g., the simplex algorithm, the interior point algorithm, etc.). Once they have been run, an optimal solution $\mathbf{x}^{*1}$ with the optimal value $\beta_1 = \mathbf{c}^{1T}\mathbf{x}^{*1}$ has been obtained. Then the preemptive scheme considers the second objective of P1 and solves another single-objective LP problem, where the domain has changed, due to the addition of the equality constraint $\mathbf{c}^{1T}\mathbf{x} = \beta_1$:

$$\begin{aligned} \text{Max} \quad & \mathbf{c}^{2T}\mathbf{x} \\ \text{s.t.} \quad & \left\{\mathbf{x} \in \mathbb{R}^n : \mathbf{A}\mathbf{x} = \mathbf{b},\ \mathbf{x} \geqslant 0,\ \mathbf{c}^{1T}\mathbf{x} = \beta_1\right\}. \end{aligned} \qquad \text{P2.2}$$

The same approach is reiterated, and the algorithm stops either when the last problem has been solved, i.e., after considering the last objective $\mathbf{c}^{rT}\mathbf{x}$ or when a unique solution has been found in the current solved LP problem. Clearly, this approach is time consuming.

2.2 The Nonpreemptive Scheme Based on Appropriate Finite Weights

It has been shown in [30] that there always exists a finite scalar $M \in \mathbb{R}$ such that the solution of the LMOLP problem P1 can be found by solving only one single-objective LP problem having the following form:

$$\begin{aligned} \text{Max} \quad & \bar{\mathbf{c}}^{T}\mathbf{x} \\ \text{s.t.} \quad & \left\{\mathbf{x} \in \mathbb{R}^n : \mathbf{A}\mathbf{x} = \mathbf{b},\ \mathbf{x} \geqslant 0\right\} \end{aligned} \qquad \text{P3}$$

where

$$\bar{\mathbf{c}} = \sum_{i=1}^{r} \mathbf{c}^i M^{-i+1}. \tag{2}$$

This is a powerful theoretical result. However, from the computational point of view, finding the value of M is not a trivial task. Finding an appropriate value of M and solving the resulting LP problem can be more time consuming than solving the original problem P1 following the preemptive approach. Indeed, the preemptive scheme requires solving r linear programming problems only in the worst case and, in addition, it does not require the computation of M.

In Sect. 3, we present a nonpreemptive approach based on infinitesimal weights (constructed by using Grossone integer powers), that overcomes the problem of computing M and still requires the solution of only one single-objective LP problem. Such an LP problem is, however, not a standard one and thus it will be called Gross-LP problem, to avoid any possible confusion with its standard formulation involving finite numbers only.

3 Grossone-Based Reformulation of the LMOLP Problem

In this Section we will introduce a nonpreemptive Grossone-based scheme for addressing LMOLP problems. In order to introduce such scheme we need the following definitions and notations. Hereinafter, an array made of Gross-scalars is called Gross-array. In particular, a vector made of Gross-scalars is called from here on Gross-vector. The definition of Gross-matrix is similar. By extension, Gross-vectors and Gross-matrices are called *purely finite* iff all of their entries are purely finite Gross-numbers (where we have defined a purely finite Gross-number as a Gross-number involving a single power of Grossone, and that power has zero as its exponent: examples are $3.56①^0$, 0.479, etc.). The used notation is summarized in Table 1.

Let us introduce now the nonpreemptive Grossone-based scheme following the lexicographic ①-based approach introduced in [5, 6, 25]. It should be stressed that it is supposed hereinafter that the original problem P1 has been stated using purely finite numbers only. This assumption is not restrictive from the practical point of view since all the LMOLP problems considered traditionally are of this kind. However, since we are now in the framework of ①-based numbers, this assumption should be explicitly stated.

Table 1 Notation used in this work

Font style	Example	Quantity
Italics lowercase	n	Purely finite real scalar
Boldface lowercase	$\mathbf{x}$	Purely finite real vector
Boldface uppercase	$\mathbf{A}$	Purely finite real matrix
Italics lowercase with tilde	$\tilde{c}$	Gross-scalar
Boldface lowercase with tilde	$\tilde{\mathbf{y}}$	Gross-vector

To state the problem P4, we reformulate P3 by making use of Gross-scalars and Gross-vectors and is defined as follows:

$$\begin{aligned} \text{Max} \quad & \tilde{\mathbf{c}}^T \mathbf{x} \\ \text{s.t.} \quad & \left\{ \mathbf{x} \in \mathbb{R}^n : \mathbf{A}\mathbf{x} = \mathbf{b}, \ \mathbf{x} \geqslant 0 \right\}, \end{aligned} \qquad \text{P4}$$

where $\tilde{\mathbf{c}}$ is a column Gross-vector having n Gross-scalar components:

$$\tilde{\mathbf{c}} = \sum_{i=1}^{r} \mathbf{c}^i ①^{-i+1} \tag{3}$$

and $\tilde{\mathbf{c}}^T \mathbf{x}$ is the Gross-scalar obtained by multiplying the Gross-vector $\tilde{\mathbf{c}}$ by purely finite vector $\mathbf{x}$

$$\tilde{\mathbf{c}}^T \mathbf{x} = (\mathbf{c}^{1T} \mathbf{x})①^0 + (\mathbf{c}^{2T} \mathbf{x})①^{-1} + ... + (\mathbf{c}^{rT} \mathbf{x})①^{-r+1}, \tag{4}$$

where (4) can be equivalently written in the extended form as:

$$\tilde{\mathbf{c}}^T \mathbf{x} = (c_1^1 x_1 + ... + c_n^1 x_n)①^0 + (c_1^2 x_1 + ... + c_n^2 x_n)①^{-1} + ... + (c_1^r x_1 + ... + c_n^r x_n)①^{-r+1}.$$

The fundamental difference between the definition of $\tilde{\mathbf{c}}$ in (3) with respect to the definition of $\bar{\mathbf{c}}$ in (2) is that $\tilde{\mathbf{c}}$ does not involve any unknown. More precisely, it does not require the specification of a real scalar value, like the value of M in P3. However, this advantage leads to the fact that standard algorithms (like the simplex algorithm or the interior point algorithm), traditionally used for solving P3, can no longer be used in this case since ①-based numbers are involved in the definition of the objective function. There will be shown in the next section that it is still possible to obtain optimality conditions and to introduce and implement a GrossSimplex algorithm (a generalization of the traditional simplex algorithm to the case of ①-based numbers) able to solve problem P4.

In the rest of this section we will prove that problems P1 and P4 are equivalent and then we will derive the optimality conditions. Before giving the equivalence theorem, the following three lemmas are required.

The first lemma states that all the optimal solutions of P4 are vertices or belong to the convex hull of vertices, where the set of all the optimal solutions for a generic problem P is denoted, from hereafter, by the symbol Ω(P).

Lemma 1 *Each* $\mathbf{x}^* \in \Omega(\text{P4})$ *is a vertex or lies on the convex hull of the optimal vertices.*

The proof of Lemma 1 can be found in the appendix.

The next lemma states that all the optimal solutions of P1 reach the same (Gross-scalar) objective value for problem P4.

Lemma 2 *For any* $\mathbf{x}^* \in \Omega(\text{P1})$ *it follows* $\tilde{\mathbf{c}}^T\mathbf{x}^* = \tilde{v}$ *(*$\tilde{v}$ being a constant Gross-scalar).

Again, the proof of Lemma 2 is given in the appendix.

The third lemma is the last step in preparing the proof of the equivalence between P1 and P4.

Lemma 3 *For all* $\mathbf{x}^* \in \Omega(\text{P1})$ *and any vertex* $\hat{\mathbf{x}}$ *of S such that* $\hat{\mathbf{x}} \notin \Omega(P1)$, *it follows*

$$\tilde{\mathbf{c}}^T\hat{\mathbf{x}} < \tilde{\mathbf{c}}^T\mathbf{x}^*.$$

As before, the proof is reported in the appendix.

We are now ready to prove that any solution to P1 is also a solution to P4, and vice-versa, as shown in next theorem.

Theorem 1 (Equivalence) $\mathbf{x}^* \in \Omega(\text{P1})$ *iff* $\mathbf{x}^* \in \Omega(P4)$.

Proof $\Rightarrow$ If $\mathbf{x}^*$ is optimal for P1, then $\tilde{\mathbf{c}}^T\mathbf{x}^* = \tilde{v}$ due to Lemma 2. Therefore, according to Lemmas 1 and 3, $\mathbf{x}^*$ is also optimal for P4.

$\Leftarrow$ If $\mathbf{x}^*$ is optimal for P4, then due to Lemma 1, it belongs to the convex hull of some vertices. But from Lemma 3 it follows that all the vertices of this kind are optimal solutions to problem P1.

4 The GrossSimplex Algorithm

The single-objective Gross-LP problem formulated in P4 using Grossone can be solved using the GrossSimplex algorithm here described, provided that the duality theory is extended to the case of Gross-scalars and Gross-vectors.

The dual problem of P4 is the following

$$\begin{aligned} &\text{Min} \quad \tilde{\mathbf{y}}^T \mathbf{b} \\ &\text{s.t.} \quad \left\{ \tilde{\mathbf{y}} : \mathbf{A}^T \tilde{\mathbf{y}} \geqslant \tilde{\mathbf{c}} \right\}, \end{aligned} \qquad \text{P5}$$

where $\tilde{\mathbf{y}} = [\tilde{y}_1, ..., \tilde{y}_m]^T$ is an m-dimensional column Gross-vector, $\mathbf{A}^T\tilde{\mathbf{y}}$ an n-dimensional column Gross-vector, and $\tilde{\mathbf{y}}^T\mathbf{b}$ a Gross-scalar.
The domain of P5 is the Gross-polyhedron $\tilde{\mathcal{D}}$ defined as:

$$\tilde{\mathcal{D}} \equiv \left\{ \tilde{\mathbf{y}} : \mathbf{A}^T \tilde{\mathbf{y}} \geqslant \tilde{\mathbf{c}} \right\}.$$

In the following we will assume that $\tilde{\mathcal{D}}$ is bounded and non-empty, as we have assumed for $\mathcal{S}$.

Since optimality conditions are the core of the simplex algorithm, we need to prove optimality conditions in our context.

Definition 1 (*Definition of optimality for the Gross-primal problem*) A point $\mathbf{x}^*$ is optimal for the problem P4 iff $\tilde{\mathbf{c}}^T\mathbf{x}^* \geqslant \tilde{\mathbf{c}}^T\mathbf{x} \quad \forall \mathbf{x} \in \mathcal{S}$.

Definition 2 (*Definition of optimality for the Gross-dual problem*) A point $\tilde{\mathbf{y}}^*$ is optimal for the problem P5 iff $\tilde{\mathbf{y}}^{*T}\mathbf{b} \leqslant \tilde{\mathbf{y}}^T\mathbf{b} \quad \forall \tilde{\mathbf{y}} \in \tilde{\mathcal{D}}$.

We need now the following lemma, taken from [6].

Lemma 4 (Weak duality) $\forall \mathbf{x} \in \mathcal{S}$ *and* $\forall\, \tilde{\mathbf{y}} \in \tilde{\mathcal{D}}$, *it follows that* $\tilde{\mathbf{y}}^T\mathbf{b} \geqslant \tilde{\mathbf{c}}^T\mathbf{x}$.

Proof Since $\mathbf{x} \in \mathcal{S}$, $\mathbf{A}\mathbf{x} = \mathbf{b}$. Pre-multiplying both by $\tilde{\mathbf{y}}^T$, we have

$$\tilde{\mathbf{y}}^T\mathbf{A}\mathbf{x} = \tilde{\mathbf{y}}^T\mathbf{b}. \tag{5}$$

Now, since $\tilde{\mathbf{y}} \in \tilde{\mathcal{D}}$, we know that $\tilde{\mathbf{y}}^T\mathbf{A} \geqslant \tilde{\mathbf{c}}^T$. By post-multiplying the latter by $\mathbf{x}$ (being $\mathbf{x} \geqslant 0$), we have:

$$\tilde{\mathbf{y}}^T\mathbf{A}\mathbf{x} \geqslant \tilde{\mathbf{c}}^T\mathbf{x}.$$

By combining it with (5) we obtain $\tilde{\mathbf{y}}^T\mathbf{b} \geqslant \tilde{\mathbf{c}}^T\mathbf{x}$.

Theorem 2 (Optimality condition) *If* $\mathbf{x}^* \in \mathcal{S}$, $\tilde{\mathbf{y}}^* \in \tilde{D}$ *and* $\tilde{\mathbf{c}}^T\mathbf{x}^* = \tilde{\mathbf{y}}^{*T}\mathbf{b}$, *then it follows that* $\mathbf{x}^* \in \Omega(\text{P1})$ *and* $\tilde{\mathbf{y}}^* \in \Omega(\text{P5})$.

Proof It is an immediate consequence of the weak duality Lemma 4.

Algorithm 1: The GrossSimplex algorithm

Step 0. The user has to provide the initial set B of basic indices.

Step 1. Solve the system $\mathbf{A}_{\mathrm{B}}^T\tilde{\mathbf{y}} = \tilde{\mathbf{c}}_{\mathrm{B}}$ (where $\mathbf{A}_{\mathrm{B}}^T$ is the sub-matrix obtained by $\mathbf{A}^T$ by considering the columns indexed by B). This can be easily calculated as $\tilde{\mathbf{y}} = \mathbf{A}_{\mathrm{B}}^{-T}\tilde{\mathbf{c}}_{\mathrm{B}}$ where $\mathbf{A}_{\mathrm{B}}^{-T}$ is an abbreviation for $(\mathbf{A}_{\mathrm{B}}^T)^{-1}$, $\tilde{\mathbf{y}}$ is a Gross-vector obtained by linearly combining the Gross-vector $\tilde{\mathbf{c}}_{\mathrm{B}}$ using the purely finite scalar elements in $\mathbf{A}_{\mathrm{B}}^{-T}$.

Step 2. Compute $\tilde{\mathbf{s}} = \tilde{\mathbf{c}}_{\mathrm{N}} - A_{\mathrm{N}}^T\tilde{\mathbf{y}}$ (where N is the complementary set of B) and then select the maximum (gradient rule). When this maximum is negative (being it finite or infinitesimal), it means that we are done (the current solution is optimal) and thus the algorithm STOPS. Otherwise, the position of the maximum in the Gross-vector $\tilde{\mathbf{s}}$ is the index k of the entering variable $\mathrm{N}(k)$.

Step 3. Solve the system $\mathbf{A}_{\mathrm{B}}\mathbf{d} = \mathbf{A}_{\mathrm{N}(k)}$.

Step 4. Find the largest t such that $\mathbf{x}_{\mathrm{B}}^* - t\mathbf{d} \geqslant 0$. If there is not such a t, then the problem is unbounded (STOP); otherwise, at least one component of $\mathbf{x}_{\mathrm{B}}^* - t\mathbf{d}$ equals zero and the corresponding variable is the leaving variable. In case of ties, use the Lexicographic Pivoting Rule [7] to break them.

Step 5. Update B and N and return to Step 1.

We are now ready to introduce the GrossSimplex algorithm (see Algorithm 1) which exploits the theoretical results presented above. Notice that the algorithm needs a first feasible basis B. It can be found by solving the standard auxiliary problem analogously to the case where the objective function is only one (no GrossSimplex required to solve it).

To illustrate the behaviour of the GrossSimplex algorithm we consider the following problem:

$$\begin{array}{rl} \text{LexMax} & 4x_1 + 2x_2, \;\; 2x_2, \;\; x_1 + x_2 \\ \text{s.t.} & 2x_1 \;\; + x_2 \;\; \leq 14 \\ & -x_1 + 2x_2 \leq 8 \\ & 2x_1 \;\;\; - x_2 \;\; \leq 10 \\ & x_1, x_2 \geq 0. \end{array}$$

The polygon $\mathcal{S}$ associated to this problem is shown in Fig. 1. It can be seen that the first objective vector $\mathbf{c}^1 = [4, 2]^T$ is orthogonal to segment $[\alpha, \beta]$ ($\alpha = (6, 2)$, $\beta = (4, 6)$) shown in the same figure. Thus all the points laying on this segment are optimal. Since the solution is not unique, there is the chance to try to improve the second objective vector ($\mathbf{c}^2 = [0, 2]^T$) without deteriorating the first one function. The point that maximizes the second objective is β, associated to solution $\mathbf{x}^* = [4, 6]^T$. Since now the solution is unique, the third objective function can not be taken into account. Thus the lexicographic optimal solution for the problem is $\mathbf{x}^* = [4, 6]^T$.

Before running the GrossSimplex algorithm we had to transform the problem into the following one, after converting the constraints into equality constraints by adding slack variables x_3, x_4, and x_5:

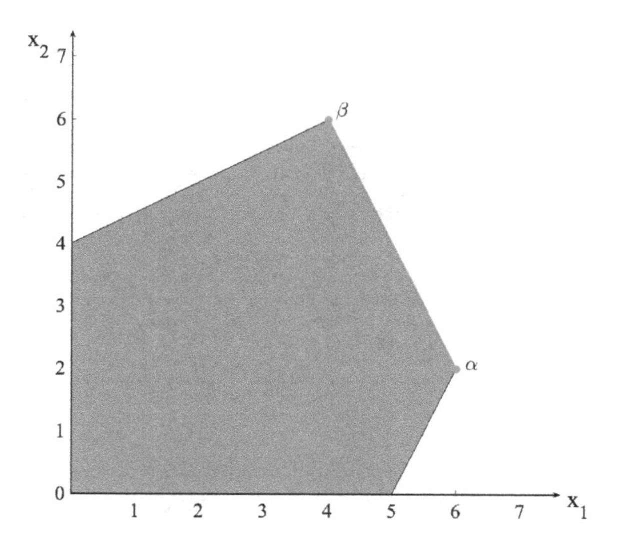

Fig. 1 An example in two dimensions with three objectives. All the points in the segment $[\alpha, \beta]$ are optimal for the first objective, while point β is the unique lexicographic optimum for the given problem ($\beta = (4, 6)$)

$$\begin{aligned}
\text{LexMax} \quad & 4x_1 + 2x_2, \quad 2x_2, \quad x_1 + x_2 \\
\text{s.t.} \quad & 2x_1 \quad + x_2 \quad + x_3 = 14 \\
& - x_1 + 2x_2 + x_4 = 8 \\
& 2x_1 \quad - x_2 \quad + x_5 = 10 \\
& x_i \geq 0 \quad i = 1, \ldots, 5.
\end{aligned} \qquad T_0$$

The GrossSimplex algorithm has been run on problem T_0, by using the initial basis $\mathrm{B} = \{3, 4, 5\}$ (N is therefore $\{1, 2\}$). The initial solution associated to the initial basis is: $\mathbf{x} = [0, 0, 14, 8, 10]^T$, which corresponds to the point (0, 0) in Fig. 1. Then the algorithm computes $\tilde{\mathbf{s}}$ as $\tilde{\mathbf{c}}_\mathrm{N} - A_\mathrm{N}^T \tilde{\mathbf{y}}$ giving

$$\tilde{\mathbf{s}} = \begin{bmatrix} 4①^0 + 0①^{-1} + 1①^{-2} \\ 2①^0 + 2①^{-1} + 1①^{-2} \end{bmatrix}.$$

Thus, according to the gradient rule for choosing the entering variable, the one in N having the first index is selected, i.e., x_1, which is the one associated to the maximum constraint violation: $4①^0 + 0①^{-1} + 1①^{-2}$. The leaving index, computed according to the lexicographic pivoting rule is the third, i.e., variable x_5. Thus the second base used is $\mathrm{B} = \{3, 4, 1\}$. The Gross-objective function is $\tilde{\mathbf{c}}^T \mathbf{x} = 20.00①^0 + 0.00①^{-1} + 5.00①^{-2}$ and, of course, coincides with that of the dual $\tilde{\mathbf{y}}^T \mathbf{b}$ (when $\tilde{\mathbf{y}}$ will be also feasible, we will be at the optimum and the algorithm will end).

The solution associated to this second base is $\mathbf{x} = [5, 0, 4, 13, 0]^T$, while the new value for $\tilde{\mathbf{s}}$ is

$$\tilde{\mathbf{s}} = \begin{bmatrix} -2①^0 + 0①^{-1} - 0.5①^{-2} \\ 4①^0 + 2①^{-1} + 1.5①^{-2} \end{bmatrix}.$$

Again, according to the gradient rule, the next entering index is the second, i.e., variable x_2. The leaving index is the first and thus the leaving variable is x_3. Therefore, the next basis (the third) is $\mathrm{B} = \{2, 4, 1\}$. The Gross-objective function is equal to $28.00①^0 + 4.00①^{-1} + 8.00①^{-2}$.

The solution associated to the third base is $\mathbf{x} = [6, 2, 0, 10, 0]^T$, while the new value for $\tilde{\mathbf{s}}$ is

$$\tilde{\mathbf{s}} = \begin{bmatrix} 0①^0 + 1①^{-1} + 0.25①^{-2} \\ -2①^0 - 1①^{-1} - 0.75①^{-2} \end{bmatrix}.$$

By following the same process, the next entering index is the first, i.e., variable x_5. It is interesting to note that in this case the constraint violation (a positive entry in $\tilde{\mathbf{s}}$) is not finite, as in previous case, but infinitesimal: $+1①^{-1} + 0.25①^{-2}$ The leaving index is the second and thus the leaving variable is x_4.

At this point, the Gross-objective function is now equal to $28.00①^0 + 12.00①^{-1} + 10.00①^{-2}$, the solution associated to the fourth base is $\mathbf{x} = [4, 6, 0, 0, 8]^T$, and the $\tilde{\mathbf{s}}$ is

$$\tilde{\mathbf{s}} = \begin{bmatrix} 0①^0 - 0.8①^{-1} - 0.2①^{-2} \\ -2①^0 - 0.4①^{-1} - 0.6①^{-2} \end{bmatrix}.$$

Since all the entries of $\tilde{\mathbf{s}}$ are negative (one is infinitesimal and one is finite), we are done. In fact, the solution found (once the slack variables are discarded), is the correct one, i.e., $\mathbf{x}^* = [4, 6]^T$. Furthermore, the Gross-objective function at the end equals to $\tilde{\mathbf{c}}^T\mathbf{x} = \tilde{\mathbf{y}}^T\mathbf{b} = 28.00①^0 + 12.00①^{-1} + 10.00①^{-2}$. This concludes the step-by-step illustration about how the GrossSimplex works in a concrete example.

As a final note, observe how in [3] a parameter-less way to obtain the initial basis is shown, based on the Big-M method with an infinitely big value for M. The idea of that method is to use a numerical, infinitely big value for M, set equal to ①.

In the next section we present the Lexicographic Multi-Objective Mixed-Integer Linear Programming problem and then, in the subsequent version, how to solve it using a Grossone-based extension of the Branch-and-Bound method.

5 Lexicographic Multi-objective Mixed-Integer Linear Programming

In this section we introduce the LMOMILP problem, which is formalized as:

$$\begin{aligned} \text{LexMin} \quad & \mathbf{c}^{1\,T}\mathbf{x},\ \mathbf{c}^{2\,T}\mathbf{x},\ \ldots,\ \mathbf{c}^{r\,T}\mathbf{x} \\ \text{s.t.} \quad & \mathbf{A}\mathbf{x} \leqslant \mathbf{b}, \\ & \mathbf{x} = \begin{bmatrix} \mathbf{p} \\ \mathbf{q} \end{bmatrix} \qquad \mathbf{p} \in \mathbb{Z}^k,\ \mathbf{q} \in \mathbb{R}^{n-k} \end{aligned} \qquad P$$

where $\mathbf{c}^i$, $i = 1, \ldots, r$, are column vectors$\in \mathbb{R}^n$, $\mathbf{x}$ is a column vector$\in \mathbb{R}^n$, $\mathbf{A}$ is a full-rank matrix$\in \mathbb{R}^{m \times n}$, $\mathbf{b}$ is a column vector$\in \mathbb{R}^m$. LexMin in P denotes the lexicographic minimum.

As in any MILP problem, from the problem P, we can define the polyhedron defined by the linear constraints alone:

$$\mathcal{S} \equiv \left\{ \mathbf{x} \in \mathbb{R}^n : \mathbf{A}\mathbf{x} \leqslant \mathbf{b} \right\}. \tag{6}$$

Thus we can define problem R, the *relaxation* of a lexicographic (mixed) integer linear problem, obtained from P by removing the integrality constraint on each variable:

$$\begin{aligned} \text{LexMin} \quad & \mathbf{c}^{1\,T}\mathbf{x},\ \mathbf{c}^{2\,T}\mathbf{x},\ \ldots,\ \mathbf{c}^{r\,T}\mathbf{x} \\ \text{s.t.} \quad & \mathbf{A}\mathbf{x} \leqslant \mathbf{b}. \end{aligned} \qquad R$$

Problem R is called LMOLP (Lexicographic Multi-Objective Linear Problem), and can be solved using the GrossSimplex algorithm shown in Sect. 4. In addition, every thing we have said above for the MILP problem is still valid for the LMOMILP.

Notice that the formulation of P makes no use of Gross-numbers or Gross-arrays involving ①, namely, it involves finite numbers only. Hereinafter we assume that $\mathcal{S}$ is bounded and non-empty. In the next we will reformulate the LMOMILP problem P using Grossone, and then we will prove their equivalence.

We present now an equivalent reformulation of the LMOMILP problem P, based on Grossone Methodology.

First of all, let us introduce the new problem $\tilde{P}$, formulated using Gross-numbers, following the same approach shown in Sect. 3:

$$\begin{aligned} \text{Min} \quad & \tilde{\mathbf{c}}^T\mathbf{x} \\ \text{s.t.} \quad & \mathbf{Ax} \leqslant \mathbf{b}, \\ & \mathbf{x} = \begin{bmatrix}\mathbf{p}\\ \mathbf{q}\end{bmatrix} \mathbf{p} \in \mathbb{Z}^k, \qquad \mathbf{q} \in \mathbb{R}^{n-k}, \end{aligned} \qquad \tilde{P}$$

where $\tilde{\mathbf{c}}$ is a column Gross-vector having n Gross-scalar components built using purely finite vectors $\mathbf{c}^i$

$$\tilde{\mathbf{c}} = \sum_{i=1}^{r} \mathbf{c}^i ①^{-i+1} \tag{7}$$

and $\tilde{\mathbf{c}^T}\mathbf{x}$ is the Gross-scalar obtained by multiplying the Gross-vector $\tilde{\mathbf{c}}$ by the purely finite vector $\mathbf{x}$:

$$\tilde{\mathbf{c}^T}\mathbf{x} = (\mathbf{c}^{1T}\mathbf{x})①^0 + (\mathbf{c}^{2T}\mathbf{x})①^{-1} + ... + (\mathbf{c}^{rT}\mathbf{x})①^{-r+1}. \tag{8}$$

Observe how Eq. (7) is identical to Eqs. (3), and (8) is the same as Eq. (4).

In Theorem 2 we will prove that problem $\tilde{P}$ is equivalent to problem P defined on Sect. 5.

What makes the new formulation $\tilde{P}$ attractive is the fact that its relaxed version (from the integrality constraint) is a Gross-LP problem, and therefore it can be effectively solved using *a single run* of the GrossSimplex algorithm introduced in Sect. 4. This means that the set of multiple objective functions is mapped into a single (Gross-) scalar function to be optimized. This opens the possibility to solve the integer-constrained variant of the problem using an adaptation of the BB algorithm (see Algorithm 2), coupled with the GrossSimplex. Of course the GrossSimplex will solve the relaxed version of $\tilde{P}$:

$$\begin{aligned} \text{Min} \quad & \tilde{\mathbf{c}}^T\mathbf{x} \\ \text{s.t.} \quad & \mathbf{Ax} \leqslant \mathbf{b}. \end{aligned} \qquad \tilde{R}$$

Theorem 3 (Equivalence of problem $\tilde{P}$and problem P) *Problem $\tilde{P}$ is equivalent to the problem P. Thus the solutions of the two are the same.*

Proof The basic observation is that the integer relaxation R of problem P is an LMOLP problem, while the integer relaxation of problem $\tilde{P}$, the $\tilde{R}$ problem defined above, is a Gross-LP problem. In Sect. 3 we have already proved the equivalence of problems R and $\tilde{R}$. Now, since problems P and $\tilde{P}$ have equivalent relaxations, the two will also share the same solutions when the same integrality constraints will be taken into account on both.

In the next subsections we will provide the pruning rules, terminating conditions and the branching rule, and then we will introduce the GrossBB algorithm, a generalization of the BB algorithm able to work with Gross-numbers.

6 A Grossone-Based Extension of the Branch-and-Bound Algorithm

In this Section we introduce a Grossone-based extension of the Branch-and-Bound algorithm, called GrossBB, to solve the Grossone-based formulation of any LMOMILP problem. As in any Branch-and-Bound algorithm, we have to provide the pruning rules (and the proof of their correctness), the termination conditions and the the branching rules.

6.1 Pruning Rules for the GrossBB

The pruning rules of a standard Branch-and-Bound algorithm (see [19]) can be adapted to the case of a Grossone reformulated version of any LMOMILP problem:

Theorem 4 (Pruning rules for the GrossBB) *Let $\mathbf{x}_{opt}$ be the best solution found so far for $\tilde{P}$, and let $\tilde{v}_S(\tilde{P}) = \tilde{\mathbf{c}}^T \mathbf{x}_{opt}$ be the current upper bound. Considering the current node $(\tilde{P}_c)$ and the associated problem $\tilde{P}_c$:*

1. *If the feasible region of problem $\tilde{P}_c$ is empty the sub-tree with root $(\tilde{P}_c)$ has no feasible solutions having values lower than $\tilde{\mathbf{c}}^T \mathbf{x}_{opt}$. So we can prune this node.*
2. *If $\tilde{v}_I(\tilde{P}_c) \geqslant \tilde{v}_S(\tilde{P})$, then we can prune at node $(\tilde{P}_c)$, since the sub-tree with root $(\tilde{P}_c)$ cannot have feasible solutions having a value lower than $\tilde{v}_S(\tilde{P})$.*
3. *If $\tilde{v}_I(\tilde{P}_c) < \tilde{v}_S(\tilde{P})$ and the optimal solution $\bar{\mathbf{x}}$ of the relaxed problem $\tilde{R}_c$ is feasible for $\tilde{P}$, then $\bar{\mathbf{x}}$ is a better candidate solution for $\tilde{P}$, and thus we can update $\mathbf{x}_{opt}$ ($\mathbf{x}_{opt} = \bar{\mathbf{x}}$) and the value of the upper bound ($\tilde{v}_S(\tilde{P}) = \tilde{v}_I(\tilde{P}_c)$). Finally, prune this node according to the second rule.*

The correctness of the above pruning rules taken from [2] and, for the convenience of the reader, is also provided below.

Proof (Pruning Rule 1) If the feasible region of the current problem $\tilde{R}_c$ is empty, then also the one of $\tilde{P}_c$ is too, since the domain of $\tilde{P}_c$ has additional constrains (the integrality constraints). Furthermore, all the domain of the problems in the leaves of the sub-tree having root in $\tilde{P}_c$ will be empty as well, since the domains of the leaves have all additional constrains with respect to $\tilde{R}_c$. This proves the correctness of the first pruning rule.

Now let us prove the second pruning rule.

Proof (Pruning Rule 2) Let us consider the generic leaf P_{leaf} of the sub-tree having root $\tilde{P}_c$. Let us indicate with $\tilde{v}(\tilde{P}_c)$ the optimal value of the current problem $\tilde{P}_c$ (the one with the integer constraints). Then the values at the leaves below $\tilde{P}_c$ must be greater than or equal to $\tilde{v}(\tilde{P}_c)$:

$$\tilde{v}(\tilde{P}_{leaf}) \geqslant \tilde{v}(\tilde{P}_c) \quad \forall \text{ leaf in SubTree}(\tilde{P}_c).$$

In fact, the domain of $\tilde{P}_c$ includes the ones of all the $\tilde{P}_{leaf}$, being each problem $\tilde{P}_{leaf}$ obtained by enriching $\tilde{P}_c$ with additional constraints. On the other hand,

$$\tilde{v}(\tilde{P}_c) \geqslant \tilde{v}_I(\tilde{P}_c)$$

since $\tilde{v}_I(\tilde{P}_c)$ is obtained as the optimal solution of $\tilde{R}_c$, a problem having a domain which includes the one of $\tilde{P}_c$. Thus the following chain of inequalities always old:

$$\tilde{v}(\tilde{P}_{leaf}) \geqslant \tilde{v}(\tilde{P}_c) \geqslant \tilde{v}_I(\tilde{P}_c) \quad \forall \text{ leaf in SubTree}(\tilde{P}_c).$$

Now, if $\tilde{v}_I(\tilde{P}_c) \geqslant \tilde{v}_S(\tilde{P})$, we can add an element to the chain:

$$\tilde{v}(\tilde{P}_{leaf}) \geqslant \tilde{v}(\tilde{P}_c) \geqslant \tilde{v}_I(\tilde{P}_c) \geqslant \tilde{v}_S(\tilde{P}) \quad \forall \text{ leaf in SubTree}(\tilde{P}_c)$$

from which we can conclude that

$$\tilde{v}(\tilde{P}_{leaf}) \geqslant \tilde{v}_S(\tilde{P}) \quad \forall \text{ leaf in SubTree}(\tilde{P}_c).$$

This means that all the leaves of the current node will contain solutions that are worse (or equivalent) than the current upper bound. Thus the sub-tree rooted in $\tilde{P}_c$ can to be pruned (i.e., not explicitly explored). This proves the correctness of the second pruning rule.

Before proving the pruning rule 3, let us observe how the pruning rule above prevents from solving multi-modal problems, in the sense that with such pruning rule we are only able to find a single optimum, not all the solutions that might have the same cost function. In other words, the proposed pruning rule does not allow to solve multi-modal problems, because we are deciding not to explore the sub-tree at a given node that could contain solutions having the same current optimal objective function value. To solve multi-modal problems, that rule must be applied only when

$$\tilde{v}_I(\tilde{P}_c) > \tilde{v}_S(\tilde{P}). \tag{9}$$

Proof (Pruning Rule 3)

If $\tilde{v}_I(\tilde{P}_c) < \tilde{v}_S(\tilde{P})$ and $\bar{\mathbf{x}}$ is feasible for $\tilde{P}$, (i.e., if all the components of $\bar{\mathbf{x}}$ that must be integer are actually ϵ-integer), we have found a better estimate for the upper bound of $\tilde{P}$, and thus we can update it:

$$\tilde{v}_S(\tilde{P}) = \tilde{v}_I(\tilde{P}_c).$$

As a result, now $\tilde{v}_I(\tilde{P}_c) = \tilde{v}_S(\tilde{P})$, and thus:

$$\tilde{v}(\tilde{P}_{leaf}) \geqslant \tilde{v}(\tilde{P}_c) \geqslant \tilde{v}_I(\tilde{P}_c) = \tilde{v}_S(\tilde{P}) \quad \forall\, leaf \text{ in SubTree}(\tilde{P}_c)$$

then again we have that the sub-tree having root $\tilde{P}_c$ cannot contain better solutions than $\bar{\mathbf{x}}$:

$$\tilde{v}(\tilde{P}_{leaf}) \geqslant \tilde{v}_S(\tilde{P}) \quad \forall \text{ leaf in SubTree}(\tilde{P}_c).$$

This proves the correctness of the third pruning rule.

6.2 Terminating Conditions, Branching Rules and the GrossBB Algorithm

Let us now discuss the terminating conditions for the GrossBB algorithm. The first two are exactly the same of the classical BB, while the third requires some attention.

The terminating conditions are:

1. **All the remaining leaves have been visited**: if all the leaves have been visited the GrossBB algorithm stops.
2. **Maximum number of iteration reached**: when a given maximum number of iterations (provided by the user at the beginning) has been reached, the GrossBB stops.

3. $\tilde{\epsilon}-$**optimality reached**: when the normalized difference between the global lower bound and the global upper bound at the i-th iteration is close enough to zero, we can stop:

$$\tilde{\Delta}^i(\tilde{P}) = \frac{\tilde{v}_S(\tilde{P}) - \tilde{v}_I(\tilde{P})}{|\tilde{v}_S(\tilde{P})|} \leqslant \tilde{\epsilon} \tag{10}$$

The last terminating condition deserves additional attention. It first involves the difference between two Gross-scalars. This intermediate result must be divided by the absolute value of $\tilde{v}_S(\tilde{P})$. While computing the absolute value of a Gross-scalar is straightforward, computing the division requires the algorithm described in [26]. The result of the Gross-division is a Gross-scalar that must be compared with the Gross-scalar $\tilde{\epsilon}$, which has the form:

$$\tilde{\epsilon} = \epsilon_0 + \epsilon_1 ①^{-1} + \epsilon_2 ①^{-2} + \ldots + \epsilon_{r-1} ①^{-r+1}.$$

Obviously, it is possible to chose $\epsilon_0 = \epsilon_1 = \epsilon_2 \ldots = \epsilon$, to simplify the presentation.

Let see the example below. Suppose to have three objectives (r=3) and to have selected $\epsilon = 10^{-6}$. Given the following $\tilde{\Delta}^i(\tilde{P})$

$$\tilde{\Delta}^i(\tilde{P}) = 1.1 \cdot 10^{-7} + 5 \cdot 10^{-3} ①^{-1} + 1.7 \cdot 10^{-8} ①^{-2}$$

it is clear that

$$\tilde{\Delta}^i(\tilde{P}) \nleqslant \tilde{\epsilon},$$

because its first-order infinitesimal component ($5 \cdot 10^{-3}$) is not less or equal to ϵ. Thus in this case the GrossBB algorithm cannot terminate: it will continue, trying to make *all* the components $\leqslant \epsilon$.

Concerning the branching rule for the GrossBB algorithm, it works as follows. When the sub-tree below $\tilde{P}_c$ cannot be pruned (because it could contain better solutions), its sub-tree must be explored. Thus we have to branch the current node into P_l and P_r and to add these two new nodes to the tail of the queue of the sub-problems to be analyzed and solved by the GrossSimplex.

We are now ready to provide the pseudo-code for the GrossBB (see Algorithm 2). The GrossBB algorithm, firstly introduced in [2], is able to solve a given $\tilde{P}$ LMOMILP problem, by internally using the GrossSimplex and the GrossBB rules provided above (pruning, terminating, and branching) .

This would allow us to create a division-free variant of the GrossBB algorithm. Anyway, we think that the use of the division is still interesting from the theoretical point of view, because it allowed us to clarify the concept of

Algorithm 2: The GrossSimplex-based GrossBB algorithm

1 Inputs: maxIter and a specific LMOMILP problem $|\tilde{P}|$, to be put within the root node $(\tilde{P})$

2 Outputs: $\mathbf{x}_{opt}$ (the optimal solution, a purely finite vector), $\tilde{f}_{opt}$ (the optimal value, a Gross-scalar)

Step 0. Insert $|\tilde{P}|$ into a queue of the sub problems that must be solved. Put $\tilde{v}_S(\tilde{P}) = ①$, $\mathbf{x}_{opt} = [\ ]$, and $\tilde{f}_{opt} = ①$ or use a greedy algorithm to get an initial feasible solution.

Step 1a. If all the remaining leaves have been visited (empty queue), or the maximum number of iterations has been reached, or the $\tilde{\epsilon}$-optimality condition holds, then goto Step 4. Otherwise extract from the head of the queue the next problem to solve and call it $\tilde{P}_c$ (*current problem*). Remark: this policy of insertion of new problems at the tail of the queue and the extraction from its head leads to a *breadth-first* visit for the binary tree of the generated problems.

Step 1b. Solve $\tilde{R}_c$, the relaxed version of the problem $\tilde{P}_c$ at hand, using the GrossSimplex and get $\bar{\mathbf{x}}$ and $\tilde{f}_c$ ($= \tilde{\mathbf{c}}^T \bar{\mathbf{x}}$):

$$[\bar{\mathbf{x}}, \tilde{f}_c, \texttt{emptyPolyhedron}] \leftarrow \texttt{GrossSimplex}(\tilde{R}_c).$$

Step 2a. If the LP solver has found that the polyhedron is empty, then prune the sub-tree of $(\tilde{P}_c)$ (according to Pruning Rule 1) by going to Step 1a (without branching $(\tilde{P}_c)$). Otherwise, we have found a new lower value for $\tilde{P}_c$:

$$\tilde{v}_I(\tilde{P}_c) = \tilde{f}_c.$$

Step 2b. If $\tilde{v}_I(\tilde{P}_c) \geqslant \tilde{v}_S(\tilde{P})$, then prune the sub-tree under $\tilde{P}_c$ (according to Pruning Rule 2), by going to Step 1a (without branching $\tilde{P}_c$).

Step 2c. If $\tilde{v}_I(\tilde{P}_c) < \tilde{v}_S(\tilde{P})$ and all components of $\bar{\mathbf{x}}$ that must be integer are actually ϵ-integer (i.e., $\bar{\mathbf{x}}$ is feasible), then we have found a better upper bound estimate. Thus we can update the value of $\tilde{v}_S(\tilde{P})$ as:

$$\tilde{v}_S(\tilde{P}) = \tilde{v}_I(\tilde{P}_c).$$

In addition we set $\mathbf{x}_{opt} = \bar{\mathbf{x}}$ and $\tilde{f}_{opt} = \tilde{v}_I(\tilde{P}_c)$. Then we also prune the sub-tree under $(\tilde{P}_c)$ (according to Pruning Rule 3) by going to Step 1a (without branching $(\tilde{P}_c)$).

Step 3. If $\tilde{v}_I(\tilde{P}_c) < \tilde{v}_S(\tilde{P})$ but *not* all components of $\bar{\mathbf{x}}$ that must be integer are actually ϵ-integer, we have to branch. Select the component $\bar{x}_t$ of $\bar{\mathbf{x}}$ having the greatest fractional part, among all the components that must be integer. Create two new nodes (i.e., problems) with a new constraint for this variable, one with a new $\leqslant$ constraint for the rounded down value of $\bar{x}_t$ and another with a new $\geqslant$ constraint for the rounded up value of $\bar{x}_t$. Let us call the two new problems $\tilde{P}_l$ and $\tilde{P}_r$ and put them at the tail of the queue of the problems to be solved, then goto Step 1a.

Step 4. End of the algorithm.

when a Gross-number is "near zero". Furthermore, the impact of the computation of the division of equation (10) on the overall computing time of the algorithm is negligible, being the overall computing time mainly affected by the time needed to compute the solutions of the relaxed LMOLP problems.

7 Experimental Results

In this section we introduce three LMOMILP test problems having known solution, then we verify that the GrossBB combined with the GrossSimplex is able to solve them. Additional examples can be found in [2].

7.1 *Test Problem 1: The "Kite" in 2D*

This problem, taken from [2], is a little variant to problem T_0 described above. The main difference is the addition of the integrality constraint. In addition, the second constraint $2x_1 + 3x_2 \leqslant 210$ has been changed into $2x_1 + 3x_2 \leqslant 210 + 2.5$). Finally observe how this formulation does not involve slack variables:

$$\begin{array}{lll} \text{LexMax} & 8x_1 + 12x_2,\ 14x_1 + 10x_2,\ x_1 + x_2 & \\ \text{s.t.} & 2x_1 + 1x_2 \leqslant 120 & \\ & 2x_1 + 3x_2 \leqslant 210 + 2.5 & \\ & 4x_1 + 3x_2 \leqslant 270 & T_1 \\ & x_1 + 2x_2 \geqslant 60 & \\ & -200 \leqslant x_1, x_2 \leqslant +200, \quad \mathbf{x} \in \mathbb{Z}^n. & \end{array}$$

The polygon $\mathcal{S}$ associated to problem T_1 is shown in Fig. 2 (left sub-figure). The integer points (feasible solutions) are shown as black spots, while in light grey we have provided the domain of the relaxed problem (without the integer constraints).
It can be seen that the first objective vector $\mathbf{c}^1 = [8,\ 12]^T$ is orthogonal to segment $[\alpha,\ \beta]$ ($\alpha = (0,\ 70.83)$, $\beta = (28.75,\ 51.67)$) shown in the same figure. All the nearest integer points parallel to this segment are optimal for the first objective (see the right sub-figure in Fig. 2). Since the solution is not unique, there is the chance to try to improve the second objective vector ($\mathbf{c}^2 = [14, 10]^T$).

Let see what happens when we solve this problem using the GrossSimplex-based GrossBB algorithm.

Since the T_1 problem is $LexMax$-formulated, we have to feed $-\tilde{\mathbf{c}}$ to the GrossSimplex-based GrossBB algorithm.

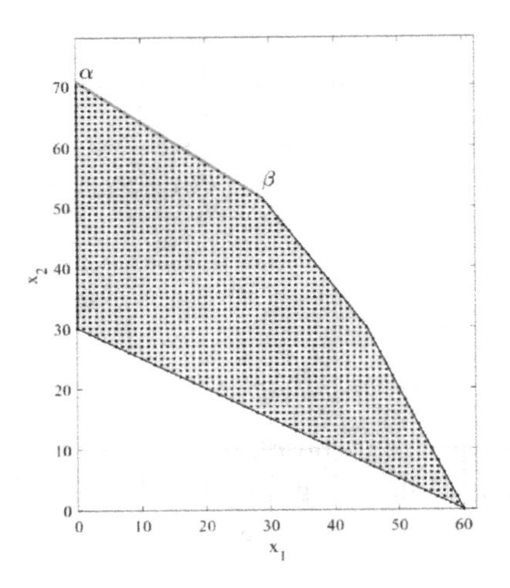

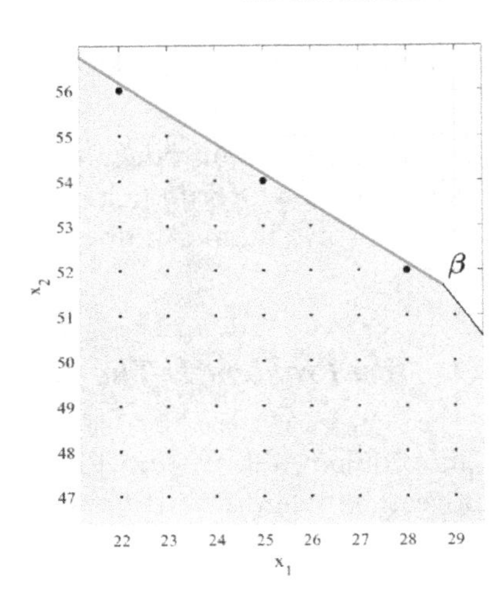

Fig. 2 An example in two dimensions with three objectives. The black points on the left figure are all the feasible solutions. All the nearest integer points parallel to the segment $[\alpha, \beta]$ (there are many), are optimal for the first objective, while point (28, 52) is the unique lexicographic optimum for the given problem (i.e., considering the second objective too). The third objective plays no role in this case. On the right, a zoom around point β is provided, with some optimal solutions for the first objective highlighted (the ones with a bigger black spot)

Initialize $\tilde{v_S}(\tilde{P}) = ①$, $\mathbf{x}_{opt} = [\ ]$, $\tilde{f}_{opt} = ①$ and insert T_1 into a queue of the sub-problems that must be solved.

Iteration 1 The GrossBB extracts from the queue of problems to be solved the only one present, and denotes it as the current problem: $\tilde{P}_c \equiv T_1$). Then the algorithm solves its relaxed version: the solution of $\tilde{R}_c$ is $\bar{\mathbf{x}} = [28.7500,\ 51.6667]^T$, with $\tilde{v_I}(\tilde{P}_c) = -850①^0 - 919.167①^{-1} - 80.4167①^{-2}$. In this case we have to branch using the component having the highest fractional part (among the variables with integer restrictions, of course). In this case, it is the first component, and thus the new sub-problem on the left $\tilde{P}_l$ will have the additional constraint $x_1 \leqslant 28$, while the new on the right $\tilde{P}_r$ will have the additional constraint $x_1 \geqslant 29$. This split makes the current solution $[28.7500,\ 51.6667]^T$ not optimal neither for problems $\tilde{P}_l$ nor for $\tilde{P}_r$ (see Fig. 3).

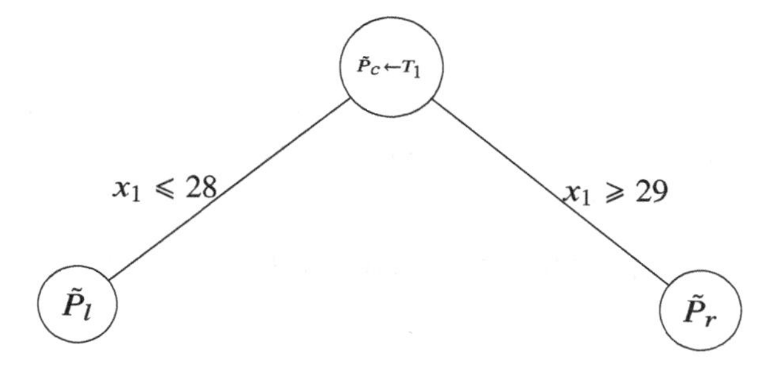

Fig. 3 Situation at the end of iteration 1, for problem T_1

Iteration 2 At this step the queue is composed by $[\tilde{P}_l, \tilde{P}_r]$, the problems generated in the previous iteration. The GrossBB extracts now the next problem from the top of the queue (*breadth-first* visit), namely $[\tilde{P}_l$ and denotes it as $\tilde{P}_c$, the current problem to solve. The optimal solution for the relaxed problem $\tilde{R}_c$ is $\bar{\mathbf{x}} = [28.25,\ 52]^T$, with $\tilde{v_I}(\tilde{P}_c) = -850①^0 - 915.5①^{-1} - 80.25①^{-2}$. We have to branch again, as we did on iteration 1, a thus a new left and right problems will be generated and added to the queue. The new left problem will have the additional constraint $x_1 \leqslant 28$, while the new right problem will have the additional constraint $x_1 \geqslant 29$. The length of the queue is now 3.

Iteration 3 Extract the next problem from the top of the queue and denote it as $\tilde{P}_c$. The optimal solution of $\tilde{R}_c$ is $\bar{\mathbf{x}} = [29.25,\ 51]^T$ and the associated $\tilde{v_I}(\tilde{P}_c) = -846①^0 - 919.5①^{-1} - 80.25①^{-2}$. We have to branch, a new left and right problem will be generated, both will be added to the queue. The left problem will have the additional constraint $x_1 \leqslant 29$, while the new on the right $\tilde{P}_r$ will have the additional constraint $x_1 \geqslant 30$. The length of the queue is now 4.

Iteration 4 Extract the next problem from the queue and indicate it as $\tilde{P}_c$. Solve $\tilde{R}_c$, the relaxation of $\tilde{P}_c$, using the GrossSimplex. Since $\tilde{R}_c$ has an empty feasible region, prune this node by applying the first pruning rule. The length of the queue is now 3.

Iteration 5 Extract the next problem from the top of the queue and denote it as $\tilde{P}_c$. The optimal solution of $\tilde{R}_c$ is $\bar{\mathbf{x}} = [28,\ 52.1667]^T$ and the associated $\tilde{v_I}(\tilde{P}_c) = -850①^0 - 913.6667①^{-1} - 80.1667①^{-2}$. We have to branch: a new left and right problems will be generated and added to the queue. The left problem will have the additional constraint $x_1 \leqslant 52$, while the right one will have the additional constraint $x_1 \geqslant 53$. The length of the queue is now

4.

Iteration 6 Extract the next problem from the queue and indicate it as $\tilde{P}_c$. This time the GrossSimplex returns an integer solution, i.e., a feasible solution for the LMOMILP initial problem T_1:

$$\bar{\mathbf{x}} = [30,\ 50]^T \quad \text{and} \quad \tilde{v_I}(\tilde{P}_c) = -840①^0 - 920①^{-1} - 80①^{-2}.$$

Since $\tilde{v}_I(\tilde{P}_c) < \tilde{v}_S(\tilde{P})$, then we can update $\tilde{v}_S(\tilde{P}) = \tilde{v}_I(\tilde{P}_c), \mathbf{x}_{opt} = \bar{\mathbf{x}}$. Finally we can prune this node, according to the third pruning rule.

Iteration 7 Extract the next problem from the queue and indicate it as $\tilde{P}_c$. Again the GrossSimplex returns an integer solution:

$$\bar{\mathbf{x}} = [29,\ 51]^T \quad \text{and} \quad \tilde{v_I}(\tilde{P}_c) = -844①^0 - 916①^{-1} - 80①^{-2}.$$

Since $\tilde{v}_I(\tilde{P}_c) < \tilde{v}_S(\tilde{P})$, then we can update $\tilde{v}_S(\tilde{P}) = \tilde{v}_I(\tilde{P}_c), \mathbf{x}_{opt} = \bar{\mathbf{x}}$. Finally we can prune this node, according to the third pruning rule.

Iteration 8 Extract the next problem from the top of the queue and indicate it as $\tilde{P}_c$. The optimal solution of $\tilde{R}_c$ is $\bar{\mathbf{x}} = [26.75,\ 53]^T$, with $\tilde{v_I}(\tilde{P}_c) = -850①^0 - 904.5①^{-1} - 79.75①^{-2}$. We have to branch: a new left and right problems will be generated and added to the queue. The left problem will have the additional constraint $x_1 \leqslant 26$, while the new on the right $\tilde{P}_r$ will have the additional constraint $x_1 \geqslant 27$. The length of the queue is now 4.

Iteration 9 Extract the next problem from the queue and designate it as the current problem $\tilde{P}_c$. Solve its relaxation, using the GrossSimplex. In this case the returned solution is feasible for the initial LMOMILP problem T_1, because it has all integral components:

$$\bar{\mathbf{x}} = [28,\ 52]^T \quad \text{and} \quad \tilde{v_I}(\tilde{P}_c) = -848①^0 - 912①^{-1} - 80①^{-2}.$$

Since $\tilde{v}_I(\tilde{P}_c) < \tilde{v}_S(\tilde{P})$, then update both $\tilde{v}_S(\tilde{P}) = \tilde{v}_I(\tilde{P}_c)$ and $\mathbf{x}_{opt} = \bar{\mathbf{x}}$. Finally prune this node by applying the third pruning rule.

Iteration 10 Extract the next problem from the queue and indicate it as $\tilde{P}_c$. Solve $\tilde{R}_c$, the relaxation of $\tilde{P}_c$, using the GrossSimplex. Since $\tilde{R}_c$ has an empty feasible region, prune this node by applying the first pruning rule.

Iterations 11–79 The GrossBB algorithm is not able to find a better solution than the $\bar{\mathbf{x}} = [28,\ 52]^T$ already found, but continues to branch and explore the tree, until only two nodes remain in the queue. The processing of the last two nodes is discussed in the last two iterations 80 and 81, below.

Iteration 80 Extract the next problem from the queue and indicate it as $\tilde{P}_c$. Solve $\tilde{R}_c$ using the GrossSimplex. Since $\tilde{R}_c$ has an empty feasible region, prune this node by applying the first pruning rule.

Iteration 81 At this point there is one last unsolved problem from the queue. Extract this problem and indicate it as $\tilde{P}_c$. The optimal solution of $\tilde{R}_c$ is:

$$\bar{\mathbf{x}} = [1,\ 70]^T, \quad \text{with} \quad \tilde{v_I}(\tilde{P}_c) = -848①^0 - 714①^{-1} - 71①^{-2}.$$

Since $\tilde{v_I}(\tilde{P}_c) \geq \tilde{v}_S(\tilde{P})$, prune this last node according to the third pruning rule. Being now the tree empty, the GrossBB algorithm stops according to the first terminating condition and returns the optimal solution found so far: $\mathbf{x}_{opt} = [28,\ 52]^T$.

The optimal value of the objective function is $\tilde{\mathbf{c}^T} x_{opt} = 848①^0 + 912①^{-1} + 80①^{-2}$.

Table 2 provides a synthesis with the most interesting iterations performed by the GrossBB.

7.2 *Test Problem 2: The Unrotated "House" in 3D*

This illustrative example is taken from [2] and is in three dimensions with three objectives:

$$\begin{aligned}
\text{LexMax} \quad & x_1, -x_2, -x_3 \\
\text{s.t.} \quad & -10.2 \leqslant x_1 \leqslant 10.2 \\
& -10 \leqslant x_2 \leqslant 10.2 \\
& -10.2 \leqslant x_3 \leqslant 10.2 \\
& -x_1 - x_2 \leqslant 2 \\
& -x_1 + x_2 \leqslant 2 \\
& -20 \leqslant x_i \leqslant 20,\ i = 1, ..., 3, \quad \mathbf{x} \in \mathbb{Z}^3,
\end{aligned} \qquad T_2$$

Table 2 Iterations performed by GrossSimplex-based GrossBB algorithm on problem T_1

Iteration	Result at node (iteration)
Initialize	– $\tilde{v_S}(\tilde{P}) = ①$ – Queue len. 1 (add the root problem to the queue)
1	$\tilde{v_I}(\tilde{P}_c)$: $-850①^0 - 919.167①^{-1} - 80.4167①^{-2}$. Queue length : 0 – No pruning rules applied, branch $\tilde{P}_c$ in two sub-problems. Queue length: 2 – $\tilde{\Delta} = 100①^0 + 100①^{-1} + 100①^{-2}$
2	$\tilde{v_I}(\tilde{P}_c)$: $-850①^0 - 915.5①^{-1} - 80.25①^{-2}$. Queue length: 1 – No pruning rules applied, branch $\tilde{P}_c$ in two sub-problems. Queue length: 3 – $\tilde{\Delta} = 100①^0 + 100①^{-1} + 100①^{-2}$
3	$\tilde{v_I}(\tilde{P}_c)$: $-846①^0 - 919.5①^{-1} - 80.25①^{-2}$. Queue length: 2 – No pruning rules applied, branch $\tilde{P}_c$ in two sub-problems. Queue length: 4 – $\tilde{\Delta} = 100①^0 + 100①^{-1} + 100①^{-2}$
4	Prune node: rule 1, empty feasible region. Queue length: 3
5	$\tilde{v_I}(\tilde{P}_c)$: $-850①^0 - 913.667①^{-1} - 80.1667①^{-2}$. Queue length: 2 – No pruning rules applied, branch $\tilde{P}_c$ in two sub-problems. Queue length: 4 – $\tilde{\Delta} = 100①^0 + 100①^{-1} + 100①^{-2}$
6	$\tilde{v_I}(\tilde{P}_c)$: $-840①^0 - 920①^{-1} - 80①^{-2}$. Queue length: 3 – A feasible solution has been found: $\mathbf{x}_{opt} = [30,\ 50]^T$ – Update $\tilde{v_S}(\tilde{P}) = \tilde{v_I}(\tilde{P}_c)$, prune node: rule 3 – $\tilde{\Delta} = 0.0119048①^0 - 0.00688406①^{-1} + 0.00208333①^{-2}$
7	$\tilde{v_I}(\tilde{P}_c)$: $-844①^0 - 916①^{-1} - 80①^{-2}$. Queue length: 2 – A feasible solution has been found: $\mathbf{x}_{opt} = [29,\ 51]^T$ – Update $\tilde{v_S}(\tilde{P}) = \tilde{v_I}(\tilde{P}_c)$, prune node: rule 3 – $\tilde{\Delta} = 0.354191①^0 - 0.127528①^{-1} + 0.104058①^{-2}$
8	$\tilde{v_I}(\tilde{P}_c)$: $-850①^0 - 904.5①^{-1} - 79.75①^{-2}$. Queue length: 1 – No pruning rules applied, branch $\tilde{P}_c$ in two sub-problems. Queue length: 3 – $\tilde{\Delta} = 0.007109①^0 - 0.00254731①^{-1} + 0.00208333①^{-2}$
9	$\tilde{v_I}(\tilde{P}_c)$: $-848①^0 - 912①^{-1} - 80①^{-2}$. Queue length: 2 – A feasible solution has been found: $\mathbf{x}_{opt} = [28,\ 52]^T$ – Update $\tilde{v_S}(\tilde{P}) = \tilde{v_I}(\tilde{P}_c)$, prune node: rule 3 – $\tilde{\Delta} = 0.00235849①^0 - 0.00822368①^{-1} - 0.003125①^{-2}$
10 ...	Prune node: rule 1, empty feasible region. Queue length: 1 ... ...
80 81	Prune node: rule 1, empty feasible region. Queue length: 1 $\tilde{v_I}(\tilde{P}_c)$: $-848①^0 - 714①^{-1} - 71①^{-2}$. Queue length: 0 – $\tilde{v_I}(\tilde{P}_c) \geqslant \tilde{v_S}(\tilde{P})$ prune node: rule 2
Result	Iteration 81. Optimization ended. Optimal solution found: $\mathbf{x}_{opt} = [28,\ 52]^T$ $\tilde{f}_{opt} = -848①^0 - 912①^{-1} - 80①^{-2}$ $\tilde{\Delta} = 0①^0 + 0①^{-1} + 0①^{-2}$

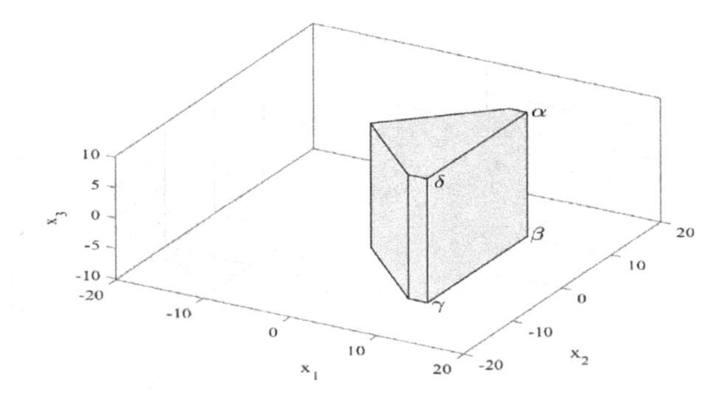

Fig. 4 A 3D unrotated "house" problem

with the domain being the cube shown in Fig. 4. It can be immediately seen that by considering the first objective alone (maximize x_1), all the nearest integer points parallel to square having vertices α, β, γ, δ (see Fig. 4) are optimal for the first objective function. Since the optimum is not unique, the second objective function can be considered in order to improve it without deteriorating the first objective. Then, all the integer points near to the segment $[\beta, \gamma]$ are all optimal for the second objective too (see Fig. 5, which provides the plant-view of Fig. 4 with $x_3 = -10$).

Again, the optimum in not unique, and thus we can consider the third objective, which allows us to select the nearest integer point to γ as the unique solution that maximizes all the three objectives. In particular, point $[10, -10, -10]$ is the lexicographic optimum to the given problem.

The problem can be solved with GrossBB algorithm, as shown in Table 3. The solution $\mathbf{x}_{opt} = [10, -10, -10]^T$ is actually found after 5 iterations. The optimal value of the objective function can be computed as $\tilde{\mathbf{c}}^T \mathbf{x}_{opt} = 10①^0 + 10①^{-1} + 10①^{-2}$.

7.3 *Test Problem 3: the Rotated "House" in 5D*

Algorithm 3 (firstly introduced in [2]) shows how to add a small rotation along the axis perpendicular to the plane containing the first two variables x_1 and x_2, for the "house" problem seen in previous example, after generalizing it to the n-dimensional case.

The problem considered here (again taken from [2]) consists of the lexicographic optimization of x_1, $-x_2$, ..., $-x_5$. The method used to generate a randomly rotated benchmark is shown on Algorithm 3 (the generated rotation

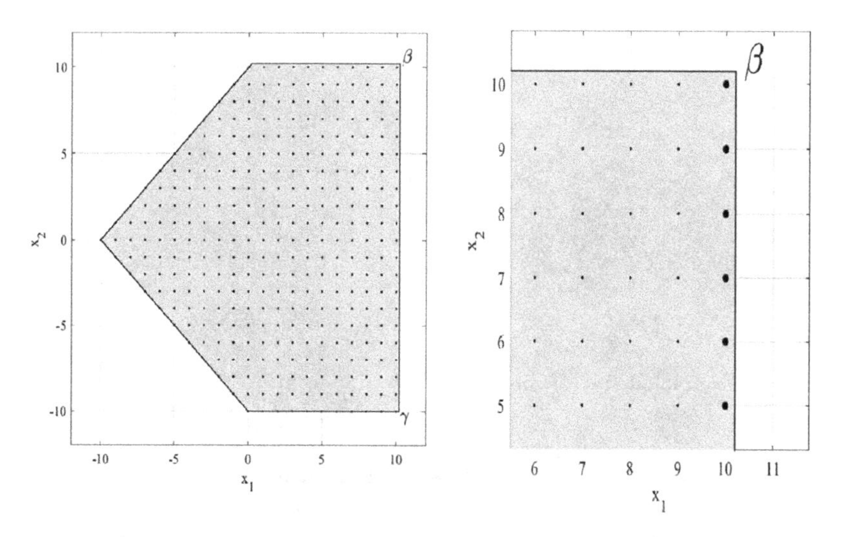

Fig. 5 Section view of Fig. 4 with $x_3 = -10$ (left) and its top-right zoom (right)

Table 3 Iterations performed by GrossBB algorithm on test problem T_2

Iteration	Result at node (iteration)
Initialize	– $\tilde{v_S}(\tilde{P}) = ①$ – Queue len. 1 (add the root problem to the queue)
1	$\tilde{v_I}(\tilde{P_c})$: $-10.2①^0 - 10①^{-1} - 10.2①^{-2}$. Queue length: 0 – no pruning rules applied, branch $\tilde{P_c}$ in two sub-problems. Queue length: 2 – $\tilde{\Delta}$ $100①^0 + 100①^{-1} + 100①^{-2}$
2	prune node: rule 1, empty feasible region. Queue length: 1
3	$\tilde{v_I}(\tilde{P_c})$: $-10①^0 - 10①^{-1} - 10.2①^{-2}$. Queue length: 0 – no pruning rules applied, branch $\tilde{P_c}$ in two sub-problems. Queue length: 2 – $\tilde{\Delta}$ $100①^0 + 100①^{-1} + 100①^{-2}$
4	$\tilde{v_I}(\tilde{P_c})$: $-10①^0 - 10①^{-1} - 10①^{-2}$. Queue length: 1 – A feasible solution has been found: $x_{opt} = [10, -10, -10]^T$ – update $\tilde{v_S}(\tilde{P}) = \tilde{v_I}(\tilde{P_c})$, prune node: rule 3 – $\tilde{\Delta}$: $0①^0 + 0①^{-1} + 0.02①^{-2}$
5	prune node: rule 1, empty feasible region. Queue length: 0
Result	Iteration 5. Optimization ended. Optimal solution found:
	$\mathbf{x}_{opt} = [10, -10, -10]^T$ $\tilde{f}_{opt} = -10①^0 - 10①^{-1} - 10①^{-2}$ $\tilde{\Delta} = 0①^0 + 0①^{-1} + 0①^{-2}$

matrix $\mathbf{Q}$ is reported in Appendix A of [2]). After the rotation, a lower bound and an upper bound were added:

$$-2\rho \leqslant x_i \leqslant 2\rho, \ i = 1, ..., 5.$$

Thus the following problem ($\mathbf{A}'$,$\mathbf{b}'$) has been generated:

$$\begin{aligned} \text{LexMax} \quad & x_1, -x_2, ..., -x_5 \\ \text{s.t.} \quad & \left\{ \mathbf{x}' \in \mathbb{Z}^5 : \mathbf{A}'\mathbf{x}' \leqslant \mathbf{b}' \right\} \end{aligned} \qquad T_3$$

where $\mathbf{C}'$, $\mathbf{A}'$ and vector $\mathbf{b}'$ are reported in Appendix A of [2].

The lexicographic optimum for this problem is:

$$\mathbf{x}_{opt} = [1000, \ -999, \ -1000, \ -1000, \ -1000]^T.$$

The new problem can be solved with GrossBB algorithm. After 11 steps, the GrossBB finds the correct lexicographic optimum (more details in [2], Table 3, Appendix B).

8 Conclusions

To conclude this chapter, let us recall that we have shown the application of Grossone to solve lexicographic multi-objective linear programming problems and their mixed-integer counterparts. We have proved that the use of Grossone allows to give an infinitely lower weights to lower-priority objectives, without the need to specify the value of the finite weight M to use. As a future work, we are planning to implement a Grossone-based Branch-and-Cut algorithm, to speedup the convergence of the GrossBB algorithm presented here.

Appendix

In this appendix we provide the proofs of Lemmas 1–3, taken from [6], and reported also here for the sake of self-completeness of this exposition.

Proof (Lemma 1) Since the objective function of problem P4 is linear in $\mathbf{x}$, the associated level sets ($\tilde{\mathbf{c}}^T\mathbf{x} = \tilde{v}$) are hyper-planes. Thus the maximum, for a bounded and non-empty polyhedron S describing the domain of the problem

Algorithm 3: Generation of a randomly rotated "house" problem in $\mathbb{R}^n$ dimensions

Step 1. Let $\{\mathbf{Ax} \leqslant \mathbf{b},\ \mathbf{x} \in \mathbb{Z}^n\}$ be the initial, unrotated problem, in n-dimensions. The problems is formulated as follow (ρ is a parameter that controls the size of the house):

$$\begin{array}{ll} \text{LexMax} & x_1,\ -x_2,\ \ldots,\ -x_n \\ \text{s.t.} & -\rho + 0.2 \leqslant x_1 \leqslant \rho + 0.2, \\ & -\rho \qquad\quad \leqslant x_2 \leqslant \rho + 0.2 \\ & -\rho + 0.2 \leqslant x_i \leqslant \rho + 0.2,\ i = 3, \ldots, n \\ & -x_1 - x_2 \leqslant 2 \\ & -x_1 + x_2 \leqslant 2 \\ & \mathbf{x} \in \mathbb{Z}^n. \end{array}$$

Step 2. Use as rotation matrix **Q** with a random little rotation:

```
rA = 0.0002;

rB = 0.0005;

ϕ = (rB-rA).*rand(1) + rA;
```

$$\mathbf{Q} = \begin{bmatrix} \cos(\phi) & \sin(\phi) & 0 & 0 & .. & 0 \\ -\sin(\phi) & \cos(\phi) & 0 & 0 & \ldots & 0 \\ 0 & 0 & 1 & 0 & \ldots & 0 \\ 0 & 0 & 0 & 1 & \ldots & 0 \\ \ldots & \ldots & \ldots & \ldots & \ldots & \ldots \\ 0 & 0 & 0 & 0 & \ldots & 1 \end{bmatrix} \in \mathbb{R}^{n \times n}$$

Step 3. Rotate the polytope: $\mathbf{A}' = \mathbf{AQ}$ ($\mathbf{b}$ and $\mathbf{C}$ does not change under rotations: $\mathbf{b}' = \mathbf{b}$ and $\mathbf{C}' = \mathbf{C}$) and then add these additional constraints as lower and upper bound for every variables to $\mathbf{A}'$ (they are twice the size of the house, in order to fully contain it):

$$-2\rho \leqslant x_i \leqslant 2\rho,\ i = 1, \ldots, n$$

Step 4 For the unrotated problem the optimal integer solution is $\mathbf{x}_{opt} = [\rho,\ -\rho,\ -\rho,\ -\rho,\ \ldots,\ -\rho]^T$.
When a rotation is applied, if the rotation is between a sufficiently small range of angles the optimal solution is: $\mathbf{x}_{opt} = [\rho,\ 1-\rho,\ -\rho,\ -\rho,\ \ldots,\ -\rho]^T$.
The optimal value is computed as: $\tilde{f}_{opt} = \tilde{\mathbf{c}}^T \mathbf{x}_{opt}$, where $\tilde{\mathbf{c}}$ is derived from $\mathbf{C}$.

---

must be on a vertex (or belong to the convex hull of the optimal vertices), for the same reasons for which the maximum of standard single-objective LP problems is located on a vertex or in the convex hull of the optimal vertices. In this case, similarly to standard LP, it follows that: (i) not all the vertices of P4 are optimal, and (ii) not all the basic solutions are vertices.

The next lemma states that all the optimal solutions of P1 reach the same (Gross-scalar) objective value for problem P4.

Proof (Lemma 2) Let $\mathbf{x}^*$ be a generic optimal solution belonging to Ω(P1), then from (4) we have that the objective value of the problem P4 associated to it is:

$$\tilde{\mathbf{c}}^T\mathbf{x}^* = (\mathbf{c}^{1T}\mathbf{x}^*)①^0 + (\mathbf{c}^{2T}\mathbf{x}^*)①^{-1} + \ldots + (\mathbf{c}^{rT}\mathbf{x}^*)①^{-r+1}.$$

We can observe that:

$$\mathbf{c}^{i^T}\mathbf{x}^* = k^i \in \mathbb{R}, \quad i \in \{1, ..., r\},$$

due to the fact that $\mathbf{x}^*$ is optimal for P1 and this problem has been formulated using purely finite numbers only. Thus we have

$$\tilde{\mathbf{c}}^T\mathbf{x}^* = k^1①^0 + k^2①^{-1} + ... + k^r①^{-r+1} = \tilde{v}.$$

□

The next lemma is the last step in preparing the proof of the equivalence between P1 and P4.

Proof (Lemma 3) Let $\hat{\mathbf{x}}$ be a vertex of S which is not optimal for P1. Thus, there exists an index $q \in \{1, ..., r\}$ such that

$$\mathbf{c}^{i^T}\hat{\mathbf{x}} = \mathbf{c}^{i^T}\mathbf{x}^*, \quad \forall i \in \{1, ..., q-1\},$$

but

$$\mathbf{c}^{q^T}\hat{\mathbf{x}} \; < \; \mathbf{c}^{q^T}\mathbf{x}^*.$$

This implies that

$$\tilde{\mathbf{c}}^T\mathbf{x}^* - \tilde{\mathbf{c}}^T\hat{\mathbf{x}} = \sum_{i=q}^{r} ①^{-i+1}(\mathbf{c}^{i^T}\mathbf{x}^* - \mathbf{c}^{i^T}\hat{\mathbf{x}}).$$

The expression above can be also expanded as follows

$$\tilde{\mathbf{c}}^T\mathbf{x}^* - \tilde{\mathbf{c}}^T\hat{\mathbf{x}} = ①^{-q+1}(\mathbf{c}^{q^T}\mathbf{x}^* - \mathbf{c}^{q^T}\hat{\mathbf{x}}) + ①^{-(q+1)+1}(\mathbf{c}^{(q+1)^T}\mathbf{x}^* - \mathbf{c}^{(q+1)^T}\hat{\mathbf{x}}) + ...$$
$$+ ①^{-r+1}(\mathbf{c}^{r^T}\mathbf{x}^* - \mathbf{c}^{r^T}\hat{\mathbf{x}}).$$

Since $①^{-q+1}(\mathbf{c}^{q^T}\mathbf{x}^* - \mathbf{c}^{q^T}\hat{\mathbf{x}}) > 0$, it follows that $\left(\tilde{\mathbf{c}}^T\mathbf{x}^* - \tilde{\mathbf{c}}^T\hat{\mathbf{x}}\right)$ is strictly positive too, since r is finite. Indeed, adding a finite number of infinitesimal contributions of orders of ① higher than $-q+1$ will keep the sum strictly positive, even when these contributions are negative in sign, due to the property of Grossone:

$$\left|①^{-q+1}(\mathbf{c}^{q^T}\mathbf{x}^* - \mathbf{c}^{q^T}\hat{\mathbf{x}})\right| \; > \; \left|①^{-q}(\mathbf{c}^{(q+1)^T}\mathbf{x}^* - \mathbf{c}^{(q+1)^T}\hat{\mathbf{x}})\right| + ... +$$
$$\left|①^{-r+1}(\mathbf{c}^{r^T}\mathbf{x}^* - \mathbf{c}^{r^T}\hat{\mathbf{x}})\right|.$$

References

1. Amodio, P., Iavernaro, F., Mazzia, F., Mukhametzhanov, M.S., Sergeyev, Y.D.: A generalized Taylor method of order three for the solution of initial value problems in standard and infinity floating-point arithmetic. Math. Comput. Simul. **141**, 24–39 (2017)
2. Cococcioni, M., Cudazzo, A., Pappalardo, M., Sergeyev, Y.D.: Solving the lexicographic multi-objective mixed-integer linear programming problem using branch-and-bound and grossone methodology. Commun. Nonlinear Sci. Numer. Simul. **84**, 105177 (2020)
3. Cococcioni, M., Fiaschi, L.: The Big-M method with the numerical infinite M. Optim. Lett. **15**, 2455–2468 (2021)
4. Cococcioni, M., Fiaschi, L., Lambertini, L.: Non-archimedean zero-sum games. J. Comput. Appl. Math. **393**, 113483 (2021)
5. Cococcioni, M., Pappalardo, M., Sergeyev, Y.D.: Towards lexicographic multi-objective linear programming using grossone methodology. In: Sergeyev, Y.D., Kvasov, D.E., Dell'Accio, F., Mukhametzhanov, M.S., (eds.), Proceedings of the 2nd International Conference "Numerical Computations: Theory and Algorithms", vol. 1776, p. 090040. AIP Publishing, New York (2016)
6. Cococcioni, M., Pappalardo, M., Sergeyev, Y.D.: Lexicographic multi-objective linear programming using grossone methodology: theory and algorithm. Appl. Math. Comput. **318**, 298–311 (2018)
7. De Cosmis, S., De Leone, R.: The use of grossone in mathematical programming and operations research. Appl. Math. Comput. **218**(16), 8029–8038 (2012)
8. De Leone, R.: Nonlinear programming and grossone: quadratic programming and the role of constraint qualifications. Appl. Math. Comput. **318**, 290–297 (2018)
9. De Leone, R., Fasano, G., Roma, M., Sergeyev, Y.D.: Iterative grossone-based computation of negative curvature directions in large-scale optimization. J. Optim. Theory Appl. **186**(2), 554–589 (2020)
10. De Leone, R., Fasano, G., Sergeyev, Y.D.: Planar methods and grossone for the conjugate gradient breakdown in nonlinear programming. Comput. Optim. Appl. **71**, 73–93 (2018)
11. Fiaschi, L., Cococcioni, M.: Numerical asymptotic results in game theory using Sergeyev's infinity computing. Int. J. Unconvent. Comput. **14**(1), 1–25 (2018)
12. Fiaschi, L., Cococcioni, M.: Non-archimedean game theory: a numerical approach. Appl. Math. Comput. **409**, 125356 (2021)
13. Gaudioso, M., Giallombardo, G., Mukhametzhanov, M.S.: Numerical infinitesimals in a variable metric method for convex nonsmooth optimization. Appl. Math. Comput. **318**, 312–320 (2018)
14. Isermann, H.: Linear lexicographic optimization. OR Spektrum **4**, 223–228 (1982)
15. Lai, L., Fiaschi, L., Cococcioni, M.: Solving mixed pareto-lexicographic many-objective optimization problems: the case of priority chains. Swarm Evol. Comput. **55**, 100687 (2020)
16. Lai, L., Fiaschi, L., Cococcioni, M., Deb, K.: Handling priority levels in mixed pareto-lexicographic many-objective optimization problems. In: Ishibuchi, H., Zhang, Q., Cheng, R., Li, K., Li, H., Wang, H., Zhou, A. (eds.) Evolutionary Multi-Criterion Optimization, pp. 362–374. Springer International Publishing, Cham (2021)
17. Lai, L., Fiaschi, L., Cococcioni, M., Deb, K.: Solving mixed pareto-lexicographic many-objective optimization problems: the case of priority levels. IEEE Trans. Evolut. Comput. (2021). http://dx.doi.org/10.1109/TEVC.2021.3068816

18. Lai, L., Fiaschi, L., Cococcioni, M., Deb, K.: Pure and mixed lexicographic-paretian many-objective evolutionary optimization: state of the art. Nat. Comput. (2022). Submitted
19. Pappalardo, M., Passacantando, M.: Ricerca Operativa. Pisa University Press (2012)
20. Pourkarimi, L., Zarepisheh, M.: A dual-based algorithm for solving lexicographic multiple objective programs. Eur. J. Oper. Res. **176**, 1348–1356 (2007)
21. Rizza, D.: Supertasks and numeral systems. In: Proceedings of the 2nd International Conference "Numerical Computations: Theory and Algorithms", vol. 1776, p. 090005. AIP Publishing, New York (2016). https://doi.org/10.1063/1.4965369
22. Rizza, D.: A study of mathematical determination through Bertrand's Paradox. Philos. Math. **26**(3), 375–395 (2018)
23. Rizza, D.: Numerical methods for infinite decision-making processes. Int. J. Unconvent. Comput. **14**(2), 139–158 (2019)
24. Sergeyev, Y.D.: Solving ordinary differential equations by working with infinitesimals numerically on the Infinity Computer. Appl. Math. Comput. **219**(22), 10668–10681 (2013)
25. Sergeyev, Y.D.: The olympic medals ranks, lexicographic ordering, and numerical infinities. Math. Intell. **37**(2), 4–8 (2015)
26. Sergeyev, Y.D.: Numerical infinities and infinitesimals: methodology, applications, and repercussions on two Hilbert problems. EMS Surv. Math. Sci. **4**, 219–320 (2017). https://doi.org/10.4171/EMSS/4-2-3
27. Sergeyev, Y.D.: Independence of the grossone-based infinity methodology from non-standard analysis and comments upon logical fallacies in some texts asserting the opposite. Found. Sci. **24**(1), 153–170 (2019)
28. Sergeyev, Y.D., Kvasov, D.E., Mukhametzhanov, M.S.: On strong homogeneity of a class of global optimization algorithms working with infinite and infinitesimal scales. Commun. Nonlinear Sci. Numer. Simul. **59**, 319–330 (2018)
29. Sergeyev, Y.D., Mukhametzhanov, M.S., Mazzia, F., Iavernaro, F., Amodio, P.: Numerical methods for solving initial value problems on the infinity computer. Int. J. Unconvent. Comput. **12**(1), 3–23 (2016)
30. Sherali, H., Soyster, A.: Preemptive and nonpreemptive multi-objective programming: relationship and counterexamples. J. Optim. Theory Appl. **39**(2), 173–186 (1983)
31. Stanimirovic, I.: Compendious lexicographic method for multi-objective optimization. Facta universitatis - Ser.: Math. Inf. **27**(1), 55–66 (2012)

The Use of Infinities and Infinitesimals for Sparse Classification Problems

Renato De Leone, Nadaniela Egidi, and Lorella Fatone

Abstract In this chapter we discuss the use of *grossone* and the new approach to infinitesimal and infinite proposed by Sergeyev in determining sparse solutions for special classes of optimization problems. In fact, in various optimization and regression problems, and in solving overdetermined systems of linear equations it is often necessary to determine a sparse solution, that is a solution with as many as possible zero components. Expanding on the results in [16], we show how continuously differentiable concave approximations of the l_0 pseudo–norm can be constructed using *grossone*, and discuss the properties of some new approximations. Finally, we will conclude discussing some applications in elastic net regularization and Sparse Support Vector Machines.

1 Introduction

In many optimization problems, in regression methods and when solving over-determined systems of equations, it is often necessary to determine a

R. De Leone (✉) · N. Egidi · L. Fatone
School of Science and Technology, University of Camerino, Camerino, MC, Italy
e-mail: renato.deleone@unicam.it

N. Egidi
e-mail: nadaniela.egidi@unicam.it

L. Fatone
e-mail: lorella.fatone@unicam.it

Y. D. Sergeyev and R. De Leone (eds.), *Numerical Infinities and Infinitesimals in Optimization*, Emergence, Complexity and Computation 43,
https://doi.org/10.1007/978-3-030-93642-6_6

sparse solution, that is, a solution with the minimum number of nonzero components. This kind of problems are known as sparse approximation problems and arise in different fields. In Machine Leaning, the Feature Extraction problem requires, for a given problem, to eliminate as many features as possible, while still maintaining a good accuracy in solving the assigned task (for example, a classification task). Sparse solutions are also required in signal/image processing problem, for example in sparse approximation of signals, image denoising, etc. [4, 12, 36].

In these cases the l_0 pseudo-norm is utilized. This pseudo-norm counts the number of nonzero elements of a vector. Problems utilizing the l_0 pseudo-norm have been considered by many researchers, but they seem "to pose many conceptual challenges that have inhibited its widespread study and application" [4]. Moreover, the resulting problem is NP–hard and, in order to construct a more tractable problem, various continuously differentiable concave approximations of the l_0 pseudo-norm are used, or the l_0 pseudo-norm is replaced by the simpler to handle 1–norm. In [27] two smooth approximations of the l_0 pseudo-norm are proposed in order to determine a vector that has the minimum l_0 pseudo-norm.

Recently, Sergeyev proposed a new approach to infinitesimals and infinities[1] based on the numeral ①, the number of elements of $\mathbb{N}$, the set of natural numbers. It is crucial to note that ① is not a symbol and is not used to perform symbolic calculations. In fact, the ① is a natural number, and it has both cardinal and ordinal properties, exactly as the "standard", finite natural numbers. Moreover, the new proposed approach is different from non–Standard Analysis, as demonstrated in [33]. A comprehensive description of the grossone–based methodology can also be found in [32].

The use of ① and the new approach to infinite and infinitesimals has been beneficial in several fields of pure and applied mathematics including optimization [6–9, 14, 15, 17–19, 23], numerical differentiation [29], ODE [1, 22, 34], hyperbolic geometry [25], infinite series and the Riemann zeta function [28, 30], biology [31], and cellular automata [13].

Moreover, this new computational methodology has been also utilized in the field of Machine Learning allowing to construct new spherical separations for classification problems [2], and novel sparse Support Vector Machines (SSVMs) [16].

In this chapter we discuss the use of ① to obtain new approximations for the l_0 pseudo-norm, and two applications are considered in detail. More specifically, the chapter is organized as follows. In Sect. 2 some of the most utilized smooth approximations of the l_0 pseudo–norm proposed in the literature are

[1] See Chap. 1 for an in–depth description of the properties of the new system and its advantages

discussed. Then, in the successive Sect. 3 it is shown how to utilize ① for constructing approximations of the l_0 pseudo-norm. Finally, in Sect. 4, mostly based on [16], two relevant applications of the newly proposed approximation scheme for the l_0 pseudo norm are discussed in detail: the elastic net regulation problem and sparse Support Vector Machines.

We briefly describe our notation now. All vectors are column vectors and will be indicated with lower case Latin letter (i.e. $x, y, \ldots$). Subscripts indicate components of a vector, while superscripts are used to identify different vectors. Matrices will be indicated with upper case Roman letter (i.e. A, B, $\ldots$). The set of natural and real numbers will be denoted, respectively, by $\mathbb{N}$ and $\mathbb{R}$. The space of the n–dimensional vectors with real components will be indicated by $\mathbb{R}^n$. Superscript ^{T} indicates transpose. The scalar product of two vectors x and y in $\mathbb{R}^n$ will be denoted by $x^T y$. Instead, for a generic Hilbert space, the scalar product of two elements x and y will be indicated by $\langle x, y\rangle$. The Euclidean norm of a vector x will be denoted by $\|x\|$. The space of the $m \times n$ matrices with real components will be indicated by $\mathbb{R}^{m\times n}$. For a $m \times n$ matrix A, A_{ij} is the element in the ith row, jth column.

In the new positional numeral system with base ①, a *gross-scalar* (or *gross–number*) C has the following representation:

$$C = C^{(p_m)}①^{p_m} + \cdots + C^{(p_1)}①^{p_1} + C^{(p_0)}①^{p_0} + C^{(p_{-1})}①^{p_{-1}} + \cdots + C^{(p_{-k})}①^{p_{-k}}, \tag{1}$$

where $m, k \in \mathbb{N}$, for $i = -k, -k+1, \ldots, -1, 0, 1, \ldots, m-1, m$, the quantities $C^{(p_i)}$ are floating-point numbers and p_i are gross-numbers such that

$$p_m > p_{m-1} > \cdots > p_1 > p_0 = 0 > p_{-1} > \cdots > p_{-k+1} > p_{-k}. \tag{2}$$

If $m = k = 0$ the gross-number C is called finite; if $m > 0$ it is called infinite; if $m = 0$, $C^{(p_0)} = 0$ and $k > 0$ it is called infinitesimal; the exponents p_i, $i = -k, -k+1, \ldots, -1, 0, 1, \ldots, m-1, m$, are called *gross-powers*.

2 The l_0 Pseudo-norm in Optimization Problems

Given a vector $x = (x_1, x_2, \ldots, x_n)^T \in \mathbb{R}^n$, the l_0 pseudo-norm of x is defined as the number of its components different from zero, that is:

$$\|x\|_0 = \text{number of nonzero components of } x = \sum_{i=1}^{n} 1_{x_i}, \tag{3}$$

where 1_a is the characteristic (indicator) function, that is, the function which is equal to 1 if $a \neq 0$ and zero otherwise.

Note that $\|\cdot\|_0$ is not a norm and hence is called, more properly, pseudo-norm. In fact, for a non zero vector $x \in \mathbb{R}^n$, and a not null constant $\lambda \in \mathbb{R}$, we have:

$$\|\lambda x\|_0 = \|x\|_0 .$$

Consequently $\|\lambda x\|_0 = |\lambda| \, \|x\|_0$, $\lambda \in \mathbb{R}$, if and only if $|\lambda| = 1$.

The l_0 pseudo-norm plays an important role in several numerical analysis and optimization problems, where it is important to get a vector with as few non-zero components as possible. For example, this pseudo-norm has important applications in elastic-net regularization, pattern recognition, machine learning, signal processing, subset selection problem in regression and portfolio optimization. For example, in signal and image processing many media types can be sparsely represented using transform-domain methods, and sparsity of the representation is fundamental in many highly used techniques of compression (see [4] and references therein). In [20, 26] the cardinality-constrained optimization problem is studied and opportunely reformulated. In [5] the general optimization problem with cardinality constraints has been reformulated as a smooth optimization problem.

The l_0 pseudo-norm is strongly related to the l_p norms. Given a vector $x = (x_1, x_2, \dots, x_n)^T \in \mathbb{R}^n$, the l_p norm of x is defined as

$$\|x\|_p := \left(\sum_{i=1}^{n} |x_i|^p \right)^{\frac{1}{p}} .$$

It is not too difficult to show that

$$\|x\|_0 = \lim_{p \to 0} \|x\|_p^p = \lim_{p \to 0} \sum_{i=1}^{n} |x_i|^p .$$

In Fig. 1 the behavior of $|\sigma|^p$ ($\sigma \in \mathbb{R}$) for different values of p is shown (see also [4]). Note that, as the figure suggests, for $0 < p < 1$, the function $\|x\|_p$ is a concave function.

It must be noted that the use of $\|x\|_0$ makes the problems extremely complicated to solve, and various approximations of the l_0 pseudo-norm have been proposed in the scientific literature, For example, in [27] two smooth approximations of the l_0 pseudo-norm are proposed in order to determine a particular vector that has the minimum l_0 pseudo-norm.

In [24], in the framework of elastic net regularization, the following approximation of $\|x\|_0$ is studied:

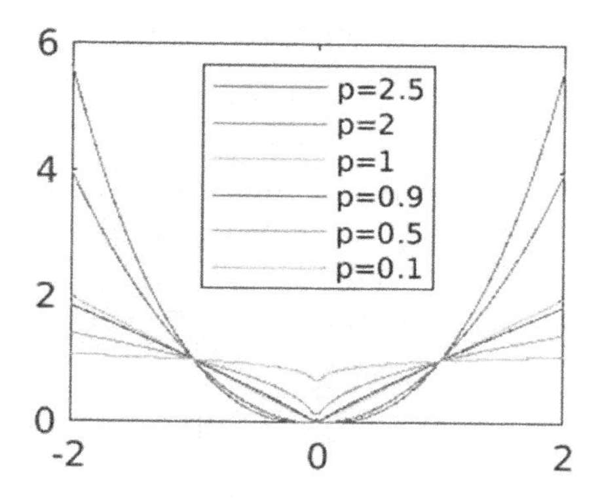

Fig. 1 The value of $|\sigma|^p$ for different values of p

$$\|x\|_{0,\delta} := \sum_{i=1}^{n} \frac{x_i^2}{x_i^2 + \delta}, \tag{4}$$

where $\delta \in \mathbb{R},\ \delta > 0$, and a small δ is suggested in order to provide a better approximation of $\|x\|_0$.

Instead, in the context of Machine Learning and Feature Selection [3], the following approximation of $\|x\|_0$:

$$\|x\|_{0,\alpha} := \sum_{i=1}^{n} \left(1 - e^{-\alpha|x_i|}\right), \tag{5}$$

where $\alpha \in \mathbb{R}$, $\alpha > 0$, is proposed, and the value $\alpha = 5$ is recommended.

By using ①, in [16] a new approximation of $\|x\|_0$ has been suggested. In the next section we discuss in detail this approximation and we also propose other approximations that use the new numeral system based on ①. Moreover, we provide the connections between $\|x\|_0$ and the new approximations.

Note that the approximation introduced in [16] has been used in connection to two different applications. The first application is an elastic net regularization. The second application concerns classification problems using sparse Support Vector Machines. These two applications are extensively reviewed in Sect. 4.

3 Some Approximations of the l_0 Pseudo-norm Using ①

The first approximation of the l_0 pseudo-norm in terms of ① was proposed in [16], where, following the idea suggested in [24] of approximating the l_0 pseudo-norm
by (4), the following approximation has been suggested:

$$\|x\|_{0,①,1} := \sum_{i=1}^{n} \frac{x_i^2}{x_i^2 + ①^{-1}}. \tag{6}$$

In this case, we have that

$$\|x\|_{0,①,1} = \|x\|_0 + C①^{-1}, \tag{7}$$

for some gross-number C which includes only finite and infinitesimal terms. Therefore, the finite parts of $\|x\|_0$ and $\|x\|_{0,①,1}$ coincide.

To this scope, let

$$\psi_1(t) = \frac{t^2}{t^2 + ①^{-1}}, \quad t \in \mathbb{R}. \tag{8}$$

We have that $\psi_1(0) = 0$ and $\psi_1(t) = 1 - ①^{-1}S$, when $t \neq 0$, where S is a gross-number such that

$$0 < S = \frac{1}{t^2 + ①^{-1}} < \frac{1}{t^2}.$$

Therefore, S has only finite and infinitesimal terms. Moreover,

$$\sum_{i=1}^{n} \frac{x_i^2}{x_i^2 + ①^{-1}} = \sum_{i=1}^{n} \psi_1(x_i) = \sum_{i=1, x_i \neq 0}^{n} \psi_1(x_i) = \|x\|_0 + C①^{-1}, \tag{9}$$

where

$$C = \begin{cases} -\displaystyle\sum_{i=1, x_i \neq 0}^{n} S_i, & \text{when } \|x\|_0 \neq 0, \\ 0, & \text{otherwise,} \end{cases}$$

and S_i is a gross-number such that

$$0 < S_i = \frac{1}{x_i^2 + ①^{-1}} < \frac{1}{x_i^2}, \quad x_i \neq 0, \quad i = 1, \dots, n.$$

Hence C is a gross-number with only finite and infinitesimal terms and the finite part of of $\|x\|_0$ and $\|x\|_{0,①,1}$ are the same.

A different proof of this result is provided in [16]. For $i = 1, \dots, n$, let assume that

$$x_i = x_i^{(0)} + R_i ①^{-1},$$

where R_i includes only finite and infinitesimal terms.

When $x_i^{(0)} = 0$:

$$\psi_1(x_i) = \frac{R_i^2 ①^{-2}}{R_i^2 ①^{-2} + ①^{-1}} = ①^{-1} \frac{R_i^2}{R_i^2 ①^{-1} + 1} = 0\,①^0 + R_i' ①^{-1},$$

where R_i' includes only finite and infinitesimal terms.

When, instead, $x_i^{(0)} \neq 0$:

$$\psi_1(x_i) = \frac{\left(x_i^{(0)} + R_i ①^{-1}\right)^2}{\left(x_i^{(0)} + R_i ①^{-1}\right)^2 + ①^{-1}} = 1 - \frac{①^{-1}}{\left(x_i^{(0)} + R_i ①^{-1}\right)^2 + ①^{-1}} = 1 + R_i' ①^{-1},$$

where, again, R_i' includes only finite and infinitesimal terms. Therefore,

$$\|x\|_{0,①,1} = \sum_{i=1}^{n} \psi_1(x_i) = \|x\|_0 + S①^{-1}$$

where S includes only finite and infinitesimal terms and hence $\|x\|_{0,①,1}$ and $\|x\|_0$ coincide in their finite part.

Following the idea suggested in [3], we now propose three novel approximation schemes of the l_0 pseudo-norm all based on the use of ①. In [3] the authors proposed to approximate the l_0 pseudo-norm using (5) and suggest to take a fixed value for α, i.e. $\alpha = 5$, or an increasing sequence of values of α.

Based on this idea, we propose the following approximation formula for $\|x\|_0$:

$$\|x\|_{0,①,2} := \sum_{i=1}^{n} \left(1 - ①^{-\alpha|x_i|}\right), \quad \alpha > 0. \tag{10}$$

Also in this case, the finite parts of $\|x\|_0$ and $\|x\|_{0,①,2}$ coincide. More precisely, let us show that

$$\|x\|_{0,①,2} = \|x\|_0 - ①^{-\alpha m_x} C, \tag{11}$$

where

$$m_x = \begin{cases} \min\Big\{|x_i| \ : \ x_i \neq 0\Big\}, & \text{when } x \neq 0, \\ 0, & \text{otherwise,} \end{cases} \tag{12}$$

and C is a gross-number which is null when $\|x\|_0 = 0$ and, otherwise, includes only finite and infinitesimal terms.

Let us define

$$\psi_2(t) = 1 - ①^{-\alpha|t|}, \quad t \in \mathbb{R}. \tag{13}$$

Since $\psi_2(0) = 0$, we have:

$$\begin{aligned} \|x\|_{0,①,2} &= \sum_{i=1}^{n} \left(1 - ①^{-\alpha|x_i|}\right) = \\ &= \sum_{i=1}^{n} \psi_2(x_i) = \sum_{i=1, x_i \neq 0}^{n} \psi_2(x_i) = \\ &= ||x||_0 - \sum_{i=1, x_i \neq 0}^{n} ①^{-\alpha|x_i|} = ||x||_0 - ①^{-\alpha m_x} C. \end{aligned} \tag{14}$$

It is easy to see that $C = 0$ when $\|x\|_0 = 0$. Instead, if $\|x\|_0 \neq 0$ then C has only finite and infinitesimal terms. This shows that $\|x\|_0$ and $\|x\|_{0,①,2}$ coincide in their finite part.

Another approximation of the l_0 pseudo–norm is given by:

$$\|x\|_0 \approx \|x\|_{0,①,3} := \sum_{i=1}^{n} \left(1 - e^{-①|x_i|}\right). \tag{15}$$

In this case, it is possible to show that

$$\|x\|_{0,①,3} = \|x\|_0 - e^{-①m_x} C, \tag{16}$$

where, as in the previous cases, C is a gross-number which includes only finite and infinitesimal terms and is null when $\|x\|_0 = 0$. Hence, also in this case we have that the finite parts of $\|x\|_0$ and $\|x\|_{0,①,3}$ coincide.

To prove the above result, let

$$\psi_3(t) = 1 - e^{-①|t|}, \quad t \in \mathbb{R}. \tag{17}$$

Since $\psi_3(0) = 0$, we have:

$$\begin{aligned} \|x\|_{0,①,3} &= \sum_{i=1}^{n} \left(1 - e^{-①|x_i|}\right) = \\ &= \sum_{i=1}^{n} \psi_3(x_i) = \sum_{i=1, x_i \neq 0}^{n} \psi_3(x_i) = \\ &= \|x\|_0 - \sum_{i=1, x_i \neq 0}^{n} e^{-①|x_i|} = \|x\|_0 - e^{-①m_x} C, \end{aligned} \tag{18}$$

where m_x is defined in (12) and C is a gross-number with only finite and infinitesimal terms. Moreover, C is null when $\|x\|_0 = 0$.

Finally, another approximation of the l_0 pseudo–norm, always in the spirit of (5), is given by:

$$\|x\|_{0,①,4} := \sum_{i=1}^{n} \left(1 - ①^{-①|x_i|}\right). \tag{19}$$

In this last case, let

$$\psi_4(t) = 1 - ①^{-①|t|}, \quad t \in \mathbb{R}. \tag{20}$$

Since even in this circumstance $\psi_4(0) = 0$, we have:

$$\begin{aligned} \|x\|_{0,①,4} &= \sum_{i=1}^{n} \left(1 - ①^{-①|x_i|}\right) = \\ &= \sum_{i=1}^{n} \psi_4(x_i) = \sum_{i=1, x_i \neq 0}^{n} \psi_4(x_i) = \\ &= \|x\|_0 - \sum_{i=1, x_i \neq 0}^{n} ①^{-①|x_i|} = \|x\|_0 - ①^{-①m_x} C, \end{aligned} \tag{21}$$

where m_x is again defined in (12) and C is a gross-number with only finite and infinitesimal terms that is null when $\|x\|_0 = 0$. As in the previous cases we have that

$$\|x\|_{0,①,4} = \|x\|_0 - ①^{-①m_x} C, \tag{22}$$

and, therefore, the finite parts of $\|x\|_0$ and $\|x\|_{0,①,4}$ coincide.

We have presented a number of different approximation schemes for the l_0 pseudo-norm. We want to stress that in all the cases the value of $\|x\|_0$ and its approximation coincide in their finite part and may only differ for infinitesimals quantities.

In the next section we will discuss some utilization of these approximating schemes in two extremely important problems: regularization and classification.

4 Applications in Regularization and Classification Problems

In this section we review some interesting uses of the proposed l_0 pseudo–norm approximations in two classes of optimization problems: elastic net regularization problems and sparse Support Vector Machine classification problems. These two applications are deeply studied in [16].

4.1 Elastic Net Regularization

There are many important applications where we want to determine a solution $x \in \mathbb{R}^n$ of a given linear system $Ax = b$, $A \in \mathbb{R}^{m\times n}$, $b \in \mathbb{R}^m$, such that x has the smallest number of nonzero components, that is

$$\begin{array}{ll} \min\limits_{x} & \|x\|_0 , \\ \text{subject to} & Ax = b. \end{array}$$

To this problem it is possible to associate the following generalized elastic net regularization:

$$\min_{x} \frac{1}{2} \|Ax - b\|_2^2 + \lambda_0 \|x\|_0 + \frac{\lambda_2}{2} \|x\|_2^2 , \tag{23}$$

where $\lambda_0 > 0$ and $\lambda_2 > 0$ are two regularization parameters (see [24] for details).

In [24] a suitable algorithm for the solution of Problem (23) with $\|x\|_{0,\delta}$ (defined in (4)) instead of $\|x\|_0$ is proposed. The corresponding solution approximates the solution of (23) and depends on the choice of $\delta > 0$ in (4).

Following the idea suggested in [24], we look for the solution of the following minimization problem:

$$\min_x f_1(x), \tag{24}$$

where

$$f_1(x) := \frac{1}{2}\|Ax - b\|_2^2 + \lambda_0 \|x\|_{0,①,1} + \frac{\lambda_2}{2}\|x\|_2^2 . \tag{25}$$

Note that Problem (24)–(25) is obtained from Problem (23) by substituting $\|x\|_{0,①,1}$ to $\|x\|_0$.

In [16] we proved that the corresponding solution coincides with the solution of the original Problem (23) apart from infinitesimal terms. In particular, in [16], the following iterative scheme for the solution of Problem (24)–(25) has been proposed: given an initial value $x^0 \in \mathbb{R}^n$, for $k = 0, 1, \ldots$, compute x^{k+1} by solving

$$\left(A^T A + \lambda_2 I + \lambda_0 D(x^k)\right) x^{k+1} = A^T b, \tag{26}$$

where $I \in \mathbb{R}^{n\times n}$ is the identity matrix and $D \in \mathbb{R}^{n\times n}$ is the following diagonal matrix:

$$D_{ii}(x) = \frac{2①^{-1}}{\left((x_i)^2 + ①^{-1}\right)^2}, \qquad D_{ij}(x) = 0,\ i \neq j. \tag{27}$$

The convergence of the sequence $\{x^k\}$ to the solution of Problem (24)–(25) is ensured by Theorem 1 in [16]. In particular, when $\mathcal{L} := \{x : f_1(x) \leq f_1(x^0)\}$ is a compact set, the above constructed sequence $\{x^k\}_k$ has at least one accumulation point, $x_k \in \mathcal{L}$ for each $k = 1, \ldots,$ and each accumulation point of $\{x^k\}_k$ belongs to $\mathcal{L}$ and is a stationary point of f_1.

In [24] a similar algorithm was proposed, where $\|x\|_0$ was substituted by (4). However, in this latter case, the quality of the final solution (in terms of being also a solution of Problem (24)–(25) strongly depends on the value of δ that is utilized. In our approach, instead, taking into account that $\|x\|_0$ and our approximation with ① only differ for infinitesimal terms, the final solution solves also Problem (24)–(25).

The results presented here are relative to the first of the four approximation schemes for $\|x\|_0$ discussed in Sect. 3.

Considering the minimization of the following functions

$$f_i(x) := \frac{1}{2}\|Ax - b\|_2^2 + \lambda_0 \|x\|_{0,①,i} + \frac{\lambda_2}{2}\|x\|_2^2, \tag{28}$$

with $i = 2$ or $i = 3$ or $i = 4$, and computing the corresponding first order optimality conditions, new iterative schemes similar to (26) can be obtained and studied.

4.2 Sparse Support Vector Machines

The grossone ① and the different approximations of l_0-pseudo norm can be also used in Sparse Support Vector Machines.

Given empirical data (training set) (x^i, y_i), $i = 1, \ldots, l$, with inputs $x^i \in \mathbb{R}^n$, and outputs $y_i \in \{-1, 1\}$, $i = 1, \ldots, l$, we want to compute a vector $w \in \mathbb{R}^n$ and a scalar θ (and hence an hyperplane) such that:

$$\begin{aligned} w^T x^i + \theta &> 0 \text{ when } y_i = 1, \\ w^T x^i + \theta &< 0 \text{ when } y_i = -1. \end{aligned}$$

The classification function is

$$h(x) = \operatorname{sign}\left(w^T x + \theta\right).$$

Given

$$\phi : \mathbb{R}^n \mapsto \mathcal{E},$$

where $\mathcal{E}$ is an Hilbert space with scalar product $\langle \cdot, \cdot \rangle$, the optimal hyperplane can be constructed by solving the following (primal) optimization problem (see [10, 11, 35] and references therein for details):

$$\begin{array}{ll} \min\limits_{w,\theta,\xi} & \frac{1}{2}\langle w, w\rangle + Ce^T\xi, \\ \text{subject to} & y_i\left(\langle w, \phi(x^i)\rangle + \theta\right) \geq 1 - \xi_i, \quad i = 1, \ldots, l, \\ & \xi_i \geq 0, \quad i = 1, \ldots, l, \end{array} \tag{29}$$

where $e \in \mathbb{R}^l$ is a vector with all elements equal to 1 and C is a positive scalar.

The dual of (29) is

$$\begin{array}{ll} \min\limits_{\alpha} & \frac{1}{2}\alpha^T Q\alpha - e^T\alpha, \\ \text{subject to} & y^T\alpha = 0, \\ & 0 \leq \alpha \leq Ce, \end{array} \tag{30}$$

where

$$Q_{ij} = y_i y_j K_{ij}, \quad K_{ij} = K(x^i, x^j) := \left\langle \phi(x^i), \phi(x^j) \right\rangle, \quad i, j = 1, \ldots, l,$$

and $K : \mathbb{R}^n \times \mathbb{R}^n \to \mathbb{R}$ is the kernel function.

We note that, this dual problem and the classification function depend only on $K_{ij} = \langle \phi(x^i), \phi(x^j) \rangle$. In fact, from the Karush–Kuhn–Tucker conditions we have

$$w = \sum_{i=1}^{l} \alpha_i y_i \phi(x^i), \tag{31}$$

and the classification function reduces to

$$h(x) = \text{sign}\Big(\langle w, \phi(x) \rangle + \theta \Big) = \text{sign}\left(\sum_{i=1}^{l} \alpha_i y_i \left\langle \phi(x^i), \phi(x) \right\rangle + \theta \right).$$

In [21], the authors consider an optimization problem based on (29) where $\frac{1}{2}\langle w, w \rangle$ is replaced with $\|\alpha\|_0$ (and, then, this term is approximated by $\frac{1}{2}\alpha \Lambda \alpha$ for opportune values of a diagonal matrix Λ) and use the expansion (31) of w in terms of α.

Furthermore, in [16] the quantity $\|\alpha\|_0$ is replaced by $\|\alpha\|_{0,①,1}$, and the following ①–Sparse SVM problem is defined:

$$\begin{aligned} \min_{\alpha,\theta,\xi} \quad & \frac{①}{2} \|\alpha\|_{0,①,1} + Ce^T \xi, \\ \text{subject to } & y_i \Big[K_{i.}{}^T \alpha + \theta \Big] \geq 1 - \xi_i, \quad i = 1, \ldots, l, \\ & \xi \geq 0, \end{aligned} \tag{32}$$

where $K_{i.}$ denotes the column vector that corresponds to the ith row of the matrix K.

The algorithmic scheme, originally proposed in [21] and revised in [16], starting from $\lambda_r^0 = 1$, $r = 1, \ldots, l$, requires, at each iteration, the solution of the following optimization problem:

$$\begin{aligned} \min_{\alpha,\theta,\xi} \quad & \frac{1}{2} \sum_{r=1}^{l} \lambda_r^k \alpha_r^2 + Ce^T \xi, \\ \text{subject to } & y_i \Big[K_{i.}{}^T \alpha + \theta \Big] \geq 1 - \xi_i, \quad i = 1, \ldots, l, \\ & \xi \geq 0, \end{aligned} \tag{33}$$

and then the update of λ^k with a suitable formula.

From the Karush–Kuhn–Tucker conditions for Problem (32), it follows that

$$\frac{1}{\left(\alpha_r^2 + ①^{-1}\right)^2}\alpha_r = \bar{K}_{r.}^T \beta, \quad r = 1, \ldots, l, \tag{34}$$

where $\bar{K}_{r.}$ is the r–th row of the matrix $\bar{K}$ with $\bar{K}_{rj} = y_j K_{jr}$, for $r, j = 1, \ldots, l$.

The Conditions (34) above suggest the more natural updating formula:

$$\lambda_r^{k+1} = \frac{1}{\left(\alpha_r^2 + ①^{-1}\right)^2}, \quad r = 1, \ldots, l. \tag{35}$$

Moreover, by considering the expansion of the gross-number α, it is easy to verify that formula (35) well mimics the updating formulas for λ^k proposed in [21], also providing a more sound justification for the updating scheme.

We note that the algorithm proposed in [16], and briefly described here, is based on the first of the approximations of $\|\alpha\|_0$ discussed in Sect. 3. Using the other different approximations introduced in the same section, new different updating formulas for λ^{k+1} can be obtained.

5 Conclusions

The use of the l_0 pseudo–norm is pervasive in optimization and numerical analysis, where a sparse solution is often required. Using the new approach to infinitesimal and infinite proposed by Sergeyev, four different approximations of the l_0 pseudo–norm are presented in this chapter. In all cases, we proved that the finite value of the l_0 pseudo–norm and and its approximation coincide, being different only for infinitesimal terms. The use of such approximations is beneficial in many applications, where the discontinuity due to the use of the l_0 pseudo–norm is easily eliminated, by using one of the four proposed approaches presented in this chapter.

References

1. Amodio, P., Iavernaro, F., Mazzia, F., Mukhametzhanov, M.S., Sergeyev, Y.D.: A generalized Taylor method of order three for the solution of initial value problems in standard and infinity floating-point arithmetic. Math. Comput. Simul. **141**, 24–39 (2017)
2. Astorino, A., Fuduli, A.: Spherical separation with infinitely far center. Soft. Comput. **24**(23), 17751–17759 (2020)
3. Bradley, P.S., Mangasarian, O.L.: Feature selection via concave minimization and support vector machines. In: Proceedings of the Fifteenth International Conference on Machine Learning, ICML '98, pp. 82–90. Morgan Kaufmann Publishers Inc., San Francisco (1998)
4. Bruckstein, A.M., Donoho, D.L., Elad, M.: From sparse solutions of systems of equations to sparse modeling of signals and images. SIAM Rev. **51**(1), 34–81 (2009)
5. Burdakov, O., Kanzow, C., Schwartz, A.: Mathematical programs with cardinality constraints: reformulation by complementarity-type conditions and a regularization method. SIAM J. Optim. **26**(1), 397–425 (2016)
6. Cococcioni, M., Cudazzo, A., Pappalardo, M., Sergeyev, Y.D.: Solving the lexicographic multi-objective mixed-integer linear programming problem using branch-and-bound and grossone methodology. Commun. Nonlinear Sci. Numer. Simul. **84**, 105177 (2020)
7. Cococcioni, M., Fiaschi, L.: The Big-M method with the numerical infinite M. Optim. Lett. **15**, 2455–2468 (2021)
8. Cococcioni, M., Pappalardo, M., Sergeyev, Y.D.: Towards lexicographic multi-objective linear programming using grossone methodology. In: Sergeyev, Y.D., Kvasov, D.E., Dell'Accio, F., Mukhametzhanov, M.S., (eds.) Proceedings of the 2nd International Conference "Numerical Computations: Theory and Algorithms", vol. 1776, p. 090040. AIP Publishing, New York (2016)
9. Cococcioni, M., Pappalardo, M., Sergeyev, Y.D.: Lexicographic multi-objective linear programming using grossone methodology: Theory and algorithm. Appl. Math. Comput. **318**, 298–311 (2018)
10. Cortes, C., Vapnik, V.: Support-vector networks. Mach. Learn. **20**(3), 273–297 (1995)
11. Cristianini, N., Shawe-Taylor, J.: An Introduction to Support Vector Machines and Other Kernel-based Learning Methods. Cambridge University Press (2000)
12. Dabov, K., Foi, A., Katkovnik, V., Egiazarian, K.: Image denoising by sparse 3-d transform-domain collaborative filtering. IEEE Trans. Image Process. **16**(8), 2080–2095 (2007)
13. D'Alotto, L.: Cellular automata using infinite computations. Appl. Math. Comput. **218**(16), 8077–8082 (2012)
14. De Cosmis, S., De Leone, R.: The use of grossone in mathematical programming and operations research. Appl. Math. Comput. **218**(16), 8029–8038 (2012)
15. De Leone, R.: Nonlinear programming and grossone: quadratic programming and the role of constraint qualifications. Appl. Math. Comput. **318**, 290–297 (2018)
16. De Leone, R., Egidi, N., Fatone, L.: The use of grossone in elastic net regularization and sparse support vector machines. Soft. Comput. **23**(24), 17669–17677 (2020)
17. De Leone, R., Fasano, G., Roma, M., Sergeyev, Y.D.: How Grossone Can Be Helpful to Iteratively Compute Negative Curvature Directions. Lecture Notes in Computer Science (including subseries Lecture Notes in Artificial Intelligence and Lecture Notes in Bioinformatics), vol. 11353, pp. 180–183 (2019)

18. De Leone, R., Fasano, G., Sergeyev, Y.D.: Planar methods and grossone for the conjugate gradient breakdown in nonlinear programming. Comput. Optim. Appl. **71**(1), 73–93 (2018)
19. Gaudioso, M., Giallombardo, G., Mukhametzhanov, M.S.: Numerical infinitesimals in a variable metric method for convex nonsmooth optimization. Appl. Math. Comput. **318**, 312–320 (2018)
20. Gotoh, J., Takeda, A., Tono, K.: DC formulations and algorithms for sparse optimization problems. Math. Program. **169**, 141–176 (2018)
21. Huang, K., Zheng, D., Sun, J., Hotta, Y., Fujimoto, K., Naoi, S.: Sparse learning for support vector classification. Pattern Recogn. Lett. **31**(13), 1944–1951 (2010)
22. Iavernaro, F., Mazzia, F., Mukhametzhanov, M.S., Sergeyev, Y.D.: Computation of higher order lie derivatives on the infinity computer. J. Computat. Appl. Math. **383** (2021)
23. Lai, L., Fiaschi, L., Cococcioni, M.: Solving mixed Pareto-Lexicographic multi-objective optimization problems: the case of priority chains. Swarm Evolut. Comput. **55**, 100687 (2020)
24. Li, S., Ye, W.: A generalized elastic net regularization with smoothed l_0 penalty. Adv. Pure Math. **7**, 66–74 (2017)
25. Margenstern, M.: An application of grossone to the study of a family of tilings of the hyperbolic plane. Appl. Math. Comput. **218**(16), 8005–8018 (2012)
26. Pham Dinh, T., Le Thi, H.A.: Recent advances in DC programming and DCA. In: Nguyen, N.T., Le Thi, H.S., (eds.) Transactions on Computational Intelligence XIII. Lecture Notes in Computer Science, vol. 8342. Springer (2014)
27. Rinaldi, F., Schoen, F., Sciandrone, M.: Concave programming for minimizing the zero-norm over polyhedral sets. Comput. Optim. Appl. **46**, 467–486 (2010)
28. Sergeyev, Y.D.: Numerical point of view on calculus for functions assuming finite, infinite, and infinitesimal values over finite, infinite, and infinitesimal domains. Nonlinear Anal. Seri. A: Theory, Methods Appl. **71**(12), e1688–e1707 (2009)
29. Sergeyev, Y.D.: Higher order numerical differentiation on the infinity computer. Optim. Lett. **5**(4), 575–585 (2011)
30. Sergeyev, Y.D.: On accuracy of mathematical languages used to deal with the Riemann zeta function and the Dirichlet eta function. p-Adic numbers. Ultrametric Anal. Appl. **3**(2), 129–148 (2011)
31. Sergeyev, Y.D.: Using blinking fractals for mathematical modelling of processes of growth in biological systems. Informatica **22**(4), 559–576 (2011)
32. Sergeyev, Y.D.: Numerical infinities and infinitesimals: methodology, applications, and repercussions on two Hilbert problems. EMS Surv. Math. Sci. **4**(2), 219–320 (2017)
33. Sergeyev, Y.D.: Independence of the grossone-based infinity methodology from non-standard analysis and comments upon logical fallacies in some texts asserting the opposite. Found. Sci. **24**(1), 153–170 (2019)
34. Sergeyev, Y.D., Mukhametzhanov, M.S., Mazzia, F., Iavernaro, F., Amodio, P.: Numerical methods for solving initial value problems on the infinity computer. Int. J. Unconv. Comput. **12**(1), 3–23 (2016)
35. Smola, A.J., Schölkopf, B.: A tutorial on support vector regression. Stat. Comput. **14**(3), 199–222 (2004)
36. Stanković, L., Sejdić, E., Stanković, S., Daković, M., Orović, I.: A tutorial on sparse signal reconstruction and its applications in signal processing. Circuits Syst. Signal Process. **38**(3), 1206–1263 (2019)

The Grossone-Based Diagonal Bundle Method

Manlio Gaudioso, Giovanni Giallombardo, and Marat S. Mukhametzhanov

Abstract We discuss the fruitful impact of the infinity computing paradigm on the practical solution of convex nonsmooth optimization problems. We consider a class of unconstrained nonsmooth optimization methods based on a variable metric approach, where the use of the infinity computing techniques allows one to numerically deal with quantities which can take arbitrarily small or large values, as a consequence of nonsmoothness. In particular, choosing a diagonal matrix with positive entries as a metric, we modify the so called Diagonal Bundle algorithm by means of matrix updates based on the infinity computing paradigm, and we provide the computational results obtained on a set of benchmark academic test-problems.

1 Introduction

We address unconstrained optimization problems of the type

$$\min_{\mathbf{x} \in \mathbb{R}^n} f(\mathbf{x}), \tag{1}$$

where $f : \mathbb{R}^n \mapsto \mathbb{R}$ is a real-valued not necessarily differentiable function of several variables. It is well known that nonsmooth (or nondifferentiable) optimization is about finding a local minimizer of f, and that even if nons-

M. Gaudioso · G. Giallombardo (✉) · M. S. Mukhametzhanov
Università della Calabria, Rende, Italy
e-mail: giovanni.giallombardo@unical.it

M. Gaudioso
e-mail: manlio.gaudioso@unical.it

M. S. Mukhametzhanov
e-mail: m.mukhametzhanov@dimes.unical.it

Y. D. Sergeyev and R. De Leone (eds.), *Numerical Infinities and Infinitesimals in Optimization*, Emergence, Complexity and Computation 43,
https://doi.org/10.1007/978-3-030-93642-6_7

moothness occurs at a zero-measure set of the function domain, convergence is not ensured when algorithms tailored to the smooth case are adopted. In fact, in-depth research activities have been developed in last decades [6, 20], and several proposals for dealing with nonsmoothness are offered in the literature (as for historical contributions, we mention here the books [14, 28, 45] and the seminal papers [8, 27]).

We particularly focus on the convex version of the nonsmooth problem (1), for which two research mainstreams are active: subgradient methods (see the classic version in [45] and, among the others, the more recent variants in [5, 17, 36]) and bundle methods. The latter stems from the seminal paper [31], and benefits from both the cutting plane model [8, 27] and the conjugate subgradient approach [46]. At each iteration of a bundle method [24] the solution of a quadratic program is required in order to find a tentative displacement from the current approximation of the minimizer. As a consequence of the nonsmoothness, no matter if line-search techniques are adopted, a sufficient reduction of the objective function may not be ensured. For such reason, bundle methods are based on the so called *null step*, whenever no progress toward the minimizer is achieved, with some additional information about the local behavior of the objective function being collected in the bundle.

Literature on bundle methods is very rich. We cite here some contributions given in [4, 13, 18, 19, 29, 33], and those (see [22, 26, 34]) where many features of the standard bundle approach are preserver, while trying to simplify the displacement search, thus avoiding solution of a too burdensome subproblem at each iteration. The latter ones exploit some ideas coming from the vast literature on the variable metric approach applied to smooth optimization, as they adopts variants of standard Quasi–Newton formulae in order to determine an approximated Hessian matrix (see [15] for a classic survey on Quasi–Newton methods and [6], with the references therein, for the applications of the variable metric approach to nonsmooth optimization).

We recall that Quasi–Newton methods for finding a local minimizer of a differentiable function $f : \mathbb{R}^n \mapsto \mathbb{R}$ are iterative techniques where, at any point $\mathbf{x}^k \in \mathbb{R}^n$, a matrix B^k is generated as an approximation of the Hessian matrix such that it satisfies the equation

$$B^k \mathbf{s}^k = \mathbf{u}^k, \tag{2}$$

where

$$\mathbf{s}^k \triangleq \mathbf{x}^k - \mathbf{x}^{k-1} \tag{3}$$

and

$$\mathbf{u}^k \triangleq \nabla f(\mathbf{x}^k) - \nabla f(\mathbf{x}^{k-1}). \tag{4}$$

Sometimes Eq. (2) is replaced by

$$H^k \mathbf{u}^k = \mathbf{s}^k, \tag{5}$$

where H^k is the inverse of B^k.

Methods extending the Quasi–Newton idea to the convex nondifferentiable case still require at each iteration the Eq. (2) be satisfied, but the definition of vector $\mathbf{u}^k$ has to be updated. In fact, denoting by $\partial f(\mathbf{x})$ the subdifferential and by $\mathbf{g} \in \partial f(\mathbf{x})$ any subgradient of f at $\mathbf{x}$, see [24], $\mathbf{u}^k$ can be restated as

$$\mathbf{u}^k \triangleq \mathbf{g}_k - \mathbf{g}_{k-1}, \tag{6}$$

with $\mathbf{g}_k$ and $\mathbf{g}_{k-1}$ being subgradients of f at the points $\mathbf{x}^k$ and $\mathbf{x}^{k-1}$, respectively. In the case of nonsmooth functions, the search for a matrix B^k satisfying (2) is a problem naturally ill-conditioned, given the possible discontinuities in the first order derivatives, which may return "large" $\mathbf{u}^k$ corresponding to "small" $\mathbf{s}^k$.

The latter remark is the main motivation underlying the approach introduced in [21], where Eq. (2) is tackled by applying the *grossone* concept [38] to handle infinite and infinitesimal numbers. Such methodology, that is not related to the well known non-standard analysis framework [41], has already been successfully applied in optimization [3, 11, 12, 16, 30, 44] and numerical differentiation [10, 48], and in a number of other theoretical and computational research areas such as cellular automata [9], Euclidean and hyperbolic geometry [35], percolation [25], fractals [40], infinite series and the Riemann zeta function [47], the first Hilbert problem [39], Turing machines [42], and supertasks [37], numerical solution of ordinary differential equations [2, 43], etc.

In the rest of the chapter, we review in Sect. 2 the relevant details of the *grossone*-based Quasi-Newton method for nonsmooth optimization presented in [21], and we report on the computational experience made on some benchmark academic test-problems in Sect. 3.

2 Grossone-Based Matrix Updates in a Diagonal Bundle Algorithm

We consider the Diagonal Bundle method (`D-Bundle`) introduced in [26] as a variable-metric method applied to the nonsmooth problem (1). `D-Bundle` is based on the limited-memory bundle approach [22], where ideas coming from the variable-metric bundle approach [32, 34] are combined with the extension to nonsmooth problems of the limited-memory approach introduced in [7]. The method adopts the diagonal update formula of the variable-metric matrix presented in [23]. Recalling that `D-Bundle` can also deal with nonconvexities, we however narrow our attention to the convex nonsmooth case. This allows one to adopt an easier structure of `D-Bundle`, thus sharpening the role of the infinity computing paradigm.

Assuming that at each point $\mathbf{x} \in \mathbb{R}^n$ it is possible to calculate $f(\mathbf{x})$ and a subgradient $\mathbf{g} \in \partial f(\mathbf{x})$, and letting $\mathbf{x}^k$ be the estimate of the minimum at iteration k, the search direction $\mathbf{d}^k$ adopted to locate the next iterate is defined as

$$\mathbf{d}^k = -H^k \boldsymbol{\xi}_a^k, \tag{7}$$

where $\boldsymbol{\xi}_a^k$ is the current aggregate subgradient, and H^k is the inverse of B^k, the positive definite variable-metric $n \times n$ matrix, resembling the classic approximation of the Hessian matrix adopted in the smooth case.

Once $\mathbf{d}^k$ is available, a line search along $\mathbf{d}^k$ is performed, which can return two possible outcomes: *serious*-step, whenever a sufficient reduction of the objective function is achieved, and *null*-step otherwise. In the serious-step case, a new approximation $\mathbf{x}^{k+1}$ of a minimizer is obtained, while in the null-step case an auxiliary point $\mathbf{y}^{k+1}$ is obtained along with a subgradient, to enrich the function model around $\mathbf{x}^k$. Further details regarding formula (7) are about the definition of the aggregate subgradient. In the serious-step case (i.e., the new iterate $\mathbf{x}^{k+1}$ has been located) the aggregate subgradient $\boldsymbol{\xi}_a^{k+1}$ is any subgradient of f at $\mathbf{x}^{k+1}$. In the null-step case, no move from the current estimate of the minimizer is made (that is $\mathbf{x}^{k+1} = \mathbf{x}^k$), hence the aggregate subgradient $\boldsymbol{\xi}_a^{k+1}$ is obtained as a convex combination of the three vectors $\mathbf{g}^k \in \partial f(\mathbf{x}^k)$, $\mathbf{g}^{k+1} \in \partial f(\mathbf{y}^{k+1})$, and $\boldsymbol{\xi}_a^k$, with multipliers λ_1^*, λ_2^*, and λ_3^*, respectively, which minimize (see [34, Sect. 4]) the following function

$$\begin{aligned}\phi(\lambda_1, \lambda_2, \lambda_3) \triangleq \frac{1}{2}\left(\lambda_1 \mathbf{g}^k + \lambda_2 \mathbf{g}^{k+1} + \lambda_3 \boldsymbol{\xi}_a^k\right)^\top H^k \left(\lambda_1 \mathbf{g}^k + \lambda_2 \mathbf{g}^{k+1} + \lambda_3 \boldsymbol{\xi}_a^k\right) + \\ + \lambda_2 \alpha^{k+1} + \lambda_3 \alpha_a^k,\end{aligned} \tag{8}$$

where α^{k+1} is the standard (nonnegative) linearization error

$$\alpha^{k+1} \triangleq f(\mathbf{x}^k) - f(\mathbf{y}^{k+1}) - (\mathbf{g}^{k+1})^\top (\mathbf{x}^k - \mathbf{y}^{k+1}),$$

and α_a^k is the aggregated linearization error. The following recursive equality

$$\alpha_a^{k+1} = \lambda_2^* \alpha^{k+1} + \lambda_3^* \alpha_a^k,$$

is adopted to update the aggregated linearization error which is initialized to zero every time a serious step takes place.

Focusing on the updating technique of matrix B^k, as in [26], matrix B^k must be kept diagonal and positive definite. We consider an easier version of the technique described in [26]. At iteration k one can only store the information about the previous iterate, and calculate two correction vectors $\mathbf{s}^k$ and $\mathbf{u}^k$, according to the definitions (3) and (6), respectively. Hence, matrix B^k is obtained by solving the following optimization problem

$$\begin{aligned} \min \quad & \|B\mathbf{s}^k - \mathbf{u}^k\| && && (9)\\ \text{s.t.} \quad & B_{ii} \geq \epsilon, && \forall i \in \{1,\ldots,n\}, && (10)\\ & B_{ij} = 0, && \forall i \neq j \in \{1,\ldots,n\}, && (11) \end{aligned}$$

for some $\epsilon > 0$, whose optimal solution can be expressed as

$$B_{ii}^k = \max\left(\epsilon, \frac{\mathbf{u}_i^k}{\mathbf{s}_i^k}\right), \quad i = 1,\ldots,n. \tag{12}$$

Ill-conditioning of problem (9)–(11) is likely induced by nonsmoothness of f, as the matrix B^k may contain arbitrarily large elements and, consequently, H^k may contain arbitrarily small elements, since

$$H_{ii}^k = (B_{ii}^k)^{-1}, \quad i = 1,\ldots,n. \tag{13}$$

Hence, adopting the infinity computing paradigm in order to control ill-conditioning [1], the correction vectors $\mathbf{s}^k$ and $\mathbf{u}^k$ are first replaced with vectors $\boldsymbol{\delta}^k$ and $\boldsymbol{\gamma}^k$, respectively, whose components are defined as follows

$$\boldsymbol{\delta}_i^k = \begin{cases} \mathbf{s}_i^k, & \text{if } |\mathbf{s}_i^k| > \epsilon, \\ ①^{-1}, & \text{otherwise,} \end{cases} \tag{14}$$

and

$$\boldsymbol{\gamma}_i^k = \begin{cases} \mathbf{u}_i^k, & \text{if } |\mathbf{u}_i^k| > \epsilon, \\ ①^{-1}, & \text{otherwise.} \end{cases} \tag{15}$$

Then, the ratio $\dfrac{\boldsymbol{\gamma}_i^k}{\boldsymbol{\delta}_i^k}$ is possibly corrected by introducing the following update rule

$$\mathbf{b}_i^k = \begin{cases} ①^{-1}, & \text{if } 0 < \dfrac{\boldsymbol{\gamma}_i^k}{\boldsymbol{\delta}_i^k} \leq \epsilon \\ \dfrac{\boldsymbol{\gamma}_i^k}{\boldsymbol{\delta}_i^k}, & \text{otherwise.} \end{cases} \tag{16}$$

Finally, similar to the solution of problem (9)–(11), the elements of the diagonal matrix B^k are set according to the rule

$$B_{ii}^k = \max\left(①^{-1}, \mathbf{b}_i^k\right), \quad i = 1,\ldots,n. \tag{17}$$

Algorithm 1: `Grossone-D-Bundle`

Input: a starting point $\mathbf{x}^0 \in \mathbb{R}^n$, parameters $\theta > 0$, $\eta > 0$, $\epsilon > 0$, $m \in (0,1)$, $\sigma \in (0,1)$
Output: an approximate local minimizer $\mathbf{x}^* \in \mathbb{R}^n$

1: Calculate $\mathbf{g}^0 \in \partial f(\mathbf{x}^0)$, and set $\xi_a^0 = \mathbf{g}^0$ ▷ Initialization
2: Set $H^0 = I$ (i.e., the $n \times n$ identity matrix) ▷
3: Set $\alpha^0 = \alpha_a^0 = 0$, and $k = 0$ ▷
4: Set $\mathbf{d}^k = -H^k \xi_a^k$ ▷ Find a search direction at $\mathbf{x}^k$
5: Set $w^k = (\xi_a^k)^\top \mathbf{d}^k - 2\alpha_a^k$ ▷ Set the desirable reduction
6: **if** $w^k \geq -\eta$ **then** ▷ Stopping test
7: set $\mathbf{x}^* = \mathbf{x}^k$ and **exit** ▷ Return $\mathbf{x}^*$ as an approximate minimizer
8: **end if**
9: Set $t_1 = 1$ and $s = 1$ ▷ Start the line-search
10: **if** $f(\mathbf{x}^k + t_s \mathbf{d}^k) \leq f(\mathbf{x}^k) + m t_s w^k$ **then** ▷ Descent test
11: Set $\mathbf{x}^{k+1} = \mathbf{x}^k + t_s \mathbf{d}^k$ ▷ Make a serious step
12: Calculate $\mathbf{g}^{k+1} \in \partial f(\mathbf{x}^{k+1})$ ▷
13: Set $\xi_a^{k+1} = \mathbf{g}^{k+1}$ ▷
14: Calculate H^{k+1} according to (7.18) ▷ Grossone matrix update
15: Set $k = k+1$ and **go to 4**
16: **end if**
17: **if** $t_s \leq \theta$ **then** ▷ Closeness test
18: Calculate $\mathbf{g}_+ \in \partial f(\mathbf{x}^k + t_s \mathbf{d}^k)$ and $\xi_a \in \text{conv}\{\mathbf{g}^k, \xi_a^k, \mathbf{g}_+\}$ ▷ Make a null step
19: Set $\alpha_+ = f(\mathbf{x}^k) - f(\mathbf{x}^k + t_s \mathbf{d}^k) + t_s g_+^\top \mathbf{d}^k$ ▷
20: Calculate $\alpha_a \in \text{conv}\{0, \alpha_a^k, \alpha_+\}$ ▷
21: Update $\xi_a^k = \xi_a$, $\alpha_a^k = \alpha_a$, and **go to 4** ▷
22: **else**
23: Set $t_{s+1} = \sigma t_s$, $s = s+1$ and **go to 10** ▷ Iterate the line-search
24: **end if**

Note that matrix B^k may contain infinite and infinitesimal numbers. However, in order to calculate $\mathbf{d}^k$ in a classical arithmetic framework, as in formula (7), we need to get rid of the dependence on ① and ①$^{-1}$ in the definition of matrix H^k. Thus, we adopt the following scheme which, as far as finite numbers are involved, reflects the standard inverse calculation scheme:

$$H_{ii}^k = \begin{cases} (B_{ii}^k)^{-1} & \text{if } B_{ii}^k \text{ neither depends on ① nor on ①}^{-1}, \\ (B_{ii}^k)^{-1} \cdot ① & \text{if } B_{ii}^k \text{ is of the type } \alpha①,\ \alpha \in \mathbb{R} \\ (B_{ii}^k)^{-1} \cdot ①^{-1} & \text{if } B_{ii}^k \text{ is of the type } \alpha①^{-1},\ \alpha \in \mathbb{R} \end{cases} \tag{18}$$

Next we report in Algorithm 1 the formal statement of the `Grossone` DBundlel method. The input parameters are: the sufficient decrease parameter $m \in (0, 1)$, the matrix updating threshold $\epsilon > 0$, the stepsize reduction parameter $\sigma \in (0, 1)$, the stopping parameter $\eta > 0$, and the null step parameter $\theta > 0$. We observe that at Step 4, a search direction $\mathbf{d}^k$ at the current point $\mathbf{x}^k$ is selected according to (7). Next, at Step 5, the desirable reduction w^k of the objective function is calculated, and the algorithm stops at Step 6 if w^k is very close to zero.

Chained LQ

$$f(\mathbf{x}) = \sum_{i=1}^{n-1} \max\left\{-x_i - x_{i+1}, -x_i - x_{i+1} + (x_i^2 + x_{i+1}^2 - 1)\right\}$$

$x_i^0 = -0.5$, for all $i = 1, \ldots, n$

$x_i^* = 1/\sqrt{2}$, for all $i = 1, \ldots, n$

$f(\mathbf{x}^*) = -(n-1)\sqrt{2}$

Chained CB3 I

$$f(\mathbf{x}) = \sum_{i=1}^{n-1} \max\left\{x_i^4 + x_{i+1}^2, (2 - x_i)^2 + (2 - x_{i+1})^2, 2e^{-x_i + x_{i+1}}\right\}$$

$x_i^0 = 2$, for all $i = 1, \ldots, n$

$x_i^* = 1$, for all $i = 1, \ldots, n$

$f(\mathbf{x}^*) = 2(n-1)$

Chained CB3 II

$$f(\mathbf{x}) = \max\left\{\sum_{i=1}^{n-1}(x_i^4 + x_{i+1}^2), \sum_{i=1}^{n-1}\left((2 - x_i)^2 + (2 - x_{i+1})^2\right), \sum_{i=1}^{n-1}(2e^{-x_i + x_{i+1}})\right\}$$

$x_i^0 = 2$, for all $i = 1, \ldots, n$

$x_i^* = 1$, for all $i = 1, \ldots, n$

$f(\mathbf{x}^*) = 2(n-1)$

Fig. 1 Test problems ($\mathbf{x}^0$ is the starting point, $\mathbf{x}^*$ is the minimizer)

Otherwise, a line-search along $\mathbf{d}^k$ is executed at Step 9, whose termination depends on the sufficient decrease condition at Step 10. In case the sufficient decrease condition is fulfilled, a serious step takes place along with the *grossone*-based update of matrix H^k. If there is no sufficient decrease at Step 10, then the line-search is iterated at Step 21, unless the step size has become very small. In the latter case a null step occurs, where the aggregate gradient $\boldsymbol{\xi}_a^k$ and the linearization error α_a^k are updated, but not the matrix H^k. For further details, especially regarding convergence properties, the definition of w^k, and the calculation of $\boldsymbol{\xi}_a^k$ and α_a^k in the null-step case, we refer the reader to [26].

Table 1 Results on Chained LQ with size $n = 50$ and $\epsilon = 10^{-2}$

N_f	f^*	e_r	N_S	$N_{①}$
50	−66.7329331	3.65E-02	8	5
100	−67.1750430	3.02E-02	11	8
200	−68.1049532	1.69E-02	16	13
300	−68.4659193	1.18E-02	21	18
400	−68.4721827	1.17E-02	22	19
500	−68.4809312	1.16E-02	24	21

Table 2 Results on Chained LQ with size $n = 100$ and $\epsilon = 10^{-2}$

N_f	f^*	e_r	N_S	$N_{①}$
50	−135.7207701	3.04E-02	8	6
100	−135.9990819	2.84E-02	10	8
200	−136.1352791	2.75E-02	12	10
300	−136.8200990	2.26E-02	15	13
400	−138.7923614	8.62E-03	20	18
500	−139.3581970	4.60E-03	24	22

3 Computational Experience

We have adopted three different classes of large-scale test problems (see Fig. 1) taken from [6] to evaluate the computational behavior of `D-Bundle`.

We have selected different values of the space dimension $n \in \{50, 100, 200\}$, and we report the results obtained by stopping the algorithm after N_f function evaluations, with $N_f \in \{50, 100, 200, 300, 400, 500\}$. In particular, we provide the objective function value f^*, the relative error $e_r = \frac{|f^* - f(\mathbf{x}^*)|}{1+|f(\mathbf{x}^*)|}$, the number of serious steps N_S, and the number of H^k updates that involved the use of grossone $N_{①}$. We report the results obtained adopting $\epsilon = 10^{-2}$, see Tables 1, 2, 3, 4, 5, 6, 7, 8 and 9, and $\epsilon = 10^{-10}$, see Tables 10, 11, 12, 13, 14, 15, 16, 17 and 18. The remaining parameters of the algorithms have been set as follows: $\sigma = 0.7$, $m = 0.1$, $\eta = 10^{-10}$ and $\theta = 10^{-4}$.

The results show that large values of the threshold ϵ for switching to the use of grossone, imply an increased number of grossone-based steps, with corresponding lack of accuracy. It can be seen that the ratio of grossone-based steps over the total number of serious steps is smaller as ϵ decreases. Hence, the use of grossone may allow reasonable treatment of ill-conditioning provided that the threshold ϵ is sufficiently small.

Table 3 Results on Chained LQ with size $n = 200$ and $\epsilon = 10^{-2}$

N_f	f^*	e_r	N_S	$N_{①}$
50	−276.9103190	1.60E-02	8	6
100	−277.4285342	1.42E-02	10	8
200	−277.4421523	1.41E-02	12	10
300	−277.5469159	1.37E-02	14	12
400	−277.5469159	1.37E-02	14	12
500	−277.5469159	1.37E-02	14	12

Table 4 Results on Chained CB3 I with size $n = 50$ and $\epsilon = 10^{-2}$

N_f	f^*	e_r	N_S	$N_{①}$
50	142.0278332	4.45E-01	7	5
100	118.9363346	2.11E-01	10	8
200	113.6568233	1.58E-01	16	14
300	104.2485194	6.31E-02	20	18
400	102.7413588	4.79E-02	25	23
500	99.2882901	1.30E-02	30	28

The `Grossone-D-Bundle` approach has been also numerically compared against its standard counterpart, namely, Algorithm 1 where at Step 14 formula (13) for calculating H_k replaces formula (18). The comparison is made for different values of ϵ in (12) and by stopping the algorithm after a given number N_f of function evaluations. We report in Table 19 the results obtained for problems with size $n = 100$, and setting $\epsilon \in \{10^{-2}, 10^{-5}, 10^{-10}\}$ and $N_f \in \{50, 100, 200\}$. In particular, we provide the relative errors $e_r^{\text{alg}} = \frac{|f^* - f(\mathbf{x}^*)|}{1+|f(\mathbf{x}^*)|}$, where alg $= D$ and alg $= G$ refer, respectively, to $B^k = B_{①}$ and $B^k = B_\epsilon$. The results show that the `Grossone-D-Bundle` approach tends to overperform the standard approach, especially as ϵ decreases.

Table 5 Results on Chained CB3 I with size $n = 100$ and $\epsilon = 10^{-2}$

N_f	f^*	e_r	N_S	$N_{①}$
50	276.0654964	3.92E-01	8	5
100	223.3313657	1.27E-01	11	8
200	218.4138518	1.03E-01	16	13
300	218.0286772	1.01E-01	20	17
400	200.6511044	1.33E-02	27	24
500	199.0866922	5.46E-03	31	28

Table 6 Results on Chained CB3 I with size $n = 200$ and $\epsilon = 10^{-2}$

N_f	f^*	e_r	N_S	$N_{①}$
50	513.8191333	2.90E-01	7	5
100	411.7056803	3.44E-02	11	9
200	403.6445105	1.41E-02	16	14
300	400.5369972	6.36E-03	20	18
400	398.4287379	1.07E-03	24	22
500	398.1830522	4.59E-04	27	25

Table 7 Results on Chained CB3 II with size $n = 50$ and $\epsilon = 10^{-2}$

N_f	f^*	e_r	N_S	$N_{①}$
50	125.5297664	2.78E-01	7	4
100	109.5757036	1.17E-01	10	7
200	105.7547185	7.83E-02	13	10
300	101.5635593	3.60E-02	17	14
400	100.9229376	2.95E-02	22	19
500	100.6930428	2.72E-02	24	21

Table 8 Results on Chained CB3 II with size $n = 100$ and $\epsilon = 10^{-2}$

N_f	f^*	e_r	N_S	$N_{①}$
50	227.4325378	1.48E-01	7	4
100	203.0976658	2.56E-02	11	8
200	202.4957777	2.26E-02	12	9
300	202.4957777	2.26E-02	12	9
400	202.4957777	2.26E-02	12	9
500	202.4957777	2.26E-02	12	9

Table 9 Results on Chained CB3 II with size $n = 200$ and $\epsilon = 10^{-2}$

N_f	f^*	e_r	N_S	$N_{①}$
50	458.0950749	1.51E-01	8	5
100	428.6046364	7.67E-02	10	7
200	424.6396931	6.68E-02	12	9
300	424.6396931	6.68E-02	12	9
400	423.8254683	6.47E-02	14	11
500	409.6297910	2.91E-02	19	16

Table 10 Results on Chained LQ with size $n = 50$ and $\epsilon = 10^{-10}$

N_f	f^*	e_r	N_S	$N_{①}$
50	−69.0981172	2.82E-03	10	4
100	−69.1800716	1.66E-03	13	7
200	−69.1801058	1.66E-03	14	8
300	−69.1801058	1.66E-03	14	8
400	−69.1801058	1.66E-03	14	8
500	−69.1801058	1.66E-03	14	8

Table 11 Results on Chained LQ with size $n = 100$ and $\epsilon = 10^{-10}$

N_f	f^*	e_r	N_S	$N_{①}$
50	−139.4816288	3.73E-03	9	4
100	−139.7674121	1.70E-03	12	6
200	−139.7711371	1.67E-03	13	7
300	−139.7711371	1.67E-03	13	7
400	−139.7711371	1.67E-03	13	7
500	−139.7711371	1.67E-03	13	7

Table 12 Results on Chained LQ with size $n = 200$ and $\epsilon = 10^{-10}$

N_f	f^*	e_r	N_S	$N_{①}$
50	−280.4361446	3.51E-03	9	4
100	−280.8968967	1.88E-03	11	5
200	−280.8968967	1.88E-03	11	5
300	−280.8968967	1.88E-03	11	5
400	−280.8968967	1.88E-03	11	5
500	−280.8968967	1.88E-03	11	5

Table 13 Results on Chained CB3 I with size $n = 50$ and $\epsilon = 10^{-10}$

N_f	f^*	e_r	N_S	$N_{①}$
50	100.6694475	2.70E-02	8	1
100	98.2901649	2.93E-03	13	3
200	98.2261786	2.28E-03	17	4
300	98.2022949	2.04E-03	18	5
400	98.2022949	2.04E-03	18	5
500	98.2022949	2.04E-03	18	5

Table 14 Results on Chained CB3 I with size $n = 100$ and $\epsilon = 10^{-10}$

N_f	f^*	e_r	N_S	$N_{①}$
50	199.9535378	9.82E-03	9	3
100	199.9446641	9.77E-03	9	3
200	199.2083426	6.07E-03	14	8
300	198.5907641	2.97E-03	19	11
400	198.5907641	2.97E-03	19	11
500	198.5907641	2.97E-03	19	11

Table 15 Results on Chained CB3 I with size $n = 200$ and $\epsilon = 10^{-10}$

N_f	f^*	e_r	N_S	$N_{①}$
50	400.6092141	6.54E-03	8	1
100	398.8440820	2.12E-03	10	3
200	398.4688649	1.18E-03	13	5
300	398.4288154	1.07E-03	17	8
400	398.2626085	6.58E-04	20	11
500	398.2282233	5.72E-04	23	14

Table 16 Results on Chained CB3 II with size $n = 50$ and $\epsilon = 10^{-10}$

N_f	f^*	e_r	N_S	$N_{①}$
50	111.4655678	1.36E-01	9	5
100	106.0858985	8.17E-02	11	7
200	106.0858985	8.17E-02	11	7
300	106.0858985	8.17E-02	11	7
400	106.0858985	8.17E-02	11	7
500	106.0858985	8.17E-02	11	7

Table 17 Results on Chained CB3 II with size $n = 100$ and $\epsilon = 10^{-10}$

N_f	f^*	e_r	N_S	$N_{①}$
50	210.9095945	6.49E-02	9	5
100	201.9139633	1.97E-02	14	10
200	201.3641400	1.69E-02	18	14
300	201.3641400	1.69E-02	18	14
400	201.3641400	1.69E-02	18	14
500	201.3641400	1.69E-02	18	14

Table 18 Results on Chained CB3 II with size $n = 200$ and $\epsilon = 10^{-10}$

N_f	f^*	e_r	N_S	$N_{①}$
50	440.0572743	1.05E-01	9	5
100	404.7495782	1.69E-02	12	8
200	404.7495782	1.69E-02	12	8
300	404.7495782	1.69E-02	12	8
400	404.7495782	1.69E-02	12	8
500	404.7495782	1.69E-02	12	8

Table 19 Comparisons between Chained LQ, Chained CB3 I, and Chained CB3 II, with size $n = 100$

	N_f	$\epsilon = 10^{-2}$		$\epsilon = 10^{-5}$		$\epsilon = 10^{-10}$	
		e_r^G	e_r^D	e_r^G	e_r^D	e_r^G	e_r^D
Chained LQ	50	3.04E-02	4.75E-01	7.64E-03	7.54E-01	3.73E-03	7.54E-01
	100	2.84E-02	2.13E-01	7.64E-03	7.49E-01	1.70E-03	7.54E-01
	200	2.75E-02	5.73E-02	7.64E-03	6.88E-01	1.67E-03	7.54E-01
Chained CB3-I	50	3.92E-01	7.43E-02	7.86E-03	1.05E-02	9.82E-03	1.05E-02
	100	1.27E-01	7.43E-02	5.76E-03	1.05E-02	9.77E-03	1.05E-02
	200	1.03E-01	1.36E-02	5.51E-04	1.05E-02	6.07E-03	1.05E-02
Chained CB3-II	50	1.48E-01	1.16E+00	6.47E-02	3.31E+00	6.49E-02	3.31E+00
	100	2.56E-02	5.47E-01	5.97E-02	1.42E+00	1.97E-02	2.98E+00
	200	2.26E-02	3.49E-01	5.97E-02	2.30E-01	1.69E-02	2.98E+00

References

1. Amodio, P., Brugnano, L., Iavernaro, F., Mazzia, F.: A dynamic precision floating-point arithmetic based on the infinity computer framework. In: Y.D. Sergeyev, D.E. Kvasov (eds.) Numerical Computations: Theory and Algorithms. NUMTA 2019, Lecture Notes in Computer Science, vol. 11974. Springer, Cham (2020)
2. Amodio, P., Iavernaro, F., Mazzia, F., Mukhametzhanov, M.S., Sergeyev, Y.D.: A generalized Taylor method of order three for the solution of initial value problems in standard and infinity floating-point arithmetic. Math. Comput. Simul. **141**, 24–39 (2017)
3. Astorino, A., Fuduli, A.: Spherical separation with infinitely far center. Soft Comput. **24**, 17751–17759 (2020)
4. Astorino, A., Gaudioso, M., Gorgone, E.: A method for convex minimization based on translated first-order approximations. Numer. Algor. **76**(3), 745–760 (2017)
5. Bagirov, A.M., Karasözen, B., Sezer, M.: Discrete gradient method: derivative-free method for nonsmooth optimization. J. Optim. Theory Appl. **137**(2), 317–334 (2008)
6. Bagirov, A.M., Karmitsa, N., Mäkelä, M.M.: Introduction to Nonsmooth Optimization: Theory, Practice and Software. Springer, Berlin (2014)
7. Byrd, R.H., Nocedal, J., Schnabel, R.B.: Representations of quasi-Newton matrices and their use in limited memory methods. Math. Program. **63**(1–3), 129–156 (1994)
8. Cheney, E.W., Goldstein, A.A.: Newton's method for convex programming and tchebycheff approximation. Numer. Math. **1**(1), 253–268 (1959)
9. D'Alotto, L.: Cellular automata using infinite computations. Appl. Math. Comput. **218**(16), 8077–8082 (2012)
10. De Cosmis, S., De Leone, R.: The use of grossone in mathematical programming and operations research. Appl. Math. Comput. **218**(16), 8029–8038 (2012)
11. De Leone, R., Egidi, N., Fatone, L.: The use of grossone in elastic net regularization and sparse support vector machines. Soft. Comput. **24**, 17669–17677 (2020)
12. De Leone, R., Fasano, G., Roma, M., Sergeyev, Y.D.: Iterative grossone-based computation of negative curvature directions in large-scale optimization. J. Optim. Theory Appl. **186**, 554–589 (2020)
13. Demyanov, A.V., Fuduli, A., Miglionico, G.: A bundle modification strategy for convex minimization. Eur. J. Oper. Res. **180**(1), 38–47 (2007)
14. Demyanov, V.F., Malozemov, V.N.: Introduction to Minimax. Wiley, New York (1974)
15. Dennis, J.E., Moré, J.J.: Quasi-Newton methods, motivation and theory. SIAM Rev. **19**(1), 46–89 (1977)
16. Fiaschi, L., Cococcioni, M.: The big-M method with the numerical infinite M. Optim. Lett. **15**, 2455–2468 (2021)
17. Frangioni, A., Gorgone, E., Gendron, B.: On the computational efficiency of subgradient methods: a case study in combinatorial optimization. Math. Progr. Comput. **9**, 573–604 (2017)
18. Fuduli, A., Gaudioso, M.: Tuning strategy for the proximity parameter in convex minimization. J. Optim. Theory Appl. **130**(1), 95–112 (2006)
19. Fuduli, A., Gaudioso, M., Giallombardo, G., Miglionico, G.: A partially inexact bundle method for convex semi-infinite minmax problems. Commun. Nonlinear Sci. Numer. Simul. **21**(1–3), 172–180 (2015)
20. Gaudioso, M., Giallombardo, G., Miglionico, G.: Essentials of numerical nonsmooth optimization. 4OR **18**(1), 1–47 (2020)

21. Gaudioso, M., Giallombardo, G., Mukhametzhanov, M.: Numerical infinitesimals in a variable metric method for convex nonsmooth optimization. Appl. Math. Comput. **318**, 312–320 (2018)
22. Haarala, N., Miettinen, K., Mäkelä, M.M.: Globally convergent limited memory bundle method for large-scale nonsmooth optimization. Math. Program. **109**(1), 181–205 (2007)
23. Herskovits, J., Goulart, R.: Sparse quasi-Newton matrices for large scale nonlinear optimization. In: Proceedings of the 6th Word Congress on Structural and Multidisciplinary Optimization (2005)
24. Hiriart-Urruty, J.B., Lemaréchal, C.: Convex Analysis and Minimization Algorithms I-II. Springer, Berlin (1993)
25. Iudin, D.I., Sergeyev, Y.D., Hayakawa, M.: Infinity computations in cellular automaton forest-fire model. Commun. Nonlinear Sci. Numer. Simul. **20**(3), 861–870 (2015)
26. Karmitsa, N.: Diagonal bundle method for nonsmooth sparse optimization. J. Optim. Theory Appl. **166**(3), 889–905 (2015)
27. Kelley, J.E.: The cutting plane method for solving convex programs. J. SIAM **8**(4), 703–712 (1960)
28. Kiwiel, K.C.: Methods of Descent for Nondifferentiable Optimization. Lecture Notes in Mathematics. Springer, Berlin (1985)
29. Kiwiel, K.C.: Proximity control in bundle methods for convex nondifferentiable minimization. Math. Program. **46**(1–3), 105–122 (1990)
30. Lai, L., Fiaschi, L., Cococcioni, M.: Solving mixed Pareto-lexicographic multi-objective optimization problems: The case of priority chains. Swarm Evol. Comput. **55**, 100687 (2020)
31. Lemaréchal, C.: An algorithm for minimizing convex functions. In: Rosenfeld, J. (ed.) Proceedings IFIP '74 Congress 17, pp. 552–556. North-Holland, Amsterdam (1974)
32. Lemaréchal, C., Sagastizábal, C.: Variable metric bundle methods: from conceptual to implementable forms. Math. Prog. Ser. B **76**(3), 393–410 (1997)
33. Lukšan, L., Vlček, J.: A bundle-Newton method for nonsmooth unconstrained minimization. Math. Prog. Ser. B **83**(3), 373–391 (1998)
34. Lukšan, L., Vlček, J.: Globally convergent variable metric method for convex nonsmooth unconstrained minimization. J. Optim. Theory Appl. **102**(3), 593–613 (1999)
35. Margenstern, M.: Fibonacci words, hyperbolic tilings and grossone. Commun. Nonlinear Sci. Numer. Simul. **21**(1–3), 3–11 (2015)
36. Nesterov, Y.: Smooth minimization of non-smooth functions. Math. Program. **103**(1), 127–152 (2005)
37. Rizza, D.: Supertasks and numeral systems. In: Y.D. Sergeyev, D.E. Kvasov, F. Dell'Accio, M.S. Mukhametzhanov (eds.) Proceedings of the 2nd International Conference on "Numerical Computations: Theory and Algorithms", vol. 1776, pp. 090005. AIP Publishing, New York (2016)
38. Sergeyev, Y.D.: Numerical point of view on calculus for functions assuming finite, infinite, and infinitesimal values over finite, infinite, and infinitesimal domains. Nonlinear Anal. Theory Methods Appl. **71**(12), 1688–1707 (2009)
39. Sergeyev, Y.D.: Counting systems and the first Hilbert problem. Nonlinear Anal. Theory Methods Appl. **72**(3–4), 1701–1708 (2010)
40. Sergeyev, Y.D.: Using blinking fractals for mathematical modelling of processes of growth in biological systems. Informatica **22**(4), 559–576 (2011)

41. Sergeyev, Y.D.: Independence of the grossone-based infinity methodology from non-standard analysis and comments upon logical fallacies in some texts asserting the opposite. Found. Sci. **24**(1), 153–170 (2019)
42. Sergeyev, Y.D., Garro, A.: The grossone methodology perspective on Turing machines. In: Adamatzky, A. (ed.) Automata, Universality, Computation, Emergence, Complexity and Computation, vol. 12, pp. 139–169. Springer, New York (2015)
43. Sergeyev, Y.D., Mukhametzhanov, M.S., Mazzia, F., Iavernaro, F., Amodio, P.: Numerical methods for solving initial value problems on the Infinity Computer. Int. J. Unconv. Comput. **12**(1), 3–23 (2016)
44. Sergeyev, Y.D., Nasso, M.C., Mukhametzhanov, M.S., Kvasov, D.E.: Novel local tuning techniques for speeding up one-dimensional algorithms in expensive global optimization using Lipschitz derivatives. J. Comput. Appl. Math. **383**, 113134 (2021)
45. Shor, N.Z.: Minimization Methods for Non-Differentiable Functions. Springer, Berlin (1985)
46. Wolfe, P.: Method of conjugate subgradients for minimizing nondifferentiable functions. In: Balinski, M., Wolfe, P. (eds.) Mathematical Programming Studies, vol. 4, pp. 145–173. North-Holland, Amsterdam (1975)
47. Zhigljavsky, A.: Computing sums of conditionally convergent and divergent series using the concept of grossone. Appl. Math. Comput. **218**(16), 8064–8076 (2012)
48. Žilinskas, A.: On strong homogeneity of two global optimization algorithms based on statistical models of multimodal objective functions. Appl. Math. Comput. **218**(16), 8131–8136 (2012)

On the Use of Grossone Methodology for Handling Priorities in Multi-objective Evolutionary Optimization

Leonardo Lai, Lorenzo Fiaschi, Marco Cococcioni, and Kalyanmoy Deb

Abstract This chapter introduces a new class of optimization problems, called Mixed Pareto-Lexicographic Multi-objective Optimization Problems (MPL-MOPs), to provide a suitable model for scenarios where some objectives have priority over some others. Specifically, this work focuses on two relevant subclasses of MPL-MOPs, namely optimization problems having the objective functions organized as *priority chains* or *priority levels*. A priority chain (PC) is a sequence of objectives ordered lexicographically by importance; conversely, a priority level (PL) is a group of objectives having the same importance in terms of optimization, but a lexicographic ordering exists between the PLs. After describing these problems and discussing why the standard algorithms are inadequate, an innovative approach to deal with them is introduced: it leverages the Grossone Methodology, a recent theory that allows handling priorities by means of infinite and infinitesimal numbers. Most interestingly, this technique can be easily embedded in most of the existing evolutionary algorithms, without altering their core logic. Three algorithms for MPL-MOPs are shown: the first two, called PC-NSGA-II and PC-MOEA/D, are the generalization of NSGA-II and MOEA/D, respectively, in the presence of PCs; the third, named PL-NSGA-II, generalizes instead NSGA-II when PLs are present. Several benchmark problems, including some from the real world, are used to evaluate the effectiveness of the proposed approach. The generalized algorithms are compared to other famous evolutionary ones, either priority-based or not, through a statistical analysis of their

L. Lai · L. Fiaschi (✉) · M. Cococcioni
Department of Information Engineering, University of Pisa, Largo Lucio Lazzarino 1, Pisa, Italy
e-mail: lorenzo.fiaschi@phd.unipi.it

M. Cococcioni
e-mail: marco.cococcioni@unipi.it

K. Deb
Department of Electrical and Computer Engineering, Michigan State University, 428 S. Shaw Lane, East Lansing, MI, USA
e-mail: kdeb@egr.msu.edu

Y. D. Sergeyev and R. De Leone (eds.), *Numerical Infinities and Infinitesimals in Optimization*, Emergence, Complexity and Computation 43,
https://doi.org/10.1007/978-3-030-93642-6_8

performances. The experiments show that the generalized algorithms are consistently able to produce more solutions and of higher quality.

1 Introduction

Evolutionary algorithms [12] are known as one of the most effective strategies to solve real problems requiring the simultaneous optimization of two or more functions, thanks to their ability to produce a group of diverse trade-off solutions rather than focusing on a single one. However, researchers have identified critical issues that arise when the number of objectives grows, often over three or four [35, 42]. The most troublesome issues when dealing with these so-called *many-objective* optimization problems include: (i) ineffectiveness of the standard Pareto dominance; (ii) need for a significantly larger population, due to the possibly increased dimensionality of the efficient set; (iii) computational inefficiency of the diversity estimation functions; and (iv) ineffectiveness of the canonical recombination operators. Many proposals attempt to address and mitigate these problems: decomposition [53], Pareto dominance alternatives [30], custom recombination operators [14], new diversity management mechanisms [1] or even many-objective specific frameworks [13, 31, 52].

While the growth in dimension raises computational challenges, the multi- or many-objective optimization problems originate from practical situations, so it's often the case that not all the objectives are of equal importance to decision makers. In other words, some objectives may be likely organized on the basis of their preference, e.g. as series of objectives ordered by priority or as multiple groups of objectives ranked by importance. One important requirement of such real world scenarios is that no objective shall be completely discarded already at the formulation stage because of its preference level. As an example, a recent problem from an automobile industry featuring six objectives clearly states—the weight objective was most important to minimize, but designers were also interested in solutions having a trade-off in other five objectives [21]. In situations like this, instead of treating all the given objectives as equally important, which is the case of the most common evolutionary multi-objective optimization (EMO) algorithms, the priority information should be taken into account during the optimization phase. This allows to guide the optimization search better, and make it more computationally efficient. Also, by focusing on the trade-off between similar priority objectives, one can obtain more fine-grained insights about the important objectives.

The domain of interest are therefore *Mixed Pareto-Lexicographic Multi-objective Optimization Problems* (abbreviated as MPL-MOPs), a broad class of problems that assume a structured precedence relationship among some objectives. Different priority structures map to different subclasses of MPL-MOPs. In this chapter, the focus is on two of them: PC-MPL-MOPs, where the objectives are organized in chains of priority [27], and PL-MPL-MOPs, where the objectives are grouped in levels of priority [28, 29]. Their models are discussed extensively in Sects. 5–7 and 8–11, respectively. Sections 2 and 4 introduce them briefly: the former is a brief recap of EMO, while the second delves further into the nature of MPL-MOPs. Conclusions are drawn in Sect. 12.

2 Multi-objective Optimization and Evolutionary Algorithms

Multi-objective optimization problems (MOPs) are typically stated as:

$$\min \begin{bmatrix} f_1(x) \\ f_2(x) \\ \vdots \\ f_m(x) \end{bmatrix} \quad s.t.\ x \in \Omega$$

where Ω represents an arbitrary input space. Finding the Pareto optimal solutions in MOPs is non-trivial, and many different methods exist, from mathematical-programming ones to EMO algorithms. Being population-based, the latter are able to deal with a large number of solutions at the same time, whereas other approaches require the same procedure to be repeated over and over (often only changing the starting point). Also, evolutionary algorithms proved to be effective to solve hard problems too, for instance those with many decision variables or local optima. Because of this, EMO has acquired popularity: noteworthy examples are *NSGA-II*, *SPEA2* and *MOEA/D* [15, 53, 55].

Recently, the research focus has shifted towards problems with a large number of objectives, known as *many-objective optimization problems (MaOPs)* [32]. This definition usually applies to any problem with more than three objectives, but it is common to have ten or more. It has been observed that traditional EMO algorithms designed for MOPs are often inadequate when facing MaOPs, for a variety of reasons. First, when the number of objectives grows, a larger fraction of the population becomes nondominated, slowing down the optimization process significantly. This is a very serious issue, which has to do with the ineffectiveness of the Pareto dominance definition; alternatives have been proposed, for instance in [19]. Second, the conflict between convergence and diversity assessment exacerbates in a large-dimensional space, making it harder to find a balance between these two pressures. Furthermore, the recombination operation may be inefficient, calculations and simulations become computationally expensive, results are harder to understand and visualize.

All these challenges have led to the development of sophisticated methods specific for MaOPs, including evolutionary algorithms that are often variations of their MOP counterpart. A remarkable example is *NSGA-III* [13], which enhances the fundamental ideas of NSGA-II to deal better with many-objective problems. Although the above issues can be mitigated by the new approaches, they can hardly be eliminated at all.

3 Metrics to Assess the Efficacy of EMO Algorithms

Several performance indicators have been proposed by researchers to evaluate the quality of a non-dominated solution set [23, 37]. One of the most popular is the *hypervolume indicator* [56], which is Pareto compliant, i.e., the better the Pareto front approximation, the better the performance indicator value. It has been shown that a Pareto non-compliant indicator may result in misleading performance results [37, 48]. However, the hypervolume indicator becomes often computationally unfeasible as the objective space dimension scales up.

Two other well known metrics are the *generational distance* (GD) and the *inverted generational distance* (IGD), described in [50] and [7] respectively. Both assess the quality of a non-dominated objective vector set $A = \{a_1, \ldots, a_n\}$ with respect to a reference Pareto set $Z = \{z_1, \ldots, z_m\}$. The analytical definition of GD is:

$$GD(A) = \frac{1}{n}\left(\sum_{i=1}^{n} d(a_i, Z)^p\right)^{\frac{1}{p}}$$

where d is the Euclidean distance from a_i to its nearest reference point in Z, and $p \in \mathbb{N}$. The IGD metrics is the inverted variation, defined as:

$$IGD(A) = \frac{1}{m}\left(\sum_{i=1}^{m} d(z_i, A)^p\right)^{\frac{1}{p}}$$

where d now represents the Euclidean distance from z_i to its nearest objective vector in A. This work uses the stabler indicator $\Delta(A) = \max\{GD(A), IGD(A)\}$, first proposed in [48]. In analogy with [26], p is set to 1, and this choice is bivariate: (i) it makes the meaning of GD and IGD interpretable (the average of Euclidean distances); (ii) it has often been used in the literature. As any other metric for EMO algorithms evaluation [33], $\Delta(\cdot)$ is not impeccable, but covers at least convergence and diversity of obtained solutions. In future works concerning PL-MPL-MOPs, we plan to include more metrics among those reviewed in [33].

To make a fair and reliable comparison of the algorithms and their performance, various non-parametric statistical tests have been carried out, specifically: (i) the Friedman's test (ii) the Iman-Davenport's test (iii) the Holm's test. Even though the resulting values are not reported here due to lack of space (they can be found in [27, 28]), it is sufficient to say that the statistical separation and hypotheses have been thoroughly tested and validated.

4 Mixed Pareto-Lexicographic Optimization

MPL-MOPs refers to a broad class of problems sharing these features:

- Multiple objectives are involved (multi-objective)
- Some objectives have precedence over some others (lexicographic)
- Some objectives cannot be compared to each other by priority (Pareto)

The mathematical formulation of a MPL-MOP is:

$$\begin{aligned} &\min\ f_1(x), f_2(x), \ldots, f_m(x), \\ &\text{s.t. } x \in \Omega \\ &\quad \wp\,(f_1, f_2, \ldots, f_m) \end{aligned}$$

where m is the number of objective functions, Ω is the feasible variables space, also called decision space or search space, and $\wp\,(\cdot)$ is a generic priority structure over the objectives. In the absence of $\wp\,(\cdot)$, the optimization is implicitly assumed to be of Paretian type, and solved with the usual EMO algorithms. At the opposite of Pareto optimization, another well-known corner case of MPL-MOPs is represented by lexicographic problems [36, 39], where $\wp\,(\cdot)$ arranges the objectives according to a total order of importance. Lexicographic problems can be denoted as:

$$\begin{aligned} &\text{lex}\min\ f_1(x), f_2(x), \ldots, f_m(x) \\ &\text{s.t. } \ x \in \Omega \end{aligned} \tag{1}$$

The expression indicates the minimization of the objectives $f_1, f_2, \ldots, f_m$ ranked lexicographically: f_1 holds absolutely priority over f_2, which in turn has absolute priority over f_3, and so on.

The more general case where $\wp\,(\cdot)$ is an arbitrary function (i.e., a generic MPL-MOP) has not received enough research attention, despite the applicability to potentially many real-world scenarios, like [21]. Some studies proposed ad-hoc approaches to deal with generic priority relations, including linear scalarization [12] or ϵ-constrained methods [12]. Other scalarization strategies can be found in [22, 38]. Despite being more effective than general purpose ones, these algorithms are not purposely designed to cope with MPL-MOPs, so they have significant shortcomings. For instance, in scalarized problems the weights must be chosen carefully, otherwise the solutions may be wrong or very inaccurate [12], sometimes, the numerical stability of the algorithm may get worse too. Moreover, scalarized approaches typically find only one solution at a time, with dramatic consequences on the overall computation time. Even the simplest priority structure in (1) hides non-trivial complexities. A common approach, called preemptive method, is to solve the single-objective subproblems sequentially: first the optimal values for the most important objective are found, then a constraint is added to force the solutions of the next subproblems to be in the optimal set of the primary objective. The same procedure is iterated over the objectives, in order of importance. However it suffers from the growing size of the set of constraints, which makes it increasingly harder to locate feasible solutions. Also the constraints may have an arbitrary from, hard to handle.

The word *mixed* in MPL-MOPs reflects the non-homogeneous nature of these problems, suggesting at the same time an hybrid approach to deal with them. How-

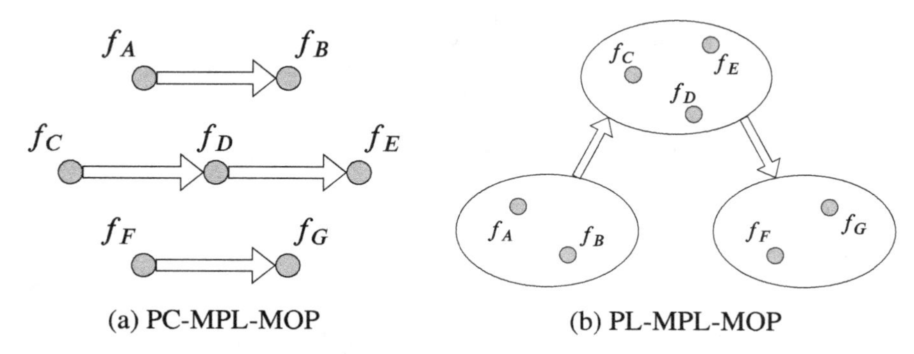

Fig. 1 Examples of two different classes of MPL-MOPs: dots represent objectives, arrows precedences, and ovals groups of objectives

ever, the relationship between Pareto optimality and priorities strongly depends on the structure of the specific problem, which in turn affects the mathematical description of the problem itself. Algorithms tailored around this model can potentially converge faster and find better solutions. The next sections introduce two of the significant families of MPL-MOPs. The first one is characterized by chain-like priority structures [27], where the objectives are indeed partitioned and a lexicographic ordering (a chain) exists within each partition. No preference ordering is defined between the chains; therefore, their simultaneous optimization is Paretian. The above problems are called Priority-Chain MPL-MOPs, or PC-MPL-MOPs in short. As opposed to them, Priority-Levels MPL-MOPs (PL-MPL-MOPs in brief) feature a $\wp(\cdot)$ that arranges the objectives into levels [28, 29]: no priorities exist within each level, and their functions are optimized concurrently in the Paretian way. On the other hand, a lexicographic priority scheme is defined across the levels themselves, that is the Pareto optimization of a level has priority over that of another level. Figure 1 illustrates two examples of PC- and PL-MPL-MOPs.

5 The Priority Chains Model

The analytical model of a PC-MPL-MOP [27] is:

$$\min \begin{bmatrix} \text{lex min } f_1^{(1)}(x), f_1^{(2)}(x), \ldots, f_1^{(p_1)}(x) \\ \text{lex min } f_2^{(1)}(x), f_2^{(2)}(x), \ldots, f_2^{(p_2)}(x) \\ \vdots \\ \text{lex min } f_m^{(1)}(x), f_m^{(2)}(x), \ldots, f_m^{(p_m)}(x) \end{bmatrix} \quad \text{s.t. } x \in \Omega. \tag{2}$$

where $\text{lex min } f_i^{(1)}, f_i^{(2)}, \ldots, f_i^{(pi)}$ indicates the minimization of the objectives $f_i^{(1)}, f_i^{(2)}, \ldots, f_i^{(pi)}$ according to the lexicographic ordering: $f_i^{(1)}$ is has priority over $f_i^{(2)}$, which in turn has priority over $f_i^{(3)}$, and so on. Each lex min block can be seen as a "container" of objectives ranked by importance: for clarity, it is referred to as *macro-objective* and denoted as $\underline{f_i}$. On the other hand, the outer min indicates the minimization (in the Pareto sense) of the macro-objectives. So (2) can be rewritten as

$$\min \begin{bmatrix} \underline{f_1}(x) \\ \underline{f_2}(x) \\ \vdots \\ \underline{f_m}(x) \end{bmatrix} \quad \text{s.t. } x \in \Omega$$

where $\underline{f_1}, \underline{f_2}, \ldots, \underline{f_m}$ are the macro-objectives.

Moving from standard objectives to macro-objectives can be fairly intuitive. When comparing the same macro-objective, say $\underline{f_i}$, of two different solutions, the one with the lowest value of the primary objective, i.e., $f_i^{(1)}$ is preferred. If the two values happen to be equal, then the preference is determined by the secondary objective, or the tertiary if they match again, and so on. In other words, secondary objectives are useful to compare solutions with equal values on the relatively more important objectives, consistently with the lexicographic ordering. Note the absence of any mutual precedence between the objectives of different macro-objectives, they are unrelated. This property implies that swapping e.g. two secondary objectives of a PC-MPL-MOPs would modify the problem itself. This is not the case for PL-MPL-MOPs.

Here is a minimal example on how to compare between macro-objectives. Consider three solutions of a minimization problem with two macro-objectives, each being a chain of two (shown horizontally):

$$A = \begin{bmatrix} [1\ 2] \\ [5\ 3] \end{bmatrix} \qquad B = \begin{bmatrix} [1\ 3] \\ [4\ 2] \end{bmatrix} \qquad C = \begin{bmatrix} [7\ 1] \\ [4\ 9] \end{bmatrix}$$

Solution A performs better than B in the first macro-objective (because $1 = 1$ and $2 < 3$), however B is better in the second macro-objective ($4 < 5$, the secondary objective does not matter this time), therefore A and B are Pareto nondominated. A and C are nondominated too, since $1 < 7$ and $5 > 4$. On the other hand, B is better than C because it dominates on both the macro-objectives: in fact, $1 < 7, 4 = 4$ and $2 < 9$.

5.1 *Need of a New Approach for PC-MPL-MOPs*

Before introducing a novel approach for PC-MPL-MOPs (Sect. 6), it is important to first highlights the limits of the standard Paretian ones. An non-priority-aware optimizer like NSGA-II can be essentially used in two ways for PC-MPL-MOPs: ignoring the priorities or the secondary objectives. Unfortunately, both alternatives have critical drawbacks.

5.1.1 Ignoring Secondary Objectives

Ignoring all the objectives except the most important ones simplifies the problem at the cost of discarding important information, hence worsening the overall quality of the found solutions. As discussed previously, the purpose of secondary objectives is essentially to help discriminating between two or more solutions when the primaries are not enough to establish a clear winner. Without secondaries, the algorithm is bound to either preserve all the nondominated solutions, that is computationally expensive, or discard one or more of them, randomly or according to some policy, still with the risk of trashing potentially good ones thus slowing the optimization. When the number of nondominated solutions is large, a common situation in many-objectives tasks, these strategy fails with a notable performance degradation.

5.1.2 Ignoring the Priorities, Possibly Filtering a Posteriori

Another possibility is to treat all the objectives as if they had the same importance, no matter what their real priorities are, and run a conventional optimization algorithm on all these objectives. Then, *after* the end of the procedure, one may post-process the final population filtering out the solutions that would be dominated by other individuals in the same set if the priorities were eventually taken into account. Although slightly less naïve than the previous option, this one shows critical flaws too. First, a problem with m chains of p objectives each would become a purely Paretian problem with $m \times p$ objectives, as a many-objective one, with all the negative consequences that this entails. Also, a secondary objective that is instead given the same importance of a primary one forces the algorithm to concurrently optimize both, or at least to try to; if those are somehow conflicting, spurious nondominated solutions or even clusters of them may emerge, in a way that is absolutely detrimental to the original optimization task. Moreover, the filtering procedure, despite being useful to pull out the very best solutions from the final set, may actually cause an overly severe skimming, sometimes preserving very few individuals out of many. Computationally speaking, ending up with two or three solutions after running the algorithm on a population of thousands is dramatically inefficient. The last two difficulties might also stack together (i.e., filtering an already mediocre set), exasperating the issue even further.

5.2 *Grossone for PC-MPL-MOPs*

A very useful application of Grossone Methodology [44] is a ploy to reformulate lexicographic problems in a way that allows performing actual numerical computations by means of macro-objectives, something that is not possible with standard model. Grossone has been successfully applied in several other optimization problems, such as in linear and non-linear programming [8, 9, 11], in global optimization [46], in large-scale unconstrained optimization [10], in machine learning [2, 3, 20] and in many other areas. In addition, the Grossone Methodology has been shown to be independent from non-standard analysis [45]. The use made of Grossone here is rather simple: given the lexicographic problem lex min $f^{(1)}(x), f^{(2)}(x), \ldots, f^{(p)}(x)$, a G-scalar is built using each objective $f^{(j)}$ as a G-digit relative to the G-power $①^{1-j}$, an idea firstly proposed in [43]. Thus that problem can be rewritten as

$$\min f^{(1)}(x) + ①^{-1} f^{(2)}(x) + \cdots + ①^{1-p} f^{(p)}(x)$$

or equivalently min $\underline{f}(x)$. The higher the priority, the higher the G-power: the most important objective, $f^{(1)}(x)$, is indeed associated the highest exponent (0), whereas $f^{(p)}(x)$ the lowest $(1 - p)$. The advantage of this approach is clear: whereas before we had objectives represented by real numbers, now we have macro-objectives represented by G-scalars. Since all the four basic operations are well-defined for G-numbers, replacing objectives with macro-objectives in an algorithm is a completely transparent operation, which does not alter the inner logic of the code. Consequently, many existing non-lexicographic optimization algorithms can potentially be extended to solve certain types of lexicographic problems too. This same technique has been recently adopted to generalize the simplex algorithm [5, 6]. So, PC-MPL-MOPs can be easily reformulated with Grossone as

$$\min \begin{bmatrix} f_1^{(1)}(x) + ①^{-1} f_1^{(2)}(x) + \cdots + ①^{1-p_1} f_1^{(p_1)}(x) \\ f_2^{(1)}(x) + ①^{-1} f_2^{(2)}(x) + \cdots + ①^{1-p_2} f_2^{(p_2)}(x) \\ \vdots \\ f_m^{(1)}(x) + ①^{-1} f_m^{(2)}(x) + \cdots + ①^{1-p_m} f_m^{(p_m)}(x) \end{bmatrix}.$$

6 PC-NSGA-II and PC-MOEA/D

This section aims at generalizing the original NSGA-II [15] and MOEA/D [53] algorithms in order to allow them to solve PC-MPL problems. As said, this result will be achieved by extending them to handle G-scalars, giving birth to *PC-NSGA-II* and *PC-MOEA/D*, respectively. Of course these new versions must be equipped with the redefinition of the four elementary operations too, in order to operate with G-scalars. Different algorithms could be chosen as well, e.g. SPEA2, but NSGA-II and MOEA/D were eventually picked because they are popular, simple and parameterless, in the sense that they do not introduce other parameters than those typically required by genetic algorithms. Notice that such extended versions inherit the very same limits and benefit of the underlying algorithms when facing MOPs or MaOPs. As it will become clear in Sect. 7.1, inheriting limits and benefits means that if an algorithm, for instance, works particularly well with binary genotypes, then its PC version will manifest that property as well. On the contrary, if an algorithm becomes less effective as the number of objectives increases, its PC counterpart will suffer MaOPs with many chains. In order to further stress the enhancement in the supported data types, he algorithms core procedures have been renamed, e.g., *PC_fast_nondominated_sort* and *PC_crowding_distance_assignment*. Again, these are aliases for the old names, as the logic behind the functions is still the same.

Consider its pseudocode in Algorithm 1:

- *PC_fast_nondominated_sort*: the nondominance operator $\underline{\prec}$ (underlined) replaces the old $\prec$.

This is very similar to the traditional definition of dominance, but the comparisons are now between macro-objectives (G-scalars) instead of single objectives (reals). Its formal definition is:

Definition 1 (*Pareto-Lexicographic dominance (PC-dominance)*) Given two solutions A and B, A dominates B (written $A \underline{\prec} B$) if:

$$A \underline{\prec} B \iff \begin{cases} \underline{f_i}(x_A) \leq \underline{f_i}(x_B) & \forall i = 1 \dots m \\ \exists j : \underline{f_j}(x_A) < \underline{f_j}(x_B) \end{cases} \tag{3}$$

The underlining notation has general validity: every time it appears, from now on, it indicates G-scalar quantities or operations between them. This remains consistent with the use made in Sect. 7. Please, notice that the lexicographic contribution is implicitly performed within the comparisons between two G-scalars.

- *PC_crowded_comparison_operator*: it works just like the old $\prec_n$ operator described in [15], but the last comparison is now performed between G-scalars because of the nature of PC_crowding_distance. It is denoted by $\underline{\prec}_n$.
- *PC_crowding_distance*: it is a G-scalar, since it is computed by means of operations (subtractions, divisions, etc.) between G-scalars. G-operations, by design, preserve the relative importance between the objectives, and such property is reflected in the crowding distance computation too. Being the finite component of the macro-objectives, primary objectives play a major role in the density estimation, while the secondaries, mostly contributing to the infinitesimal part, are still very useful in those situations where it is hard to choose, e.g. when sorting two elements that have the same finite crowding distance but different infinitesimal values. The definition also makes sense from the perspective of the decision-maker, as it is reasonable to prefer having a larger variety of options in the domain of the most relevant aspects, rather than for less important objectives. Notice that the crowding distance of the extreme points can be initialized to ①, similarly to NSGA-II where it is set to ∞. If we represent the priorities with non-positive G-powers only (a non-restrictive assumption) ① is always guaranteed to be greater than any other crowding distance value.

Algorithm 1: PC-NSGA-II algorithm

1: **procedure** PC_NSGA_II

2: $R_t = P_t \cup Q_t$

3: /* *fast_nondominated_sort with gross-scalar objectives* */

4: $F = PC\_fast\_nondominated\_sort(R_t, 0)$

5: $P_{t+1} = \emptyset$

6: $i = 1$

7: **while** $|P_{t+1}| + |F_i| \leq N$ **do**

8: /* *Crowding distance with gross-scalars* */

9: $PC\_crowding\_distance\_assignment(F_i)$

10: $P_{t+1} = P_{t+1} \cup F_i$

11: $i = i + 1$

12: /* *Sort by PC_crowded_comparison_operator* */

13: $sort(F_i, \preceq_n)$

14: $P_{t+1} = P_{t+1} \cup F_i[1 : (N - [P_{t+1}])]$

15: $Q_{t+1} = make\_new\_pop(P_{t+1})$

16: $t = t + 1$

1: **procedure** PC_FAST_NONDOMINATED_SORT(P)

2: **for all** $p \in P$ **do**

3: $S_p = \emptyset$, $n_p = 0$

4: **for all** $q \in P$ **do**

5: **if** $p \preceq q$ **then** ▹ PC-dominance being used here

6: $S_p = S_p \cup \{q\}$

7: **else if** $q \preceq p$ **then** ▹ PC-dominance used also here

8: $n_p = n_p + 1$

9: **if** $n_p = 0$ **then**

10: $p_{rank} = 1$

11: $F_1 = F_1 \cup \{p\}$

12: $i = 1$

13: **while** $F_i \neq \emptyset$ **do**

14: $Q = \emptyset$

15: **for all** $p \in F_i$ **do**

16: **for all** $q \in S_p$ **do**

17: $n_q = n_q - 1$

18: **if** $n_q = 0$ **then**

19: $q_{rank} = i + 1$

20: $Q = Q \cup \{q\}$

21: $i = i + 1$

22: $F_i = Q$

1: **procedure** PC_CROWDING_DISTANCE_ASSIGNMENT(F)

2: $n = |F|$ ▹ number of solution in the current front F

3: **for all** $i \in F$ **do**

4: $F[i]_{\underline{dist}} = 0$

5: **for** $j = 1 \ldots m$ **do** ▹ m is the number of macro-objectives

6: $F = sort(F, f_j)$ ▹ sort F according to j-th macro-objective

7: $F[1]_{\underline{dist}} = F[n]_{\underline{dist}} = \text{①}$

8: **for** $i = 2 \ldots (n-1)$ **do**

9: /* *the PC_crowding_distance is a gross-scalar* */

10: $F[i]_{\underline{dist}} = F[i]_{\underline{dist}} + \frac{\underline{f_j}(F[i+1]) - \underline{f_j}(F[i-1])}{\underline{f_j}^{max} - \underline{f_j}^{min}}$

Similarly, MOEA/D can be extended by means of G-scalars to obtain PC-MOEA/D (see Algorithm 2). For brevity reasons we do not go into the algorithm details, but please note that the PC extension of MOEA/D is even easier than the NSGA-II one.

Algorithm 2: PC-MOEA/D algorithm

```
1: procedure PC_MOEA/D
2:     EP = ∅
3:     for i = 1 . . . N do
4:         λ^{i_1}, . . . , λ^{i_T} = find_closest_weights(λ_i, T)
5:         B_i = {i_1, . . . , i_T}
6:     x^1, . . . , x^N = initialize_population(N)
7:     for i = 1 . . . N do
8:         FV^i = F(x^i)
9:     z_1, . . . , z_m = initialize_ref_point(m)
10:    while stop_criteria() = False do
11:        for i = 1 . . . N do
12:            k, l = rand(B_i, 2)
13:            y = make_new_sol(x^k, x^l)
14:            y' = mutate(y)
15:            for j = 1 . . . m do
16:                if z_j < f_j(y') then
17:                    z_j = f_j(y')
18:            for all j ∈ B_i do
19:                if g^te(y'|λ^j, z) ≤ g^te(x^j|λ^j, z) then
20:                    x^j = y'
21:                    FV^j = F(y')
22:            DD = PC_find_dominated_sol(EP, F(y'))
23:            DG = PC_find_dominating_sol(EP, F(y'))
24:            EP = EP \ DD
25:            if DG = ∅ then
26:                EP = EP ∪ F(y')
```

Indeed, the function to compute the dominance relation is the only one which needs to be extended to the G-scalar case. The new function works similarly to its old version except for the dominance operator, which becomes the $\preceq$ described in (3) instead of the standard $\prec$.

7 Test Cases for PC-MPL MOPs

Many test problems were proposed to assess the performance of multi-objective optimization algorithms [16, 50, 54]. However, very few articles include problems where priorities exist among the objectives, and none specifically for the MPL class. The following sections contain some new PC-MPL benchmarks to verify the effectiveness of the PC-generalizations of Sect. 6. More problems are also discussed in [27]. Note that this study aims only at validating the conceptual superiority of priority-structure-aware algorithms design versus their native counterparts or other priority-based algorithms; instead, it is out of the scope of this chapter to analyze the raw performance of the PC-algorithms, which ultimately depends on the backbone MOP algorithm (NSGA-II or MOEA/D here). A detailed analysis concerning the identification and construction of interesting and challenging PC-MPL benchmarks is left for future studies.

All the experiments use SBX crossover and polynomial mutation for real-encoded genotypes, and two-point crossover and flip-bit mutation with probability $\frac{1}{l}$ (bit-string length l) for binary-encoded ones. The experiments are repeated 50 times for each benchmark, with 500 epochs per run before halting the algorithm. The initial population is made of 100 random individuals. In addition to PC-NSGA-II and PC-MOEA/D, six more algorithms compete on each benchmark: NSGA-II and MOEA/D ignoring the secondary objectives (Sect. 5.1.1), NSGA-II and MOEA/D ignoring the priorities and a posteriori filtering (Sect. 5.1.2), Tan et al. [49], Chang et al. [4].

7.1 Test Problem: PC-1

PC-1 is a PC-MPL 01-knapsack problem, featuring 6 objectives (2 chains of 3 objectives):

$$\begin{aligned}
&\max \begin{bmatrix} \underline{f}_1(x) \\ \underline{f}_2(x) \end{bmatrix} \\
&\text{s.t.} \quad Wx \leq C \\
&\underline{f}_j(x) = \underline{V}_j x \\
&\underline{V}_j \;\; = V_j^{(1)} + ①^{-1} V_j^{(2)} + ①^{-2} V_j^{(3)} \\
&V_1^{(1)} = [7, 1, 5, 2, 9, 4, 4, 2, 8, 2, 1, 6, 9, 5, 4, 4, 9, 5, 3, 3]^T \\
&V_1^{(2)} = [3, 1, 6, 6, 7, 2, 5, 2, 1, 9, 1, 8, 9, 6, 7, 4, 4, 5, 9, 4]^T \\
&V_1^{(3)} = [2, 1, 3, 5, 8, 9, 5, 7, 1, 6, 4, 7, 9, 5, 1, 5, 5, 4, 4, 2]^T \\
&V_2^{(1)} = [1, 7, 3, 6, 2, 3, 7, 7, 9, 9, 5, 3, 5, 3, 1, 8, 6, 1, 9, 3]^T \\
&V_2^{(2)} = [2, 6, 4, 7, 7, 4, 2, 4, 9, 4, 3, 3, 6, 8, 2, 8, 1, 6, 8, 8]^T \\
&V_2^{(3)} = [8, 9, 5, 9, 8, 9, 4, 9, 7, 5, 4, 5, 3, 4, 4, 7, 3, 8, 2, 4]^T \\
&W \;\;\; = [4, 8, 6, 5, 5, 5, 7, 4, 8, 4, 4, 1, 8, 7, 4, 3, 5, 1, 9, 5]^T \\
&C \;\;\;\; = 50, \quad x_i \in \{0, 1\} \;\; i = 1, \ldots, 20
\end{aligned} \tag{4}$$

To give an idea of how important the secondaries are in the optimization process, consider what happens when running PC-NSGA-II and NSGA-II (primary-only) on this problem. PC-NSGA-II gives the result shown in Fig. 2a, which includes 19 distinct solutions. The True Pareto front (computed through brute-force enumeration), consists of 22 solutions (see [27]). A comparison between the two highlights that, out of the 19 found solutions, 16 are true optima, 2 differ from their closest true optimum for one secondary, 1 for a primary. As for the 6 missing true optima, another solution with the same primary values was found for 2 of them, while for the other 4 the primaries error is <= 3.

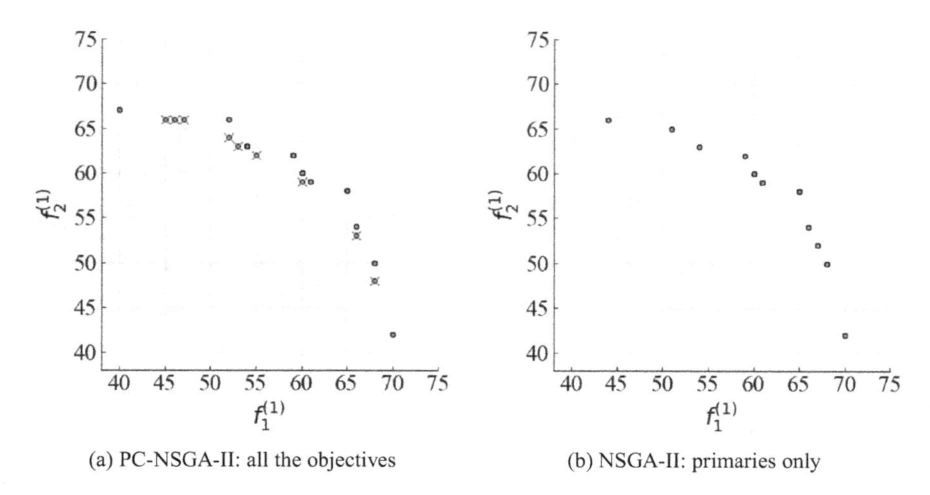

Fig. 2 Primary objectives of PC-1, after 300 generations with 400 individuals. PC-NSGA-II solutions that would be instead discarded by NSGA-II (i.e. missing on the right plot) are marked with a cross

With NSGA-II, the secondary objectives are computed only at the end of the optimization process, and unused during the evolution phase. Not only it finds fewer solutions (Fig. 2b), but also of lower quality. Out of 11, 8 are true optima, 2 are sub-optimal with respect to primaries, 1 to secondaries. This shortage of solutions from NSGA-II is the consequence of its inability to preserve "apparently horizontal/vertical" solutions, which differ in their infinitesimal components only.

The results are analyzed quantitatively with the $\Delta(\cdot)$ metric, already introduced in Sect. 3. Table 1 reports the mean and standard deviation of the performance measurements collected over 50 runs. It turns out that the PC algorithms do better than their standard counterparts. Specifically, PC-MOEA/D turns out to be the best performing algorithm, immediately followed by MOEA/D without secondaries. This confirms that PC algorithms inherit benefits and some limitations from their underlying optimizer too. To clarify, based on the empirical evidences, it is reasonable to say that MOEA/D works consistently better than NSGA-II on this specific benchmark (PC-1). Indeed, not only PC-MOEA/D gets a lower score (i.e., mean) than PC-NSGA-II, but also MOEA/D without secondaries beats NSGA-II without secondaries, and the same is true even for the third version of MOEA/D versus NSGA-II. Of course, the property of MOEA/D being better than NSGA-II is not a general one: as we will see, there exist other problems where the opposite is true, that is NSGA-II beats MOEA/D. Moreover, the algorithms without secondaries usually perform similarly to their PC variation, at least on the primary objectives, since both try to optimize them in a Pareto manner. These considerations justify why PC-MOEA/D is at the top of the ranking and why MOEA/D without secondaries is second.

Table 1 Mean and Std of metric $\Delta(\cdot)$ on PC-1 after 50 repetitions

Algorithm	Mean	Std
PC-MOEA/D	$\mathbf{0.36 + 0.25①^{-1} + 0.99①^{-2}}$	$\mathbf{0.31 - 0.07①^{-1} + 1.11①^{-2}}$
MOEA/D pre	$0.54 + 0.43①^{-1} + 3.22①^{-2}$	$0.19 - 0.22①^{-1} + 1.37①^{-2}$
MOEA/D post	$1.44 - 0.22①^{-1} + 0.88①^{-2}$	$0.54 + 0.15①^{-1} + 0.59①^{-2}$
PC-NSGA-II	$\mathbf{1.6 + 0.17①^{-1} + 1.2①^{-2}}$	$\mathbf{0.74 + 0.19①^{-1} + 0.44①^{-2}}$
NSGA-II pre	$1.95 + 0.57①^{-1} + 2.37①^{-2}$	$0.91 + 0.21①^{-1} + 0.53①^{-2}$
NSGA-II post	$1.79 + 0.08①^{-1} - 0.77①^{-2}$	$0.33 - 0.02①^{-1} + 0.88①^{-2}$
Tan et al.	$12.83 + 2.71①^{-1} + 4.03①^{-2}$	$0.73 + 0.04①^{-1} + 1.2①^{-2}$
Chang et al.	$18.01 + 4.45①^{-1} - 4.42①^{-2}$	$2.53 + 2.31①^{-1} + 3.48①^{-2}$

7.2 *Test Problem: PC-3*

The PC-3 test case is a modified version of POL, a well-known benchmark based on the Poloni's study [41], part of the Van Veldhuizen's suite [50]:

$$\begin{aligned}
&\min \begin{bmatrix} f_1^{(1)}(x) + ①^{-1} f_1^{(2)}(x) \\ f_2^{(1)}(x) \end{bmatrix} \\
&f_1^{(1)}(x) = \left\lfloor \alpha \left(1 + (A_1 - B_1)^2 + (A_2 - B_2)^2\right) \right\rfloor / \alpha \\
&f_2^{(1)}(x) = \left\lfloor \alpha \left((x_1 + 3)^2 + (x_2 + 1)^2\right) \right\rfloor / \alpha \\
&f_1^{(2)}(x) = x_1^2 \\
&A_1 = 0.5 \sin(1) - 2\cos(1) + \sin(2) - 1.5\cos(2) \\
&A_2 = 1.5 \sin(1) - \cos(1) + 2\sin(2) - 0.5\cos(2) \\
&B_1 = 0.5 \sin(x_1) - 2\cos(x_1) + \sin(x_2) - 1.5\cos(x_2) \\
&B_2 = 1.5 \sin(x_1) - \cos(x_1) + 2\sin(x_2) - 0.5\cos(x_2) \\
&x_1, x_2 \in [-\pi, \pi], \quad \alpha = 3
\end{aligned}$$

The original structure of POL remains unchanged, except for the insertion of the *floor* function (to discretize the problem) and the addition of a third objective, $f_1^{(2)}$, which is to be considered less important than $f_1^{(1)}$.

The parameter α can be tuned to adjust the granularity of the discretization. In the nondominance ranking procedure, the secondary objective $f_1^{(2)}$ intervenes only when two solutions have the same value of $f_1^{(1)}$, favoring the one whose coordinate x_1 is closer to 0. Therefore, it is reasonable to expect the new optimal solutions to be very similar to those of original POL, at most slightly shifted towards the line $x_1 = 0$; the magnitude of such effect depends on α.

Figure 4 shows the results of PC-NSGA-II for PC-3. As expected, the shape of the Pareto front is similar to the POL one, the most noticeable difference being that it appears as a set of disconnected points due to the discretization (floor function). As

Table 2 Mean and Std of metric $\Delta(\cdot)$ on PC-3 after 50 repetitions

Algorithm	Mean	Std
PC-NSGA-II	$\mathbf{0.01}$	$\mathbf{0.02 + 0.01①^{-1}}$
NSGA-II pre	$0.34 + 0.09①^{-1}$	$0.09 + 0.02①^{-1}$
PC-MOEA/D	$\mathbf{0.59 - 0.02①^{-1}}$	$\mathbf{0.58 - 0.02①^{-1}}$
NSGA-II post	$0.65 - 0.08①^{-1}$	$0.16 - 0.07①^{-1}$
MOEA/D pre	$0.89 + 0.05①^{-1}$	$1.4 - 0.01①^{-1}$
MOEA/D post	$1.23 - 0.03①^{-1}$	1.48
Tan et al.	$1.34 - 0.82①^{-1}$	$0.28 - 0.16①^{-1}$
Chang et al.	$38.06 - 7.83①^{-1}$	$3.0 - 0.09①^{-1}$

it is often the case, there are some solutions in the Pareto frontier of primary objectives with equal values of $f_1^{(1)}$ but different values of $f_2^{(1)}$, which graphically appear as vertically-aligned points. While the picture seems to show only a few dozens of solutions, there are in fact 700 fighting for optimality, but many of them are so close to each other that they overlap. Using the standard approach of removing priorities, running NSGA-II and eventually filtering by priority-aware nondominance, the obtained results are those of Fig. 3. What at first glance seems a richer front is actually noise. Treating the third objective as important as the other two causes a dramatic loss of accuracy in the search for optimal solutions, because part of the optimization focus is shifted towards the minimization of this objective and dragged away from the minimization of the other (most important) two. Indeed, after filtering the final set of NSGA-II (Fig. 3), only 7 out of 700 solutions are left, a very small fraction of the total population. Also, the solutions found by NSGA-II turn out to be farther away from the real Pareto front (thus worse) than those found by PC-NSGA-II. Essentially, it happens because NSGA-II completely ignores the priorities of the objectives, not taking advantage of that valuable information. On the other hand, PC-NSGA-II is able to exploit it, succeeding to find more accurate solutions.

Table 2 reports the mean and the standard deviation of the algorithms performances on the PC-3 benchmark. We observe once again that the PC algorithms perform better than their standard counterparts and the other priority-based approaches. Moreover, PC-NSGA-II turns out to be the overall best performing algorithm this time; NSGA-II appears to work better than MOEA/D on the current benchmark. Indeed, the second best performing algorithm is NSGA-II without secondaries. However, PC-MOEA/D still gets a good score, being third and very close to the second.

8 The Priority Levels Model

PL-MPL-MOPs [28, 29] allow one to consider groups of conflicting objectives ordered by priority. For instance, among the seven objectives of Fig. 1b, a prior-

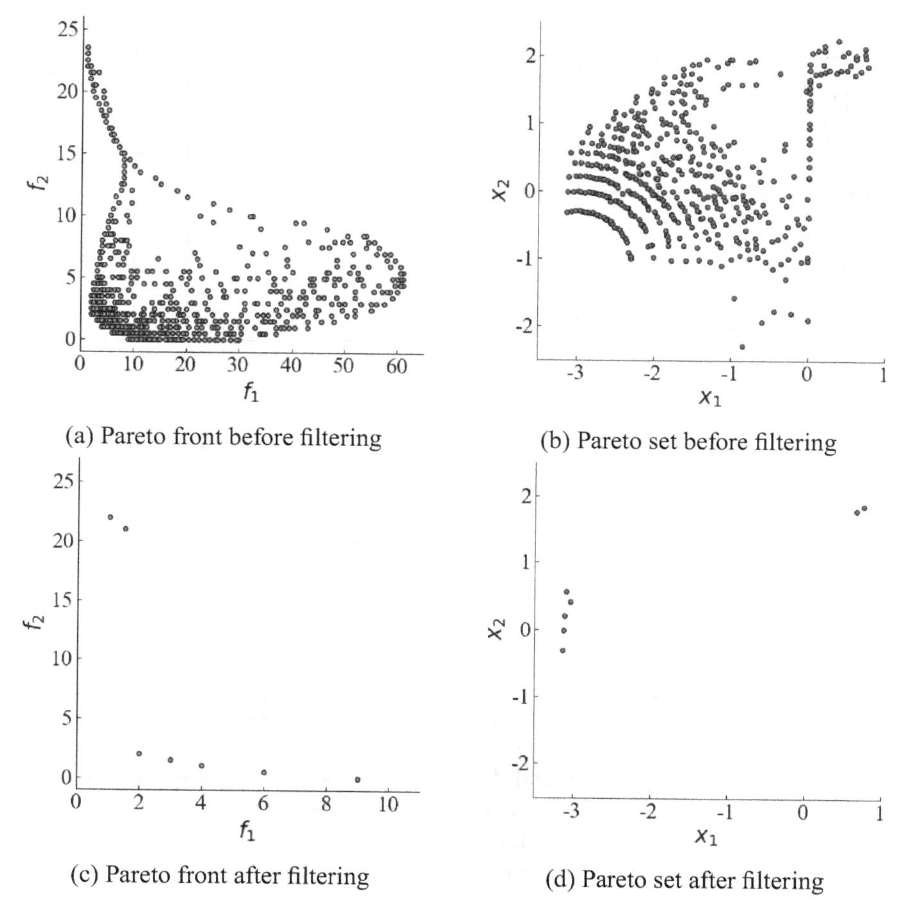

Fig. 3 Obtained solutions by NSGA-II for PC-3

ity structure can be defined as illustrated: f_A and f_B are in the same priority level and f_C, f_D and f_E are also in the same priority level, but the former is lexicographically more important than the latter. The objectives f_F and f_G are in the same priority level, which has lesser priority than the other ones.

Definition 2 (*PL-MPL-MOP*) Any MPL-MOP satisfying the two conditions below is a PL-MPL-MOP:

- Some groups of objectives have clear precedence over some others and they can be totally ordered according to it
- Within each group, objectives cannot be compared to each other on the basis of importance

Finally, the mathematical formulation of a PL-MPL-MOP is:

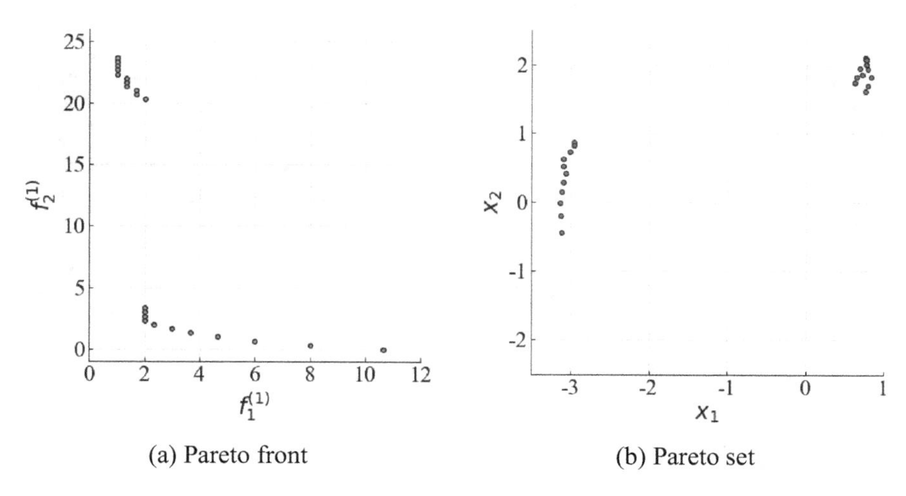

(a) Pareto front (b) Pareto set

Fig. 4 Solutions obtained by PC-NSGA-II for PC-3

$$\text{lex}\min\left[\min\begin{pmatrix} f_1^{(1)}(x) \\ f_2^{(1)}(x) \\ \vdots \\ f_{m_1}^{(1)}(x) \end{pmatrix}, \min\begin{pmatrix} f_1^{(2)}(x) \\ f_2^{(2)}(x) \\ \vdots \\ f_{m_2}^{(2)}(x) \end{pmatrix}, \ldots, \min\begin{pmatrix} f_1^{(p)}(x) \\ f_2^{(p)}(x) \\ \vdots \\ f_{m_p}^{(p)}(x) \end{pmatrix}\right], \tag{5}$$

where p is the number of levels and $f_i^{(j)}$ is the i_{th} objective in the j_{th} level.

In a PL-MPL-MOP, the Pareto-optimal solutions of the objectives in the first level form the decision space for the Pareto optimization of the objectives in the second level, and then again the latter solutions become the domain for the third one, and so on for every level. PL-MPL-MOPs are structurally quite similar to purely lexicographic problems, except that each element of the sequence is now a multi-objective problem on its own, instead of single-objective function in the original lexicographic sense [40]. Of course, the formulation above covers the original case as well, since a purely lexicographic problem is nothing but a particular case with one objective function in each class. Similarly, the classical formulation of Pareto MOPs occurs when there is only one priority level.

8.1 A First Example of a Real-World PL-MPL-MO Problem

The automotive crashworthiness problem [34] consists in finding the best car designs balancing performance and safety. The original benchmark aims at maximizing the vehicle acceleration while minimizing mass and toe-board intrusion. Even if the three objectives are all necessary for a consistent design, one may assume that they are not all equally important for a firm, also their mutual relation may vary across manufacturers. For instance, a company may be interested in designing high performing cars, in which case it would put more emphasis on the first two objectives (mass and acceleration) than on the safety one. Therefore, during the optimization the toe-board intrusion should play a role in discriminating among the high speed cars, rather than on the whole pool of vehicles. A possible approach is to reformulate the problem in a PL-fashion, as illustrated in (6). The model splits the original 3-objective optimization task in two PLs: the first (high priority) contains the mass and the acceleration, while the second (lower priority) focuses on safety while still preserving the goal of maximizing the acceleration.

$$\begin{aligned} &\operatorname{lex\,min} \left[\min \begin{pmatrix} \texttt{mass}(x) \\ -\texttt{accel}(x) \end{pmatrix}, \min \begin{pmatrix} \texttt{toe}(x) \\ -\texttt{accel}(x) \end{pmatrix} \right] \\ &\text{s.t. } x \in [1,\ 3]^5. \end{aligned} \tag{6}$$

Conversely, other firms may prefer to focus on safe cars. Such problem will be analyzed in particular in the experimental section:

$$\begin{aligned} &\operatorname{lex\,min} \left[\min \begin{pmatrix} \texttt{mass}(x) \\ \texttt{toe}(x) \end{pmatrix}, \min \begin{pmatrix} \texttt{toe}(x) \\ -\texttt{accel}(x) \end{pmatrix} \right] \\ &\text{s.t. } x \in [1,\ 3]^5. \end{aligned}$$

8.2 *Need of a New Approach for PL-MPL-MOPs*

In this subsection, we aim to point out the limitations of strategies based on standard optimization techniques when solving PL-MPL-MOPs. In the absence of tools that are able to concurrently perform Pareto and lexicographic optimization, any viable approach necessarily requires these two to run separately in distinct stages. How these steps are interleaved may vary, leading to different options. We identify here two strategies that are worth of consideration and which recall a lot those in Sect. 5.1.

The first way is to simply ignore the precedence relations during the multi-objective search, thus Pareto optimizing all the objectives together, indistinctly. Only afterwards the priorities are considered, by filtering the solutions that are group-lexicographically optimal. We refer to this approach as *postfiltering*. Clearly, such an approach is computationally inefficient, as many more solutions need to be found by the original EMO algorithm in order to have a sizable number of solutions at the end.

The second approach uses a multi-objective optimizer solving the problem considering only the first level. Only at the end of the process, the objectives from the other levels are exploited to rank the obtained solutions. This scheme, called here *prefiltering*, is motivated by the fact that primary objectives have the highest impact during the optimization. The weakness, however, is the little to no influence of the less important objectives in the result. Unlike the postfiltering method, it puts indeed a lot of responsibility on the primary objectives.

Section 10 presents a new algorithm that is able to overcome these difficulties, that does not split the optimization into sub-tasks, uses all the objectives concurrently, yet still privileges those belonging to higher levels. Again, the approach uses the Grossone Methodology.

8.3 Enhancing PL-MPL Problems by Means of Grossone

The idea to implement PL-aware algorithms is the similar to the one in Sect. 5.2 in the way it uses ① for lexicographic ranking. However, here the ① powers represent scaling factors for groups of objectives, i.e., multi-objective optimization tasks rather than single functions. PL-MPL-MOPs can be thus reformulated as:

$$\min \left[\begin{pmatrix} f_1^{(1)}(x) \\ f_2^{(1)}(x) \\ \vdots \\ f_{m1}^{(1)}(x) \end{pmatrix} + ①^{-1} \begin{pmatrix} f_1^{(2)}(x) \\ f_2^{(2)}(x) \\ \vdots \\ f_{m2}^{(2)}(x) \end{pmatrix} + \cdots + ①^{1-p} \begin{pmatrix} f_1^{(p)}(x) \\ f_2^{(p)}(x) \\ \vdots \\ f_{mp}^{(p)}(x) \end{pmatrix} \right].$$

Without going into the mathematical details, the optimization problem can be thought has a minimization problem (and not a lexicographic one) of the PLs, whose contribution is weighted in accordance to their priority. The level with the highest priority is weighted by $①^0$ (=1, omitted), the second one by $①^{-1}$, and so on until no more PLs remain, thereby giving infinitely more importance to the higher levels. As a practical example, the Grossone-based rewriting of the problem in (6), called PL-Crash, is

$$\min \left[\begin{pmatrix} \texttt{mass}(x) \\ -\texttt{accel}(x) \end{pmatrix} + ①^{-1} \begin{pmatrix} \texttt{toe}(x) \\ -\texttt{accel}(x) \end{pmatrix} \right]$$
$$\text{s.t. } x \in [1,\ 3]^5.$$

With such reformulation of the problem, only one more ingredient is needed to propose a PL extension of an EMO algorithm for solving PL-MPL-MOPs. Such element is a novel, Grossone-based definition of dominance, that will be presented in the next section. After that, the improvement of some fundamental routines of the a standard EMO algorithm in order to handle PL-MPL-MOPs in an effective way is discussed.

9 Handling Precedence in PL-MPL-MOPs

9.1 A New Definition of Dominance: PL-Dominance

As the original Pareto-optimality definition is not enough to cope with PL-MPL-MOPs, a new one is proposed to take the priority structure into account and guide the search towards better optima. Unlike PC-dominance, creating a good *PL-dominance* is not trivial [28], as it must be well-defined (the transitive property requires special attention) and be effective in practice. The idea behind the PL-dominance proposed here is to extend the concept of *non-dominated fronts*. According to the Pareto dominance, a front is defined as a set of solutions which do not dominate each other, but are dominated by those of previous fronts and, in turn, dominate those in the subsequent fronts. The new definition introduces the concept of "subfronts", which generalize "fronts" and consist in nested fronts determined by the priorities. The procedure to partition the population into subfronts, described below, may resemble in some aspects the preference-based ranking scheme by Fonseca and Fleming [17].

First, considering only the objectives with the highest priority, a set of non-dominated fronts is determined as usual, according to Pareto dominance. Then, these fronts are taken individually (one by one), and each is further split on the basis of the objectives of the second highest priority level, determining new groups of (sub)fronts. The same procedure is iterated recursively on the newly defined subfronts, until there are no more priority levels left. The whole population is eventually partitioned in a hierarchy of subfronts, where the term "subfronts" indicates the groups obtained after every splitting. Grossone Methodology comes in handy, letting one uniquely identify any subfront within the hierarchy by means of a G-scalar index. For instance, F_i where $i = 2①^0 + 7①^{-1} + 3①^{-2}$ denotes all the solutions in the front 2 for the primary objectives, front 7 for the secondary objectives (within the individuals in the front 2 for the primaries) and front 3 for the tertiaries (among those with primary front 2 and secondary front 7). The formal definition is:

Definition 3 (*PL-Dominance*) Given a PL-MPL-MOP and two solutions A and B belonging to the subfronts F_i and F_j, respectively, A "PL-dominates" B if and only if i is strictly smaller than j: $A \prec^* B \iff i < j$.

It is worth noting the *a posteriori* nature of the definition: it is not always possible to tell whether two elements are dominated or not before assigning a fitness rank to the whole population. In other words, the non-dominance relation is not just a function of two arguments (solutions), but has a global dependency on all the other individuals.

The transitivity of the proposed *PL-Dominance* can be proven:

Proposition 1 *PL-dominance in Definition 3 is transitive.*

Proof Let $\mathcal{S}$ be a set of distinct solutions already partitioned in subfronts. Following the notation above, indicate the partitioning in subfronts as follows: $\forall\, X \in \mathcal{S},\ X \in$

F_x. In order to demonstrate that $\prec^*$ is transitive, the following must hold: $\forall\, X, Y, Z \in \mathcal{S}, X \prec^* Y \wedge Y \prec^* Z \Rightarrow X \prec^* Z$. One can show that it is always true by repeatedly applying Definition 3. Indeed, $X \prec^* Y \wedge Y \prec^* Z \Longrightarrow x < y \wedge y < z \Longrightarrow x < y < z$.

9.2 PL-Dominance in PL-Crash Problem

For a better comprehension of how PL-Dominance works, this subsection illustrates its step-by-step application in the case of PL-Crash problem (6). In accordance with Definition 3, the optimal Pareto front shall consists of all the solution falling in the subfront indexed by the G-scalar $1 + 1①^{-1}$. This means that among those solutions which are Pareto efficient considering the objective in the first PL (mass and acceleration), the ones that are truly optimal are also Pareto efficient considering acceleration and toe-board intrusion only.

This request affects the optimal three-dimensional front of the original problem (given in [13]) as described in the next lines. First, all the solutions are projected on the mass-acceleration plane and only the Pareto efficient survives. Figure 5 reports this process: the green dots indicate the individuals which are preserved and therefore form the parent front, the one indexed by the G-scalar 1. Then, the survived solutions are projected on the toe-acceleration plane (second priority level, Fig. 6), and only those which are non-dominated also there can be considered truly optimal. Such individuals, highlighted in red in Fig. 7, are the only ones which belong to the subfront $1 + 1①^{-1}$, since they are the non-dominated solutions (including secondaries) of the non-dominated ones in the first PL.

If one wants to retrieve the sub-optimal solutions belonging to the $1 + 2①^{-1}$ subfront (assuming it is not empty), s/he should discard the solutions just found and repeat the procedure above. On the other hand, if one wants the solutions in the $2 + ①^{-1}$ subfront, s/he should eliminate all the solutions in the front 1 (regardless the infinitesimals), and again repeat the procedure just illustrated.

Figure 8 shows the solutions of the PL-MPL-MOP problem (red points) in the entire three-dimensional objective space (marked with blue points). No known domination check on the entire objective space is able to promote the red points only. This further highlights the novelty in the PL-dominance definition as well as its effectiveness in PL-MPL-MOPs. Note that the postponed filtering only works when the whole front of the non-prioritized problem is known, which may not be the case for high-dimensional problems Sect. 8.2. To overcome these issues, next section introduces a generalization of NSGA-II, enhanced with Grossone Methodology and PL-Dominance in order to transparently and directly cope with any PL-MPL-MOP.

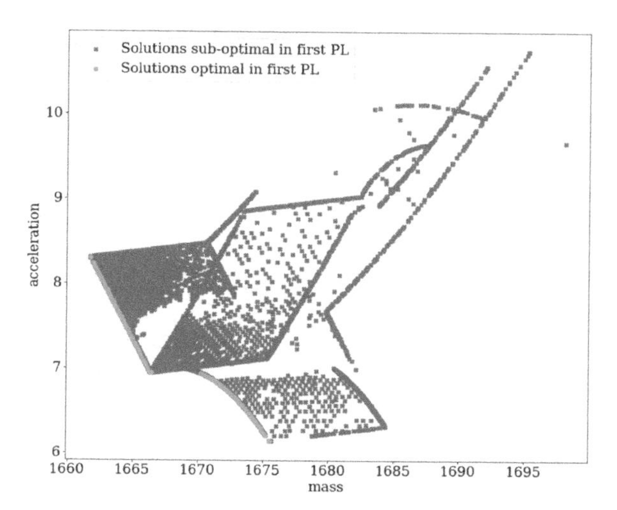

Fig. 5 Original optimal solutions on the mass-acceleration plan. First PL in green

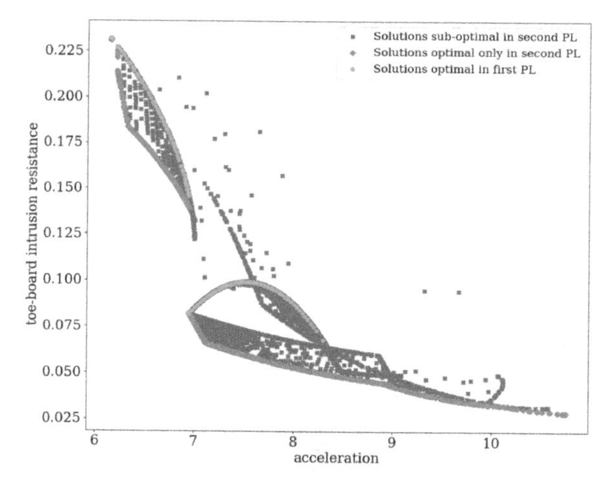

Fig. 6 Original optima on the toe-acceleration plan. First PL optima in green, second in magenta

10 Algorithm for PL: PL-NSGA-II

The new definition of PL-dominance paves the way for a generalization of ranking-based EMO algorithms to solve PL-MPL problems. Here, it is presented for NSGA-II only, yet nothing prevents the very same approach to be exploited in other frameworks as well. NSGA-II core operations are already illustrated Sect. 6 and are reported here too to simplify the reading and help the reader understanding: (i) assignment of a non-dominance rank to each solution; (ii) sorting the population by rank; (iii) best-to-worst accommodation of the fronts in the next population, until a too large to fit one is found; (iv) filling the next population with the solutions of the last front according to their crowding distance.

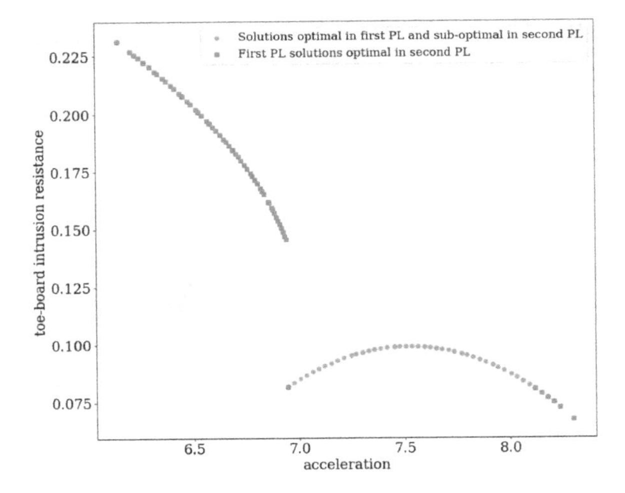

Fig. 7 Original optimal solutions on the toe-acceleration plan: in red the optimal solutions for both PLs

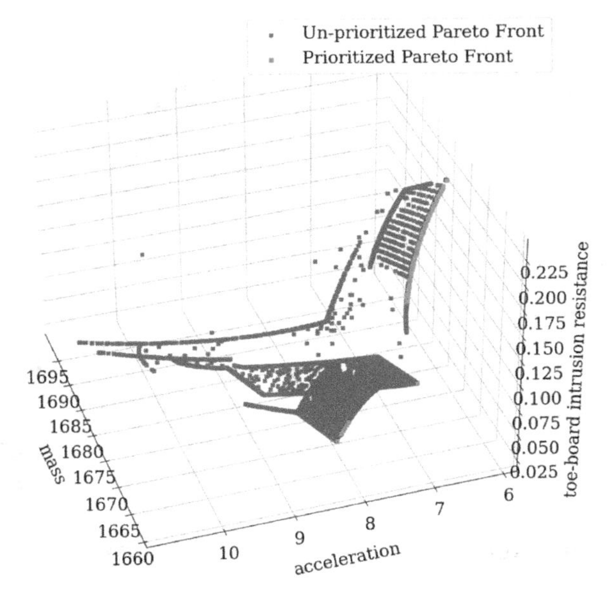

Fig. 8 Effect of the priority on the 3D Pareto front. Only the red solutions are truly optimal

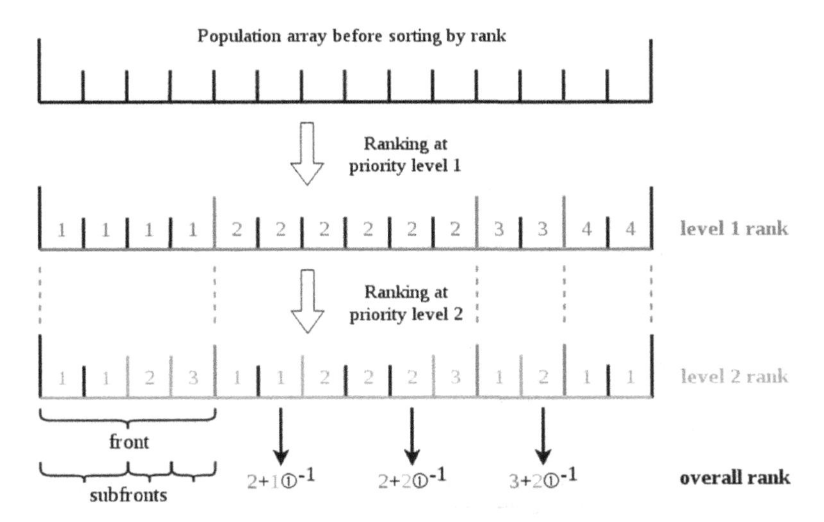

Fig. 9 An example of ranking procedure based on Definition 3

The first two steps, together, are efficiently carried out by the NSGA-II subprocedure called *fast_nondominated_sort*, while the fourth mainly leverages on the routine *crowding_distance_assignment*. Since they are not designed to do anything special when some objectives have priority over others, we improved them to increase their effectiveness when solving PL-MPL-MOPs, obtaining their PL extension. The pieces are eventually put together, creating *PL-NSGA-II*, our variant of NSGA-II specifically adapted for PL-MPL-MOPs.

10.1 PL Fast Non-dominated Sort and PL Crowding Distance

In PL-NSGA-II, the solutions are split into subfronts on the basis of the PL-Dominance, which by definition makes two solutions less likely to be non-dominated each other than NSGA-II does. As a consequence, *PL_fast_nondominated_sort* is expected to return a reasonably larger number of smaller-sized fronts, that is a positive thing considering that the main weakness of many-objective algorithms is indeed having large sets of non-dominated solutions, as mentioned earlier. This difficulty is alleviated by taking into account the priority levels of the objectives. A visual example of the rank assignment procedure is outlined in Fig. 9. Refer to Algorithms 3 and 4 for the corresponding pseudocode.

Algorithm 3: Priority Levels fast non-dominated sort

1: /* *This function is recursive* */
2: /* *P is the population, lvl is the current level to consider* */
3: **procedure** PL_FAST_NONDOMINATED_SORT(P, lvl)
4: /* *Base case of recursion* */
5: **if** $lvl < min\_lvl$ **then return**
6: /* *The first iteration also initializes all ranks to 0* */
7: **if** $lvl == 0$ **then**
8: **for all** $p \in P$ **do**
9: $\underline{p_{rank}} = 0$
10: /* *Ranking within the priority level to determine subfronts* */
11: $F^{(lvl)} = fast\_nondom\_sort\_in\_level(P, lvl)$
12: /* *Repeat for every subfront found* */
13: **for all** $F_{\underline{i}} \in F^{(lvl)}$ **do**
14: /* *In the next priority level* */
15: $F_{\underline{i}} = PL\_fast\_nondominated\_sort(F_{\underline{i}}, lvl\text{-}1)$

Algorithm 4: Fast non-dominated sort in each level

1: **procedure** FAST_NONDOM_SORT_IN_LEVEL(P, lvl)
2: **for all** $p \in P$ **do**
3: $S_p = \emptyset$
4: $n_p = 0$
5: **for all** $q \in P$ **do**
6: /* *non-dominance at specified priority level* */
7: **if** $p \prec^{lvl} q$ **then**
8: $S_p = S_p \cup \{q\}$
9: **else if** $q \prec^{lvl} p$ **then**
10: $n_p = n_p + 1$
11: **if** $n_p == 0$ **then**
12: /* *First subfront within P* */
13: $\underline{p_{rank}} = \underline{p_{rank}} + ①^{lvl}$
14: $F_1 = F_1 \cup \{p\}$
15: /* *Start from the first subfront* */
16: $\underline{i} = ①^{lvl}$
17: **while** $F_{\underline{i}} \neq \emptyset$ **do**
18: $Q = \emptyset$
19: **for all** $p \in F_i$ **do**
20: **for all** $q \in S_p$ **do**
21: $n_q = n_q - 1$
22: **if** $n_q == 0$ **then**
23: /* *q belongs to the i–th subfront* */
24: $\underline{q_{rank}} = \underline{q_{rank}} + \underline{i} + ①^{lvl}$
25: $Q = Q \cup \{q\}$
26: /* *Index of the next subfront* */
27: $\underline{i} = \underline{i} + ①^{lvl}$
28: $F_{\underline{i}} = Q$

The proposed *PL_crowding_distance_assignment* is described in Algorithm 5, and it basically consists in a G-weighted sum of the standard crowding distances computed within every priority level. In practice, it follows these steps: first, the standard crowding distance is computed for the primary (read most important) objectives and weighted by $①^0$; then, the same happens considering the second level objectives only, with weight $①^{-1}$; this is repeated for every level, decreasing the G-weight after each step; finally, all the partial results are added together to form a G-scalar, representing the PL crowding distance.

Since *PL_crowding_distance_assignment* is computed within each subfront, and subfronts tend to be generally smaller with respect to *NSGA-II* fronts because of the new ranking system, we can observe that in *PL-NSGA-II* the non-dominated sorting algorithm carries more weight than the crowding distance one. A shift in the delicate balance between convergence and diversity preservation may look questionable, but it can be justified by the following consideration. Diversity preservation mechanisms exist to keep a certain degree of spread among the best solutions. With the addition of extra information to the problem, namely the priorities, we have more means to tell which solutions are better than which, therefore it is easier to circumscribe a smaller set of best solutions. Such an increase in the ability to estimate the quality of solutions partially alleviates the need to maintain a large diversity between them because, from the perspective of a decision maker, it may be preferable to have a narrow set of truly optimal solutions rather than a larger group of diversified but potentially sub-optimal ones. That being said, the PL crowding distance still plays a crucial role after all, and does a decent job at spreading the solutions across the efficient set, especially nearer to the convergence.

Algorithm 5: Priority Levels crowding distance assignment

1: /* *F is a leaf-front in the hierarchy of population fronts partitioning* */
2: **procedure** PL_CROWDING_DIST_ASSIGNMENT(F)
3: $n = |F|$
4: **for all** $i \in F$ **do**
5: $F[i]_{dist} = 0$
6: /* *For each level of priority q; p is the index of the last level* */
7: **for** $q = 1 \ldots p$ **do**
8: **for** $j = 1 \ldots m_q$ **do**
9: $F = sort\left(F, f_j^{(q)}\right)$
10: /* +Inf *means "full-scale" (IEEE 754 standard)* */
11: $F[1]_{dist}$ += $+\text{Inf}\ ①^{1-q}$
12: $F[n]_{dist}$ += $+\text{Inf}\ ①^{1-q}$
13: **for** $i = 2 \ldots (n-1)$ **do**
14: /* *PL_crowd_dist_ass. has infinitesimal parts* */
15: $F[i]_{dist}$ += $①^{1-q} \frac{f_j^{(q)}(F[i+1]) - f_j^{(q)}(F[i-1])}{f_j^{(q)max} - f_j^{(q)min}}$

10.2 PL-NSGA-II

By incorporating both *PL_fast_nondominated_sort* and *PL_crowding_distance _assignment* into NSGA-II, one obtains an enhanced version that is also capable of handling PL-MPL-MOPs in a proper way; the name of this algorithm is *PL-NSGA-II*. Now, look how the two new functions affect the *make_new_pop* procedure, that is drafting the parent population for the next generation. Similarly to what happens in NSGA-II, the subfronts are sequentially included in the next population in order of rank; when there remains no enough room to insert the next subfront in its entirety, the latter is sorted by PL crowding distance and the fittest individuals eventually survive. PL-NSGA-II could in theory support custom recombination and mutation operators too, which is another very promising research direction. The pseudocode of PL-NSGA-II is specified in Algorithm 6.

Algorithm 6: PL-NSGA-II algorithm

1: /* P_0 *is the starting population,* T *is the total number of iterations* */
2: /* *The function returns the approximated Pareto front* P_T */
3: **procedure** PL_NSGA_II (P_0, T)
4: **for** $t = 0 \ldots T - 1$ **do**
5: $Q_t = make\_new\_pop(P_t)$
6: $R_t = P_t \cup Q_t$
7: /* F *is the set of all the non-dominated subfronts* */
8: $F = PL\_fast\_nondom\_sort(R_t, 0)$
9: $P_{t+1} = \emptyset$
10: /* i *is a G-index, a.k.a. G-scalar* */
11: $\underline{i} = 1①^0$
12: **while** $|P_{t+1}| + |F_{\underline{i}}| \leq N$ **do**
13: /* *Crowding distance is computed within subfront* $F_{\underline{i}}$ */
14: $PL\_crowding\_dist\_assignment(F_{\underline{i}})$
15: $P_{t+1} = P_{t+1} \cup F_{\underline{i}}$
16: /* *Move to the next subfront* */
17: $\underline{i} = next\_index(F, \underline{i})$
18: $Sort(F_{\underline{i}}, <_n^*)$
19: $P_{t+1} = P_{t+1} \cup F_{\underline{i}}[1 : (N - [P_{t+1}])]$
return P_T

Of course, there may be other ways to implement these functionalities, and the same ideas can also be ported to other EMO algorithms as well. This algorithm was chosen for its simplicity and similarity to the Grossone-based generalization of NSGA-II.

11 Test Cases for PL-MPL-MOPs

As for PC-MPL-MOPs, there is a lack of existing PL-MPL-MOPs test problems too. The following paragraphs deal with both synthetic problems and a real-world one, which is an adaptation of the 10-objective GAA problem [51]. The latter allows to introduce of levels of priority quite naturally (given the insights of a domain expert, and it has thus been renamed PL-GAA. Finally, the PL-Crash problem already introduced in Sect. 8.1 is revisited.

The performance of PL-NSGA-II is compared against seven classic EMO algorithms: (i) postfiltered NSGA-II (NSGA-II-post), (ii) prefiltered NSGA-II (NSGA-II-pre), (iii) postfiltered NSGA-III (NSGA-III-post), (iv) postfiltered MOEA/D (MOEA/D-post), (v) prefiltered MOEA/D (MOEA/D-pre), (vi) the favoring graph-based approach proposed in [47] (Schmiedle et al.), and (vii) the dynamic prioritized goal-based approach proposed in [49] (Tan et al.), already considered in Sect. 7 for priority chains. NSGA-III prefiltered is not considered as an interesting option because the number of objectives at the first level is too low to justify the use of a many-objective optimizer. Note that the performances of direction-based algorithms such as NSGA-III and MOEA/D would be better if the priority information was somehow used to guide the search: this idea is left for a future study. As in Sect. 7, SBX crossover and polynomial mutation operators are adopted, each algorithm run 50 times per benchmark.

11.1 Test Problem: PL-1

PL-1 has five objectives, two of which are less important than the others.

$$
\begin{aligned}
&\min \left[\begin{pmatrix} f_1^{(1)}(x) \\ f_2^{(1)}(x) \\ f_3^{(1)}(x) \end{pmatrix} + ①^{-1} \begin{pmatrix} f_1^{(2)}(x) \\ f_2^{(2)}(x) \end{pmatrix} \right], \\
&f_1^{(1)}(x) = \cos\left(\frac{\pi}{10}x_1\right)\cos\left(\frac{\pi}{10}x_2\right), \\
&f_2^{(1)}(x) = \cos\left(\frac{\pi}{10}x_1\right)\sin\left(\frac{\pi}{10}x_2\right), \\
&f_3^{(1)}(x) = \sin\left(\frac{\pi}{10}x_1\right) + g(x), \\
&f_1^{(2)}(x) = ((x_1-2)^2 + (x_2-2)^2 - 1)^2, \\
&f_2^{(2)}(x) = ((x_1-2)^2 + (x_2-2)^2 - 3)^2, \\
&g(x) = \left(x_3 - \frac{45}{9 + (x_1-2)^2 + (x_2-2)^2}\right)^2, \\
&x_1, x_2, x_3 \in [0, 5].
\end{aligned}
$$

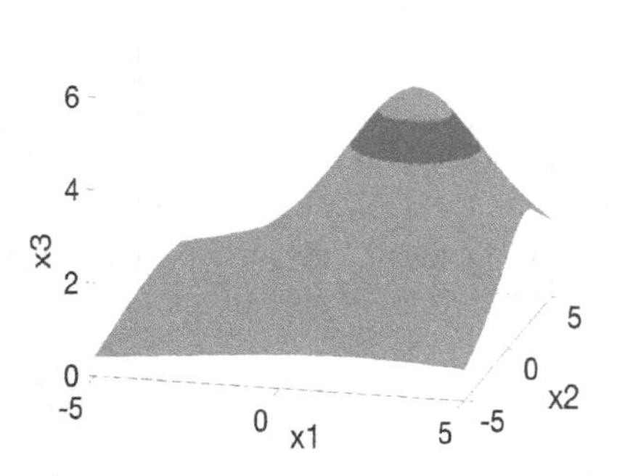

(a) Bell-shaped surface satisfying $g(x) = 0$.

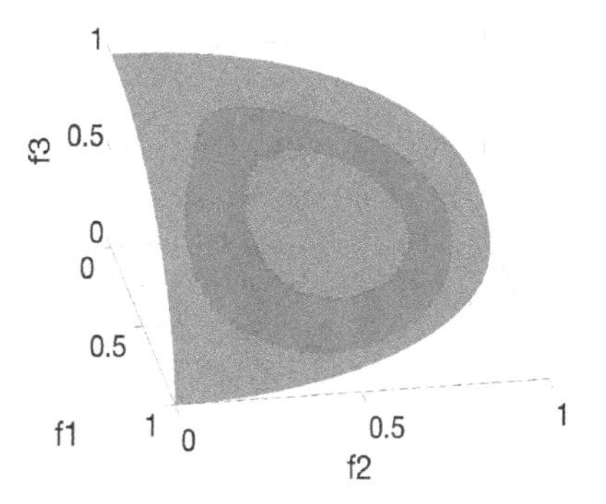

(b) Resulting efficient set in objective space.

Fig. 10 PL-1 theoretical Pareto set (blue) and efficient set (red)

The easiest way to grasp this problem is to decompose it into smaller parts. Considering only the first three objectives, namely those whose priority is maximum, it can be recognized that the Pareto surface (on which $g(x) = 0$) is an octant of a unit sphere. In the variable space, it is generated by the following set of points:

$$\mathcal{S} := \left\{(x_1, x_2, x_3) \in \mathbb{R}^3 \mid x_1 \in [0, 5],\ x_2 \in [0, 5],\ x_3 = 45/\left(9 + (x_1 - 2)^2 + (x_2 - 2)^2\right)\right\},$$

which corresponds to a bell-shaped curve on the x_3-axis. Figure 10a reports it in gray. Among all Pareto-optimal solutions for the primary objectives, i.e., the points in $\mathcal{S}$, secondary objectives $f_1^{(2)}$ and $f_2^{(2)}$ make non-dominated only the points within the blue annular region. Moving to the objective space, the entire positive octant represents the efficient set in the primary objective space, while only the red region is it also for the whole problem, i.e., considering also the secondaries (see Fig. 10b).

The solution found by PL-NSGA-II for PL-1 is the one in Fig. 11: the quality of the solutions found by PL-NSGA-II is evident. Indeed, the set of optimal points both in the decision (Fig. 11a) and in the objective space (Fig. 11b) seems to closely fit the theoretical expectations of Fig. 10.

To obtain it, PL-NSGA-II run for 500 generations with a population of 100 individuals, and the same holds for all the other algorithms. Table 3 reports the mean and the standard deviation of their performance. PL-NSGA-II appears to be the best algorithm, outperforming the second one both in terms of mean and standard deviation. In other words, not only PL-NSGA-II achieves better results, but it is also stabler.

11.2 PL-GAA: A Real-World Problem

The GAA benchmark [51] is a difficult many-objective problem from a real-world scenario. It challenges the decision maker to tune 27 constrained variables in order to design a family of three slightly different aircrafts for general aviation. The crucial

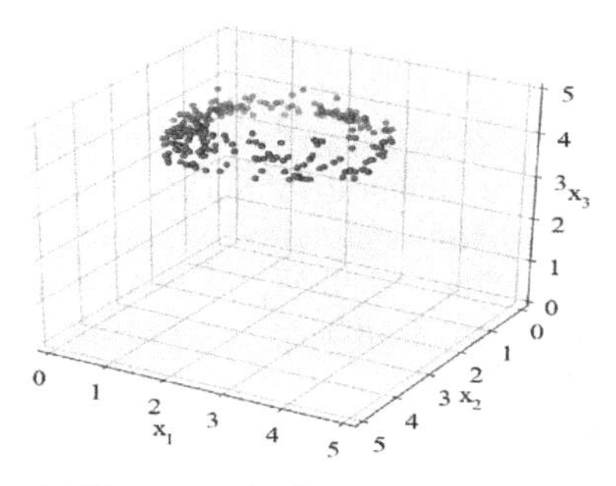

(a) Pareto set in decision space.

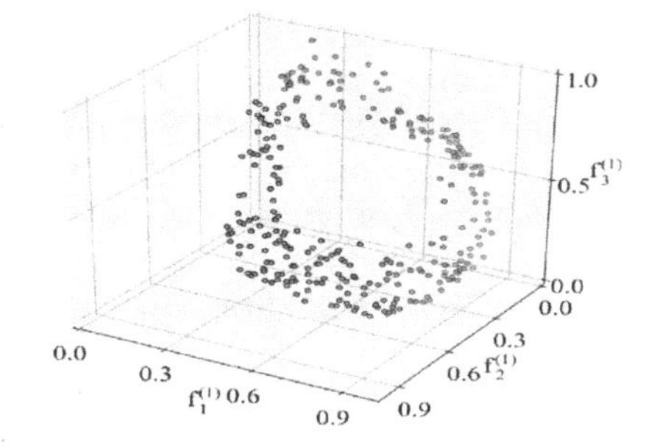

(b) Efficient set in obj. space.

Fig. 11 Obtained solutions by PL-NSGA-II for PL-1

Table 3 PL-1: Mean and Std. Dev. for metric $\Delta(\cdot)$

Algorithm	Mean	Std
PL-NSGA-II	$\mathbf{0.002 + 0.43①^{-1}}$	$\mathbf{0.002 - 0.09①^{-1}}$
MOEA/D-post	$0.01 + 2.18①^{-1}$	$0.01 + 0.33①^{-1}$
MOEA/D-pre	$0.01 + 2.94①^{-1}$	$0.01 + 0.90①^{-1}$
NSGA-II-pre	$0.01 + 3.53①^{-1}$	$0.01 + 0.53①^{-1}$
NSGA-II-post	$0.15 + 2.42e4①^{-1}$	$0.03 - 6.94e3①^{-1}$
NSGA-III-post	$0.49 + 1.47e3①^{-1}$	$0.23 + 8.11e3①^{-1}$
Tan et al.	$6.63 + 6.62e4①^{-1}$	$7.16 - 3.42e4①^{-1}$
Schmiedle et al.	$8.92 + 4.10e4①^{-1}$	$6.8 - 7.12e3①^{-1}$

point is that not only the optimization procedure must maximize the aircraft performances over conflicting parameters, but also the resulting airplanes must be similar enough not to force industrial facilities diversification. The objective space has 10 dimensions, 9 of which describe the aircraft performances, while the last one measures their diversity (henceforth called PFPF). The 9 performance metrics are: the takeoff noise (NOISE), empty weight (WEMP), direct operating costs (DOC), ride roughness (ROUGH), fuel weight (WFUEL), flight range (RANGE), purchase price (PURCH), maximum lift/drag ratio (LDMAX), maximum cruise speed (VCMAX). Also, PFPF and the first 6 performance metrics are constrained objectives.

With the help of a domain expert, the problem has been reformulate dividing the 10 objectives in 4 PLs, obtaining the new problem PL-GAA:

$$\min \left[(\lfloor \text{PFPF} \rfloor) + ①^{-1} \begin{pmatrix} \text{DOC} \\ \text{-LDMAX} \\ \text{-RANGE} \end{pmatrix} + ①^{-2} \begin{pmatrix} \text{NOISE} \\ \text{PURCH} \\ \text{WEMP} \end{pmatrix} + ①^{-3} \begin{pmatrix} \text{ROUGH} \\ \text{WFUEL} \\ \text{-VCMAX} \end{pmatrix} \right].$$

The PL with the highest importance consists of only one objective: PFPF. Since the aircraft similarity is the decision-maker's main concern, this choice comes as a natural option. As stated in [51], the decision-maker can tolerate a diversity level not greater than 0.05. Thus, the PFPF output space is discretized in intervals of length 0.05, by means of the floor operator: $\lfloor \cdot \rfloor$. In this way, all the aircraft configurations within the tolerance constraint (PFPF $\in [0, 0.05]$) are considered maximally optimized for the first PL. This will produce a number of feasible designs which will be considered equally good for PFPF objective. Subsequent levels will then determine a single or more preferred solutions. With regard to the second PL, the choice fell on the DOC objective because its importance is already pointed out in [51], while the other two (LDMAX and RANGE) are two critical factors in aircraft performance and usability, respectively. A detailed discussion about the remaining two PLs is omitted for brevity; nevertheless, it should be noted that the third PL still contains some highly impacting aircraft features, like NOISE (which can limit the possibility to land close to an urban center) or PURCH (of course, the price plays a relevant role in the decision process). Conversely, the least important level contains not so crucial properties (e.g., the maximum speed VCMAX), or aspects for which relatively cheap and simple customer solutions already exist (e.g., the roughness ROUGH).

In the experiments, due to the complexity of the problem, each algorithm run with 200 individuals for 2,000 generations. Moreover, due to the presence of constraints also on the objectives, the pre-filtered approaches have been excluded from the performance comparison, since they necessarily perform poorly. Table 4 reports the algorithms average performance (the performance of NSGA-II-pre and MOEA/D-pre have not been reported, being their use on this problem clearly meaningless). As before, PL-NSGA-II finds noticeably better solutions than the other algorithms, even for this challenging problem. It is worth noting that the finite component of the $\Delta(\cdot)$ metric has a special meaning: it is zero only when PFPF falls within $[0, 0.05]$, which indicates that all the algorithms with a non-infinitesimal mean were not always able to fully satisfy the PFPF constraint. Also, the higher the mean value is, the higher the average decision maker dissatisfaction is; the higher the variance is, the higher the randomness of the performance quality is.

The numerical evidences tell us that only two algorithms are able to respect such a constraint, namely PL-NSGA-II (always) and Schmiedle et al. (most of the cases). This fact is quite interesting indeed, while PL-NSGA-II is designed to properly master the Pareto dominance also in the presence of priority levels (PL-dominance), Schmiedle et al. does not use the Pareto dominance at all. Instead, it leverages on the *favoring* binary relation (see [47]), suggesting to address to this aspect the reasons of its high performances and capability to satisfy the PFPF's constraint. All in all, the only algorithm with an infinitesimal mean, i.e., able to always guarantee the PFPF constraint satisfaction is PL-NSGA-II, proving its strength.

Table 4 PL-GAA: Mean and Std. Dev. for metric $\Delta(\cdot)$ (NSGA-II-pre and MOEA/D-pre not reported, being their use meaningless on this problem)

Algorithm	Mean	Std. Dev.
PL-NSGA-II	$\mathbf{4.29e3①^{-1} + 1.91e4①^{-2} + 21.82①^{-3}}$	$\mathbf{1.65e4①^{-1} + 3.23e4①^{-2} + 4.04e4①^{-3}}$
Schmiedle et al.	$0.06 + 1.00e4①^{-1} + 7.74e4①^{-2} + 19.7①^{-3}$	$0.24 - 1.30e3①^{-1} + 5.63e8①^{-2} + 3.08e12①^{-3}$
NSGA-II-post	$1.24 + 1.12e5①^{-1} + 6.66e5①^{-2} + 482.83①^{-3}$	$1.36 - 1.12e4①^{-1} + 2.55e9①^{-2} + 2.08e13①^{-3}$
NSGA-III-post	$3.9 + 3.24e4①^{-1} + 4.13e5①^{-2} + 214.36①^{-3}$	$4.99 + 1.13e4①^{-1} + 8.48e7①^{-2} - 1.91e11①^{-3}$
Tan et al.	$102.55 + 8.90e4①^{-1} + 5.48e5①^{-2} + 421.65①^{-3}$	$1.22e2 + 1.23e4①^{-1} + 1.92e7①^{-2} - 1.99e9①^{-3}$
MOEA/D-post	$113.08 + 8.23e4①^{-1} + 8.66e5①^{-2} + 1.47e3①^{-3}$	$26.11 - 5.64e3①^{-1} + 9.56e7①^{-2} + 2.01e10①^{-3}$

Table 5 PL-Crash: Mean and Std. Dev. for metric $\Delta(\cdot)$

Algorithm	Mean	Std
PL-NSGA-II	$\mathbf{0.002 + 2.99e - 5①^{-1}}$	$\mathbf{0.02 + 2.00e - 4①^{-1}}$
NSGA-II-post	$1.00 + 0.52①^{-1}$	$0.22 - 0.001①^{-1}$
NSGA-II-pre	$1.12 + 0.57①^{-1}$	$0.23 - 0.01①^{-1}$
MOEA/D-post	$14.29 + 1.80①^{-1}$	$11.76 - 0.37①^{-1}$
MOEA/D-pre	$19.50 + 1.43①^{-1}$	$12.12 - 0.30①^{-1}$
Tan et al.	$29.96 + 1.73①^{-1}$	$16.97 + 0.48①^{-1}$
NSGA-III-post	$94.51 + 4.51①^{-1}$	$44.14 + 1.55①^{-1}$
Schmiedle et al.	$143.14 + 10.76①^{-1}$	$32.06 + 2.23①^{-1}$

11.3 Test Problem: PL-Crash

This section provides brief experimental results on the PL-Crash problem already discussed in the previous sections of this work. Table 5 contains the mean and standard deviation values of the $\Delta(\cdot)$ metric. Again, PL-NSGA-II significantly outperforms the competitors.

12 Conclusions

In this chapter the novel and very general class of Mixed Pareto-Lexicographic Multi-Objective Problems has been introduced and discussed. The peculiarity of these problems is to have the objective functions organized according to a priority structure, and different structures lead to different models. In particular, two models have been deeply investigated: PC-MPL-MOPs, where the objective are organized in chains of priority; and PL-MPL-MOPs, where the functions are grouped in levels which are ordered by priority. In order to solve the intrinsic limits of standard approaches when dealing with these two classes of MPL-MOPs, the Grossone Methodology has been leveraged to achieve two results: (i) a reformulation of the models which is easier to be processed by a computer; (ii) the generalization of standard EMO algorithms in order to make them able to cope with these more general problems. The effectiveness of the generalizations has been tested on several benchmark and compared with several well-known algorithms (Grossone-based algorithms have been evaluated on an Infinity Computer simulator implemented by the authors). The results testify that the additional information about priority relations among objectives can be exploited to significantly improve the algorithms search, guiding them towards higher quality solutions than those provided by conventional multi-objective algorithms.

References

1. Adra, S.F., Fleming, P.J.: Diversity management in evolutionary many-objective optimization. IEEE Trans. Evolut. Comput. **15**, 183–195 (2011)
2. Astorino, A., Fuduli, A.: Spherical separation with infinitely far center. Soft. Comput. **24**(23), 17751–17759 (2020)
3. Cavoretto, R., De Rossi, A., Mukhametzhanov, M.S., Sergeyev, Y.D.: On the search of the shape parameter in radial basis functions using univariate global optimization methods. J. Global Optim. **79**(2), 305–327 (2021)
4. Chang, P.-C., Hsieh, J.-C., Lin, S.-G.: The development of gradual-priority weighting approach for the multi-objective flowshop scheduling problem. Int. J. Prod. Econ. **79**, 171–183 (2002)
5. Cococcioni, M., Pappalardo, M., Sergeyev, Y.D.: Towards lexicographic multi-objective linear programming using grossone methodology. In: Proceedings of the 2nd International Conference "Numerical Computations: Theory and Algorithms". AIP Conference Proceedings, vol. 1776, p. 90040 (2016)
6. Cococcioni, M., Pappalardo, M., Sergeyev, Y.D.: Lexicographic multi-objective linear programming using grossone methodology: theory and algorithm. Appl. Math. Comput. **318**, 298–311 (2018)
7. Coello Coello, C.A., Sierra, M.R.: A study of the parallelization of a coevolutionary multi-objective evolutionary algorithm. In: Mexican International Conference on Artificial Intelligence, pp. 688–697 (2004)
8. De Cosmis, S., De Leone, R.: The use of grossone in mathematical programming and operations research. Appl. Math. Comput. **218**(16), 8029–8038 (2012)
9. De Leone, R.: Nonlinear programming and grossone: quadratic programming and the role of constraint qualifications. Appl. Math. Comput. **218**(16), 290–297 (2018)
10. De Leone, R., Fasano, G., Roma, M., Sergeyev, Y.D.: Iterative grossone-based computation of negative curvature directions in large-scale optimization. J. Optim. Theory Appl. **186**, 554–589 (2020)
11. De Leone, R., Fasano, G., Sergeyev, Y.D.: Planar methods and grossone for the conjugate gradient breakdown in nonlinear programming. Comput. Optim. Appl. **71**(1), 73–93 (2018)
12. Deb, K.: Multi-objective Optimization Using Evolutionary Algorithms, vol. 16. Wiley, New York (2001)
13. Deb, K., Jain, H.: An evolutionary many-objective optimization algorithm using reference-point-based nondominated sorting approach, part I: Solving problems with box constraints. IEEE Trans. Evolut. Comput. **18**, 577–601 (2014)

14. Deb, K., Joshi, D., Anand, A.: Real-coded evolutionary algorithms with parent-centric recombination. In: Proceedings of the 2002 Congress on Evolutionary Computation, vol. 1, pp. 61–66 (2002)
15. Deb, K., Pratap, A., Agarwal, S., Meyarivan, T.: A fast and elitist multiobjective genetic algorithm: NSGA-II. IEEE Trans. Evolut. Comput. **6**, 182–197 (2002)
16. Deb, K., Thiele, L., Laumanns, M., Zitzler, E.: Scalable test problems for evolutionary multiobjective optimization. Evolut. Multiobjective Optim. 105–145 (2005)
17. Fonseca, C.M., Fleming, P.J.: Multiobjective optimization and multiple constraint handling with evolutionary algorithms – Part I: A unified formulation. IEEE Trans. Syst Man Cybern.-Part A: Syst. Humans **28**, 26–37 (1998)
18. García, S., Molina, D., Lozano, M., Herrera, F.: A study on the use of non-parametric tests for analyzing the evolutionary algorithms' behaviour: a case study on the CEC' 2005 special session on real parameter optimization. J. Heuristics **15**, 617 (2009)
19. Garza-Fabre, M., Pulido, G. T., Coello Coello, C.A.: Ranking methods for many-objective optimization. In: Mexican International Conference on Artificial Intelligence, pp. 633–645 (2009)
20. Gaudioso, M., Giallombardo, G., Mukhametzhanov, M.: Numerical infinitesimals in a variable metric method for convex nonsmooth optimization. Appl. Math. Comput. **318**, 312–320 (2018)
21. Gaur, A., Khaled Talukder, A., Deb, K., Tiwari, S., Xu, S., Jones, D.: Unconventional optimization for achieving well-informed design solutions for the automobile industry. Eng. Optim. **52**, 1542–1560 (2020)
22. Gergel, V, Grishagin, V., Israfilov, R.: Adaptive dimensionality reduction in multiobjective optimization with multiextremal criteria. In: Machine Learning, Optimization, and Data Science, pp. 129–140 (2019)
23. Hansen, M.P., Jaszkiewicz, A.: Evaluating the quality of approximations to the non-dominated set. IMM, Department of Mathematical Modelling, TU Denmark (1994)
24. Holm, S.: A simple sequentially rejective multiple test procedure. Scand. J. Stat. 65–70 (1979)
25. Iman, R.L., Davenport, J.M.: Approximations of the critical region of the fbietkan statistic. Commun. Stat.-Theory Methods **9**, 571–595 (1980)
26. Ishibuchi, H., Masuda, H., Tanigaki, Y., Nojima, Y.: Modified distance calculation in generational distance and inverted generational distance. In: International Conference on Evolutionary Multi-Criterion Optimization, pp. 110–125 (2015)
27. Lai, L., Fiaschi, L., Cococcioni, M.: Solving mixed Pareto-Lexicographic multi-objective optimization problems: the case of priority chains. Swarm Evolut. Comput. **55**, 100687 (2020)
28. Lai, L., Fiaschi, L., Cococcioni, M., Deb, K.: Solving mixed pareto-lexicographic multi-objective optimization problems: the case of priority levels. IEEE Trans. Evolut. Comput. (2021)
29. Lai, L., Fiaschi, L., Cococcioni, M., Deb, K.: Handling priority levels in mixed pareto-lexicographic many-objective optimization problems. In: Proceedings of the 2021 International Conference on Evolutionary Multi-Criterion Optimization, Shenzhen, China, pp. 362–374 (2021)
30. Laumanns, M., Thiele, L., Deb, K., Zitzler, E.: Combining convergence and diversity in evolutionary multiobjective optimization. Evolut. Comput. **10**, 263–282 (2002)
31. Li, K., Deb, K., Kwong, S.: An evolutionary many-objective optimization algorithm based on dominance and decomposition. IEEE Trans. Evol. Comp. **19**, 694–716 (2015)
32. Li, H., Deb, K., Zhang, Q., Suganthan, P.N., Chen, L.: Comparison between MOEA/D and NSGA-III on a set of novel many and multi-objective benchmark problems with challenging difficulties. Swarm Evol. Comput. **46**, 104–117 (2019)
33. Li, M., Yao, X.: Quality evaluation of solution sets in multiobjective optimisation: a survey. ACM Comput. Surv. (CSUR) **52**, 1–38 (2019)
34. Liao, X., Li, Q., Yang, X., Zhang, W., Li, W.: Multiobjective optimization for crash safety design of vehicles using stepwise regression model. Struct. Multidiscip. Optim. **35**, 561–569 (2008)

35. Khare, V., Yao, X., Deb, K.: Performance scaling of multi-objective evolutionary algorithms. In: International Conference on Evolutionary Multi-Criterion Optimization, pp. 376–390 (2003)
36. Khosravani, S., Jalali, M., Khajepour, A., Kasaiezadeh, A., Chen, S.K., Litkouhi, B.: Application of Lexicographic optimization method to integrated vehicle control systems. IEEE Trans. Ind. Electron. **65**, 9677–9686 (2018)
37. Knowles, J., Corne, D.: On metrics for comparing nondominated sets. In: IEEE Proceedings of the 2002 Congress on Evolutionary Computation, vol. 1, pp. 711–716 (2002)
38. Marler, R.T., Arora, J.S.: Survey of multi-objective optimization methods for engineering. Struct. Multidiscip. Optim. **26**, 369–395 (2004)
39. Marques-Silva, J., Argelich, J., Graça, A., Lynce, I.: Boolean lexicographic optimization: algorithms & applications. Ann. Math. Art. Int. **62**, 317–343 (2011)
40. Miettinen, K.: Nonlinear Multiobjective Optimization. Springer Science, New York (1999)
41. Poloni, C.: Hybrid GA for multi objective aerodynamic shape optimisation. Genetic Algorithms in Engineering and Computer Science, 397–415 (1995)
42. Purshouse, R.C., Fleming, P.J.: Evolutionary many-objective optimisation: an exploratory analysis. In: The 2003 IEEE Congress on Evolutionary Computation, vol. 3, pp. 2066–2073 (2003)
43. Sergeyev, Y.D.: The Olympic medals ranks, lexicographic ordering, and numerical infinities. Math. Intell. **37**(2), 4–8 (2015)
44. Sergeyev, Y.D.: Numerical infinities and infinitesimals: methodology, applications, and repercussions on two Hilbert problems. EMS Surv. Math. Sci. **4**(2), 219–320 (2017)
45. Sergeyev, Y.D.: Independence of the grossone-based infinity methodology from non-standard analysis and comments upon logical fallacies in some texts asserting the opposite. Found. Sci. **24**(1), 153–170 (2019)
46. Sergeyev, Y.D., Nasso, M.C., Mukhametzhanov, M.S., Kvasov, D.E.: Novel local tuning techniques for speeding up one-dimensional algorithms in expensive global optimization using Lipschitz derivatives. J. Comput. Appl. Math. **383**, 113134 (2021)
47. Schmiedle, F., Drechsler, N., Große, D., Drechsler, R.: Priorities in multi-objective optimization for genetic programming. In: Proceedings of the 3rd Annual Conference on Genetic and Evolutionary Computation, pp. 129–136 (2001)
48. Schutze, O., Esquivel, X., Lara, A., Coello Coello, C.A.: Using the averaged Hausdorff distance as a performance measure in evolutionary multiobjective optimization. IEEE Trans. Evol. Comput. **16**, 504–522 (2012)
49. Tan, K.C., Khor, E.F., Lee, T.H., Sathikannan, R.: An evolutionary algorithm with advanced goal and priority specification for multi-objective optimization. J. Artif. Intell. Res. **18**, 183–215 (2003)
50. Van Veldhuizen, D.A.: Multiobjective evolutionary algorithms: classifications, analyses, and new innovations. Air Force Institute of Technology Wright Patterson AFB, OH, USA (1999)
51. Wang, L., Ng, A.H.C., Deb, K.: Multi-objective Evolutionary Optimisation for Product Design and Manufacturing. Springer Nature, Berlin (2011)
52. Yuan, Y., Xu, H., Wang, B., Yao, X.: A new dominance relation-based evolutionary algorithm for many-objective optimization. IEEE Trans. Evolut. Comput. **20**, 16–37 (2016)
53. Zhang, Q., Li, H.: MOEA/D: a multiobjective evolutionary algorithm based on decomposition. IEEE Trans. Evolut. Comput. **11**, 712–731 (2007)
54. Zhang, Q., Zhou, A., Zhao, S., Suganthan, P.N., Liu, W., Tiwari, S.: Multiobjective optimization test instances for the CEC 2009 special session and competition. University of Essex, Colchester, and Nanyang technological University, Singapore, TR, vol. 264 (2008)
55. Zitzler, E., Laumanns, M., Thiele, L.: SPEA2: improving the strength Pareto evolutionary algorithm. TIK-report 103 (2001)
56. Zitzler, E., Thiele, L.: Multiobjective optimization using evolutionary algorithms: a comparative case study. In: International Conference on Parallel Problem Solving from Nature, pp. 292–301 (1998)

Applications and Implementations

Exact Numerical Differentiation on the Infinity Computer and Applications in Global Optimization

Maria Chiara Nasso and Yaroslav D. Sergeyev

Abstract There exist many applications where it is necessary to approximate numerically derivatives of a function $f(x)$ which is given by a computer procedure. A novel way to efficiently compute *exact* derivatives (the word "exact" means here with respect to the accuracy of the implementation of $f(x)$) is presented in this Chapter. It uses a new kind of a supercomputer—the Infinity Computer—able to work numerically with different finite, infinite, and infinitesimal numbers. Numerical examples illustrating these concepts and numerical tools are given. In particular, the field of Lipschitz global optimization having a special interest in exact numerical differentiation is considered in cases where there exists a code for computing $f(x)$ but a code for its derivative $f'(x)$ is not available. In addition, it is supposed that the first derivative $f'(x)$ satisfies the Lipschitz condition. Algorithms using smooth piece-wise quadratic support functions and their convergence conditions are discussed. All the methods are implemented both in the traditional floating-point arithmetic and in the Infinity Computing framework.

M. C. Nasso (✉) · Y. D. Sergeyev
University of Calabria, Rende, CS, Italy
e-mail: mc.nasso@dimes.unical.it

Y. D. Sergeyev
e-mail: yaro@dimes.unical.it

Y. D. Sergeyev
Lobachevsky State University, Nizhny Novgorod, Russia

Y. D. Sergeyev and R. De Leone (eds.), *Numerical Infinities and Infinitesimals in Optimization*, Emergence, Complexity and Computation 43,
https://doi.org/10.1007/978-3-030-93642-6_9

1 Introduction

The capabilities of the Infinity Computer framework cover a wide range of actions. In fact, this methodology, working with finite, infinite and infinitesimal numbers, was successfully applied in several areas of mathematics and computer science. In [8, 12, 16, 37], the Infinity Computer framework was applied in game theory and probability, in hyperbolic geometry and percolation (see [26, 32]), fractals (see [3, 7]), infinite series (see [45, 51]), Turing machines, cellular automata and supertasks (see [11, 38, 47]), local, global, and multiple criteria optimization (see [14, 17, 52]), numerical differentiation and numerical solution of ordinary differential equations (see [1, 15, 24]), etc. The methodology is taught nowadays in several countries (see, e.g., [2, 23, 25]), in particular, the University of East Anglia, UK developed a web page ([22]) in which a teaching manual can be found. In the following pages we will focus our attention on exact numerical differentiation on the Infinity Computer and describe how it can be used in Global Optimization.

In many applications (e.g. numerical simulation, differentiation, and integration) it is required to calculate derivatives of a function $g(x)$ which is given by a computer procedure calculating its approximation $f(x)$. Usually, procedures for evaluating the exact values of derivatives of $f(x)$ are not available and numerical approximations or automatic differentiation techniques are often used for this purpose (see, e.g. [4, 9, 33] and references given therein for a detailed discussion).

The simplest formulae used on traditional computer to approximate the first derivative $f'(x)$ which require evaluation of $f(x)$ at two points and use forward, backward, and central differences are the following:

$$f'(x) \approx \frac{f(x+h) - f(x)}{h}, \; f'(x) \approx \frac{f(x) - f(x-h)}{h}, \tag{1}$$

$$f'(x) \approx \frac{f(x+h) - f(x-h)}{2h}. \tag{2}$$

However, due to the finiteness of digits in the mantissa of floating-point numbers, round-off errors in these procedures dominate calculation when $h \to 0$. Both $f(x+h)$ and $f(x-h)$ tend to $f(x)$, so that their difference tends to the difference of two almost equal quantities and thus contains fewer and fewer significant digits that provokes an explosion of the computational error. As an example, let us consider a computer procedure $f(x)$ implementing the function $g(x) = \frac{x+3}{x+1}$ and study errors provided by formulae (1), (2) during approximation of the value $f'(y)$ at the point $y = 1$ in dependence of the

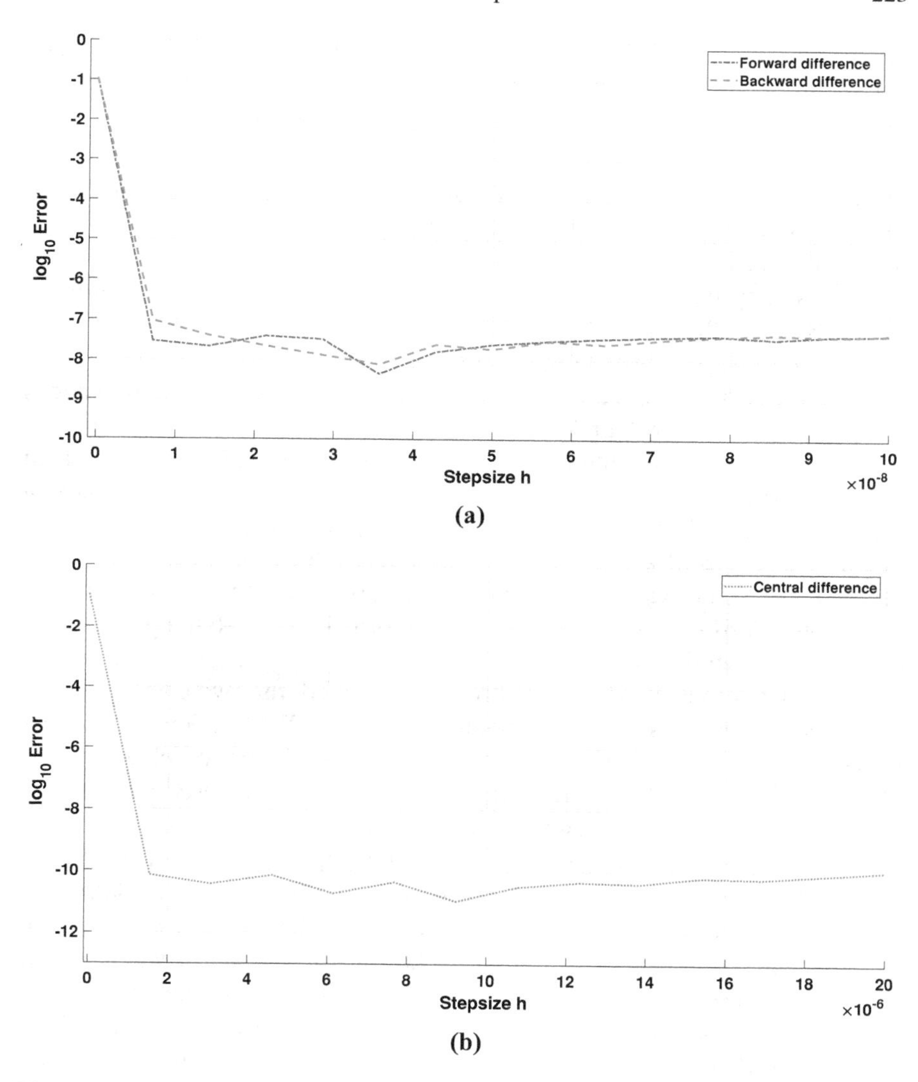

Fig. 1 When h becomes sufficiently small the error of approximation increases drastically

step h. It can be seen from Fig. 1 that when h becomes sufficiently small the error of approximation increases drastically. Thus, it is meaningless to carry out these computations beyond a certain treshold value of h. Calculations of higher derivatives suffer from the same problems.

Let us consider some other techniques. The complex step method (see [31]) allows one to improve approximations of $f'(x)$ avoiding subtractive cancellation errors present in (1), (2) by using the following formula to approximate $f'(x)$

$$f'(x) \approx \frac{Im[f(x+ih)]}{h}, \tag{3}$$

where $Im(u)$ is the imaginary part of u. Though this estimate does not involve the dangerous difference operation, it is still an approximation of $f'(x)$ because it depends on the choice of the step h.

Another approach consists of the usage of symbolic (algebraic) computations (see, e.g., [9]) where $f(x)$ is differentiated as an expression in symbolic form in contrast to manipulating numerical quantities used to express $f(x)$. Unfortunately this approach can be too slow when it is applied to long codes coming from real world applications.

There is an extensive literature (see, e.g., [4, 5, 10] and references given therein) dedicated to automatic (algorithmic) differentiation (AD) that is a set of techniques based on the mechanical application of the chain rule to obtain derivatives of a function given as a computer program. By applying the chain rule of derivation to elementary operations this approach allows one to compute derivatives of arbitrary order automatically with the precision of the code representing $f(x)$.

Implementations of AD can be broadly classified into two categories that have their advantages and disadvantages (see [5, 10] for a detailed discussion): (i) AD tools based on source-to-source transformation changing the semantics by explicitly rewriting the code; (ii) AD tools based on operator overloading using the fact that modern programming languages offer the possibility to redefine the semantics of elementary operators. In particular, the dual numbers extending the real numbers by adjoining one new element d with the property $d^2 = 0$ (i.e., d is nilpotent) can be used for this purpose (see, e.g., [4]). Every dual number has the form $v = a + db$, where a and b are real numbers and v can be represented as the ordered pair (a, b). On the one hand, dual numbers have a clear similarity with complex numbers $z = a + ib$ where $i^2 = -1$. On the other hand, speaking informally it can be said that the imaginary unit d of dual numbers is a close relative to infinitesimals (we mean here a general non formalized idea about infinitesimals) since the square (or any higher power) of d is exactly zero and the square of an infinitesimal is 'almost zero'.

All the methods described above use traditional computers as computational devices and propose a number of techniques to calculate derivatives on them. In this Chapter, a new way to calculate derivatives numerically is described. It is made by using a new kind of a supercomputer—the Infinity Computer—introduced in [43, 45] and able to work *numerically* with different finite, infinite, and infinitesimal quantities. This computer is based on a new applied point of view on infinite and infinitesimal numbers (see [42]) that, as shown in [46], is not related to non-standard analysis. The new approach does not use Cantor's ideas and works with infinite and infinitesimal numbers being in accordance with the principle 'The part is less than the whole'. Since this framework allows one to efficiently compute exact derivatives, it is applied in one-dimensional algorithms in expensive global optimization using derivatives in the case where the optimized function is given as a black box.

2 Numerical Differentiation on the Infinity Computer

In order to start, let us recall the basics of the positional numeral system with the base grossone used at the Infinity Computer (grossone is expressed by the numeral ①, see [45] and Chap. 1 of this book for detailed descriptions of ①). To express an infinite, finite, or infinitesimal number C at the Infinity Computer we subdivide C into groups corresponding to different powers of ①:

$$C = c_{p_m} ①^{p_m} + \dots + c_{p_1} ①^{p_1} + c_{p_0} ①^{p_0} + c_{p_{-1}} ①^{p_{-1}} + \dots + c_{p_{-k}} ①^{p_{-k}}. \tag{4}$$

Then, the record

$$C = c_{p_m} ①^{p_m} \dots c_{p_1} ①^{p_1} c_{p_0} ①^{p_0} c_{p_{-1}} ①^{p_{-1}} \dots c_{p_{-k}} ①^{p_{-k}} \tag{5}$$

represents the number C, where finite numbers c_i called *grossdigits* can be both positive and negative (in case during computations a grossdigit $c_{p_i} = 0$ is obtained, the corresponding term $c_{p_i} ①^{p_i}$ is excluded from C). Grossdigits show how many corresponding units should be added or subtracted in order to form the number C. Grossdigits can be expressed by several symbols. Numbers p_i in (5) called *grosspowers* can be finite, infinite, and infinitesimal, they are sorted in the decreasing order with $p_0 = 0$.

Purely finite numbers in this numeral system are represented by numerals having only one grosspower $p_0 = 0$. In fact, if we have a number C such that $m = k = 0$ in representation (5), then, since $①^0 = 1$ (see [45]), we have $C = c_0 ①^0 = c_0$. Thus, the number C in this case does not contain grossone and is equal to the grossdigit c_0 being a conventional finite number expressed in a traditional finite numeral system. Terms having negative finite grosspowers represent the simplest infinitesimal parts of C. For instance, the number $2①^{-1} = \frac{2}{①}$ is infinitesimal. If a number contains a term $c_0 ①^0$ and some infinitesimal terms, then it is finite but not purely finite. Terms having finite positive grosspowers represent the simplest infinite parts of C.

Let us now suppose that we have a set of basic functions ($\sin(x)$, $\cos(x)$, a^x etc.) represented at the Infinity Computer by their truncated Taylor series or other kinds of approximations (see [34]) using the argument x and finite constants that are connected by four arithmetical operations. We consider a function $g(x)$ and a computer procedure calculating its approximation $f(x)$ that is constructed using the said implementations of basic functions, the argument x, and finite constants connected by four arithmetical operations.

We suppose that $f(x)$ approximates $g(x)$ sufficiently well with respect to some criteria and we shall not discuss the goodness of this approximation in this Chapter. Our attention will be attracted to numerical calculations of derivatives $f'(x), \dots f^{(k)}(x)$ and the information that can be obtained from the computer procedure $f(x)$ for this purpose. The following theorem holds (see [44]).

Theorem 1 *Suppose that:*
(i) for a function $f(x)$ calculated by a procedure implemented at the Infinity Computer there exists an unknown Taylor expansion over an interval $[a, b]$, with a and b purely finite numbers, containing a purely finite point y;
(ii) $f(x)$, $f'(x)$, $f^{(2)}(x), \dots f^{(k)}(x)$ assume purely finite values or are equal to zero for purely finite points $x \in [a, b]$;
(iii) $f(x)$ has been evaluated at a point $y + ①^{-1} \in [a, b]$. Then the Infinity Computer returns the result of this evaluation in the positional numeral system with the infinite radix ① in the following form

$$f(y + ①^{-1}) = c_0①^0 c_{-1}①^{-1} c_{-2}①^{-2} \dots c_{-(k-1)}①^{-(k-1)} c_{-k}①^{-k}, \quad (6)$$

where

$$f(y) = c_0, \;\; f'(y) = c_{-1}, \;\; f^{(2)}(y) = 2! \cdot c_{-2}, \;\; \dots \;\; f^{(k)}(y) = k! \cdot c_{-k}. \quad (7)$$

Proof Let us consider the Taylor expansion for $f(x)$ with $h > 0$

$$f(x + h) = f(x) + f'(x)h + f^{(2)}(x)\frac{h^2}{2} + \cdots$$

by assuming $x = y$ and $h = ①^{-1}$. Obviously, the Taylor expansion of $f(x)$ is unknown for the Infinity Computer. Due to the rules of its operation, while calculating $f(y + ①^{-1})$, different exponents of ① are simply collected in independent groups with finite grossdigits. Since functions $f(x)$, $f'(x)$, $f^{(2)}(x), \dots f^{(k)}(x)$ assume purely finite values or are equal to zero over the interval $[a, b]$ with a, b which are also purely finite, the highest grosspower in the number (6) is necessary less or equal to zero. Thus, the number that the Infinity Computer returns can have only a finite and infinitesimal parts. The fact that four arithmetical operations (see [45]) executed by the Infinity Computer with the operands having finite integer grosspowers in the form (5) produce only results with finite integer grosspowers concludes the proof. □

Let us comment upon the theorem. It describes a situation where we need to evaluate $f(x)$ and its derivatives at a point $x = y$ but analytic expressions of $f(x), f'(x), f^{(2)}(x), \dots f^{(k)}(x)$ are unknown and computer procedures for calculating $f'(x), f^{(2)}(x), \dots f^{(k)}(x)$ are unavailable. Moreover, the internal structure of the procedure $f(x)$ can also be unknown for us.

Instead of the usage of traditional formulae (1), (2), we evaluate $f(x)$ at the point $y + ①^{-1}$. The Infinity Computer will return the number in the form (6) from where we can easily obtain exact values of $f(y)$ and $f'(y), f^{(2)}(x), \dots f^{(k)}(y)$ as shown in (7) without any knowledge of the formula and/or computer procedure for evaluating derivatives. Due to the fact that the Infinity Computer is able to work with infinite and infinitesimal numbers numerically, the values $f'(y), \dots f^{(k)}(y)$ are calculated exactly at the point $x = y$ without introduction of dangerous operations (1), (2) related to the necessity to use finite values of h when one works with traditional computers.

If we come back to the function $g(x) = \frac{x+3}{x+1}$ and calculate the value $f(1 + ①^{-1})$ of its implementation $f(x)$ obtained using the Infinity Computer framework, we have that $f(1 + ①^{-1}) = 2①^0 - \frac{1}{2}①^{-1}$ from where one can obtain that $f(1) = 2$, $f'(1) = -\frac{1}{2}$ which coincide with the exact values of the functions $g(x)$, $g'(x)$ in the point $x = 1$. Let us now illustrate the theorem by other examples.

Example 1 Suppose that we have a computer procedure implementing the following function $f(x) = 2x^2$ and we want to evaluate the values $f(y), f'(y), f^{(2)}(y)$, and $f^{(3)}(y)$ at a point $x = y$. If we evaluate now $f(x)$ at a point $y + ①^{-1}$ we obtain

$$f(y + ①^{-1}) = 2(y + ①^{-1})^2 = 2y^2 + 4y①^{-1} + 2①^{-2} = \tag{8}$$

$$2y^2①^0 4y①^{-1} 2①^{-2}. \tag{9}$$

By applying (7) we immediately calculate the required values

$$f(y) = 2y^2,\ f'(y) = 4y,\ f^{(2)}(y) = 4.$$

That, obviously, coincide with the respective derivatives

$$f'(x) = 4x,\ f^{(2)}(x) = 4,$$

at the point $x = y$. The Infinity Computer executes (8), (9) numerically using a given value of y. For instance, for $y = 36$ it executes the following operations

$$f(36 + ①^{-1}) = 2 \cdot 36①^{0}1①^{-1} \cdot 36①^{0}1①^{-1} =$$
$$2592①^{0}144①^{-1}2①^{-2}. \tag{10}$$

From (10) by applying (7) we obtain that

$$f(36) = 2592,\ f'(36) = 144,\ f^{(2)}(36) = 2! \cdot 2 = 4,$$

that are correct values of $f(x)$ and the derivatives at the point $y = 36$. □

Example 2 Suppose that we have two functions $g_1(x) = \sin(3x)$ and $g_2(x) = e^x$ and they are represented in the Infinity Computer as

$$f_1(x) = 3x - \frac{9x^3}{2},\ f_2(x) = 1 + x + \frac{x^2}{2} \tag{11}$$

being respectively the first two and three items in the corresponding Taylor expansions. If we want to evaluate $f_1(x)$ and $f_1'(x)$ at a point y, we apply the first formula from (11) at the Infinity Computer as follows

$$f_1(y + ①^{-1}) = 3(y + ①^{-1}) - \frac{9(y + ①^{-1})^3}{2} = 3y + 3①^{-1} - \frac{9(y^3 + 3y^2①^{-1} + 3y①^{-2} + ①^{-3})}{2} =$$
$$3y - \frac{9}{2}y^3 + \left(3 - \frac{27}{2}y^2\right)①^{-1} - \frac{27}{2}y①^{-2} - \frac{9}{2}①^{-3}.$$

Thus, the Infinity Computer returns

$$f_1(y) = 3y - \frac{9}{2}y^3,\ f_1'(y) = 3 - \frac{27}{2}y^2.$$

For istance, for $y = 2$ we obtain

$$f_1(2 + ①^{-1}) = 3(2 + ①^{-1}) - \frac{9(2 + ①^{-1})^3}{2} =$$
$$-30 - 51①^{-1} - 27①^{-2} - \frac{9}{2}①^{-3}.$$

Therefore

$$f_1(2) = -30,\ f_1'(y) = -51.$$

If we want to calculate $f_2(x)$, $f_2'(x)$ and $f_2^{(2)}(x)$ at a point y, we obtain

$$f_2(y + ①^{-1}) = 1 + y + ①^{-1} + \frac{(y + ①^{-1})^2}{2} =$$

$$(1 + y + \frac{y^2}{2})①^0(1 + y)①^{-1}\frac{1}{2}①^{-2}.$$

Therefore

$$f_2(y) = 1 + y + \frac{y^2}{2},\ f_2'(y) = 1 + y,\ f_2^{(2)}(y) = 1.$$

Thus for $y = -4$, the Infinity Computer returns

$$f_2(-4 + ①^{-1}) = 1 - 4 + ①^{-1} + \frac{(-4 + ①^{-1})^2}{2} =$$

$$5①^0 - 3①^{-1}\frac{1}{2}①^{-2}.$$

Then we have

$$f_2(-4) = 5,\ f_2'(-4) = -3,\ f_2^{(2)}(-4) = 1.$$

Notice that the Infinity Computer just uses formulae in (11) and works with the accuracy of approximation of $g_1(x)$ and $g_2(x)$ bounded by the accuracy of $f_1(x)$ and $f_2(x)$. □

Example 3 Suppose that we have a computer procedure implementing the following function $f(x) = \frac{x+5}{x^2}$ and we want to calculate the values $f(y)$, $f'(y)$, $f^{(2)}(y)$, and $f^{(3)}(y)$ at the point $y = 2$. As in the previous examples, we evaluate $f(x)$ at a point $y + ①^{-1}$. We consider the Infinity Computer that returns grossdigits corresponding (as it was in the previous examples automatically) to the exponents of grossone from 0 to -3. Then we have

$$f(2 + ①^{-1}) = \frac{2 + ①^{-1} + 5}{(2 + ①^{-1})^2} = \frac{7①^0①^{-1}}{4①^0 4①^{-1}①^{-2}} =$$

$$1.7500①^0 - 1.500①^{-1}1.0625①^{-2} - 0.6875①^{-3}.$$

By applying (7) we obtain that

$$f(2) = 1.7500,\ f'(2) = -1.500,$$

$$f^{(2)}(2) = 2! \cdot 1.0625 = 2.1250,\ f^{(3)}(2) = 3! \cdot -0.6875 = -4.125,$$

that are values which one obtains by using explicit formulae

$$f(x) = \frac{x+5}{x^2},\ f'(x) = -\frac{x+10}{x^3},\ f^{(2)}(x) = \frac{2(x+15)}{x^4},\ f^{(3)}(x) = -\frac{6(x+20)}{x^5}$$

for $f(x)$ and its derivatives at the point $x = 2$. □

3 Application in Lipschitz Global Optimization

Global optimization is an important research field with numerous applications in engineering, electronics, machine learning, optimal decision making, etc. In many of these applications, even in the univariate case, evaluations of the objective functions and derivatives are often time-consuming and the number of function evaluations executed by algorithms is extremely high due to the presence of multiple local extrema. As a result, the problems of an acceleration of the global search and computing efficiently exact derivatives in the case where the optimized function is given as a black box arise inevitably.

The necessity to find the best (in other words, global) solution in the situation where a high number of local extrema is present explains the continuously increasing interest of researchers to global optimization algorithms looking for global minimum (or maximum). One of the important methodologies developed to attack this problem is Lipschitz global optimization (see, e.g., [29, 35]). It uses a natural assumption on the global optimization problem supposing that the objective function under consideration has bounded slopes, in other words, it satisfies the Lipschitz property and at the same moment it can be multiextremal and each evaluation can be a very time-consuming operation. An important subclass in Lipschitz global optimization consists of functions with the first derivative satisfying the Lipschitz condition (see [13, 28], etc.).

Problems belonging to Lipschitz global optimization are extremely difficult even in the one-dimensional case and are under an intensive study at least for two reasons. First, there exists a huge number of applications where problems of this kind arise. The second reason is that one-dimensional schemes are broadly used for constructing multi-dimensional global optimization methods.

In this Chapter, our attention is devoted to the global optimization problem

$$f^* = f(x^*) = \min f(x), \quad x \in D, \tag{12}$$

where $f(x)$, with $x \in D = [a, b]$, is the objective black-box function and its first derivative $f'(x)$ satisfies the Lipschitz condition with an unknown Lipschitz constant $0 < M < \infty$, i.e.

$$|f'(x) - f'(y)| \leq M|x - y|, \quad x, y \in D. \tag{13}$$

An additional practical difficulty consists in the fact that it is supposed that there exists a code for computing $f(x)$ only and a code for the derivative $f'(x)$ is not available. The problem (12), (13) is under a close scrutiny since the nineties of the last century. Breiman and Cutler (see [6]) have proposed a method to solve the problem (12) where the constant M from (13) is a priori known whereas Gergel in [18] has proposed a global optimization method that estimates M in the course of the search. Both methods use in their work auxiliary non-smooth support functions that are less or equal to $f(x)$, $x \in D$. Since the objective function $f(x)$ is differentiable over the search region D, in [30, 40] there have been introduced methods constructing smooth support functions that are closer to the objective function $f(x)$ with respect to non-smooth ones providing so a notable acceleration in comparison with the methods [6, 18].

Moreover, it is well known in Lipschitz global optimization that the usage of the global (i.e., the same for the whole search region D) Lipschitz constant or its global estimate can slow down the search. In order to overcome this difficulty, a number of local tuning techniques automatically controlling the exploitation-exploration trade-off have been proposed in [19, 27, 40, 49]. These techniques adaptively estimate local Lipschitz constants over different subregions of D constructing auxiliary functions that are closer to $f(x)$ with respect to those using the global Lipschitz constant (or its estimates). As a result, methods using local tuning techniques are significantly faster than algorithms working with the global Lipschitz constant.

Let us present now a theoretical background required for construction of smooth support functions for global optimization algorithms for solving the problem (12), (13). Global optimization methods introduced in [40] construct a smooth support function $\psi_i(x)$ at each subinterval $[x_{i-1}, x_i]$, $2 \leq i \leq k$, of the search region D where the points x_i, $1 \leq i \leq k$, are the so-called *trial points*, i.e., points where $f(x)$ and $f'(x)$ have been evaluated. The function $\psi_i(x)$ is constructed using the fact that the maximal possible curvature of $f(x)$ is determined by the Lipschitz constant M from (13). The functions

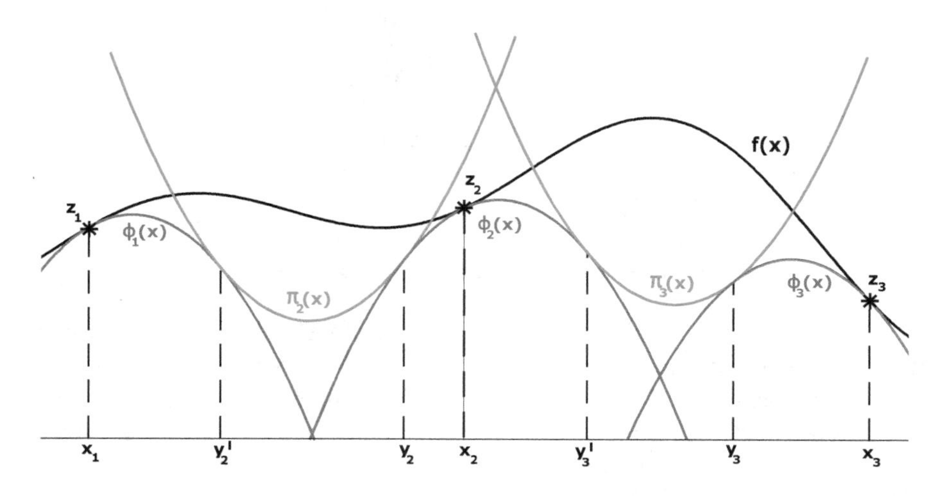

Fig. 2 Constructing smooth support functions $\psi_i(x)$ for $f(x)$

$$\psi_i(x) \leq f(x), \quad x \in [x_{i-1}, x_i], \; 2 \leq i \leq k, \tag{14}$$

can be built using the following three functions

$$\pi_i(x) = 0.5Mx^2 + b_i x + c_i, \tag{15}$$

$$\phi_{i-1}(x) = z_{i-1} + z'_{i-1}(x - x_{i-1}) - 0.5M(x - x_{i-1})^2, \tag{16}$$

$$\phi_i(x) = z_i - z'_i(x_i - x) - 0.5M(x_i - x)^2, \tag{17}$$

where $z_i = f(x_i)$, $z'_i = f'(x_i)$, and b_i, c_i are two parameters to be determined.

Notice that $\phi_{i-1}(x)$ and $\phi_i(x)$ are support functions for $f(x)$ and are obtained from the Taylor formulae based at the points x_{i-1} and x_i, respectively. The meaning of the parabola $\pi_i(x)$ will be explained in a minute. It has been shown in [6, 18] that

$$\max\{\phi_{i-1}(x), \phi_i(x)\}, \quad x \in [x_{i-1}, x_i], \; 2 \leq i \leq k,$$

is a non-smooth minorant for $f(x)$ over $[x_{i-1}, x_i]$. Then, the key observation made in [40] in order to construct a smooth minorant consists in the fact that due to the boundedness of the curvature of $f(x)$, it cannot be below the parabola $\pi_i(x)$ over the interval $[y'_i, y_i]$ that can be found by asking that the piece-wise quadratic function

$$\psi_i(x) = \begin{cases} \phi_{i-1}(x), & x \in [x_{i-1}, y_i'], \\ \pi_i(x), & x \in [y_i', y_i], \\ \phi_i(x), & x \in [y_i, x_i], \end{cases} \tag{18}$$

is a smooth support function for $f(x)$ over $[x_{i-1}, x_i]$ (see Fig. 2). The values y_i, y_i' and b_i, c_i from (15) can be determined by "gluing" the functions $\phi_{i-1}(x)$, $\pi_i(x)$, $\phi_i(x)$ and their first derivatives. In other words, the following system of equations should be solved:

$$\begin{cases} \phi_{i-1}(y_i') = \pi_i(y_i') \\ \phi_i(y_i) = \pi_i(y_i) \\ \phi_{i-1}'(y_i') = \pi_i'(y_i') \\ \phi_i'(y_i) = \pi_i'(y_i) \end{cases} \tag{19}$$

As was shown in [40], its solution is:

$$y_i = \frac{x_i - x_{i-1}}{4} + \frac{z_i' - z_{i-1}'}{4M} + \frac{z_{i-1} - z_i + z_i'x_i - z_{i-1}'x_{i-1} + 0.5M(x_i^2 - x_{i-1}^2)}{M(x_i - x_{i-1}) + z_i' - z_{i-1}'}, \tag{20}$$

$$y_i' = -\frac{x_i - x_{i-1}}{4} - \frac{z_i' - z_{i-1}'}{4M} + \frac{z_{i-1} - z_i + z_i'x_i - z_{i-1}'x_{i-1} + 0.5M(x_i^2 - x_{i-1}^2)}{M(x_i - x_{i-1}) + z_i' - z_{i-1}'}, \tag{21}$$

$$b_i = z_i' - 2My_i + Mx_i, \tag{22}$$

$$c_i = z_i - z_i'x_i - 0.5Mx_i^2 + My_i^2. \tag{23}$$

Then we construct $\psi_i(x)$, searching for the point

$$p_i = \min\{\psi_i(x) : x \in [x_{i-1}, x_i]\}$$

and the respective value $R(i) = \psi_i(p_i)$ called *characteristic* of the interval $[x_{i-1}, x_i]$. The term characteristic is due to the class of characteristic methods introduced in [20] for solving Lipschitz global optimization problems. The same terminology is used in Divide the Best methods (see [41]) to determine a goodness of each subregion of the search domain. In fact, the methods under consideration belong to the class of Divide the Best algorithms.

Since in practice it is difficult to know a priori the global constant M and its usage over the whole region D can slow down the search, Local Tuning techniques can be applied in order to automatically balance global and local information during the global search and allow one to obtain a significant acceleration of the algorithms.

4 Local Tuning Techniques

The first local tuning technique automatically balancing global and local information during the global search has been initially introduced in [39] for Lipschitz functions that can be non-differentiable. It has been shown that strategies of this kind allow one to obtain a significant acceleration of the search. As a result, an intensive research activity focused on developing new local tuning strategies and finding new classes of problems where they can be applied has begun (see, for example, [21, 27, 48]).

We start this section by presenting two traditional strategies for choosing an estimate for the global value M from (13). After that a known local tuning from [40] and two recent local tuning techniques from [49] will be introduced. In order to describe these procedures let us denote as $\{x_i\}^k$ the ordered trial points (the operation of evaluation of $f(x)$ and $f'(x)$ at a point x is called *trial*), where $k \geq 2$ is the number of iterations of the algorithm (for $k = 2$: $x_1 = a$ and $x_2 = b$). Let $r > 1$ be the reliability parameter of the methods.

For each interval $[x_{i-1}, x_i]$, $2 \leq i \leq k$, its local Lipschitz constant for $f'(x)$ is estimated by values m_i in one of the following five ways:

1. A priori given Lipschitz constant. Set

$$m_i = M, \quad 2 \leq i \leq k. \tag{24}$$

In this case, an algorithm uses the same exact a priori known value M from (13) of the Lipschitz constant of $f'(x)$ for each subinterval $[x_{i-1}, x_i]$, $2 \leq i \leq k$.

2. Global estimate. Compute estimates

$$m_i = r \cdot \max\{\xi, H^k\}, \quad 2 \leq i \leq k, \tag{25}$$

where $\xi > 0$ is a technical parameter (a small number greater than 0) reflecting the supposition that $f'(x)$ is not constant over the interval $[x_{i-1}, x_i]$. Then,

$$H^k = \max\{v_i : 2 \leq i \leq k\}, \tag{26}$$

with

$$v_i = \frac{|2(z_{i-1} - z_i) + (z'_{i-1} + z'_i)(x_i - x_{i-1})| + d_i}{(x_i - x_{i-1})^2} \tag{27}$$

and

$$d_i = \sqrt{|2(z_{i-1} - z_i) + (z'_{i-1} + z'_i)(x_i - x_{i-1})|^2 + (z'_i - z'_{i-1})^2(x_i - x_{i-1})^2}. \tag{28}$$

Notice that this adaptive estimate of the global Lipschitz constant is obtained by imposing that the upper bound

$$\phi^+_{i-1}(x) = z_{i-1} + z'_{i-1}(x - x_{i-1}) + \frac{1}{2}m_i(x - x_{i-1})^2,$$

based on the point x_{i-1} is equal or greater than the lower bound

$$\phi^-_i(x) = z_i + z'_i(x - x_i) - \frac{1}{2}m_i(x - x_i)^2,$$

based on the point x_i, over the interval $[x_{i-1}, x_i]$, i.e.,

$$\phi^+_{i-1}(x) \geq \phi^-_i(x), \quad x \in [x_{i-1}, x_i], \tag{29}$$

and the upper bound

$$\phi^+_i(x) = z_i + z'_i(x - x_i) + \frac{1}{2}m_i(x - x_i)^2,$$

based on the point x_i is equal or greater than the lower bound

$$\phi^-_{i-1}(x) = z_{i-1} + z'_{i-1}(x - x_{i-1}) - \frac{1}{2}m_i(x - x_{i-1})^2,$$

based on the point x_{i-1}, over the interval $[x_{i-1}, x_i]$ i.e.,

$$\phi^+_i(x) \geq \phi^-_{i-1}(x), \quad x \in [x_{i-1}, x_i], \tag{30}$$

As shown in [50], in order to satisfy (29), (30) it is sufficient that the following inequality holds for m_i:

$$m_i \geq \tau(x), \quad x \in [x_{i-1}, x_i],$$

where

$$\tau(x) = 2\frac{|z_i - z_{i-1} + z'_i(x - x_i) - z'_{i-1}(x - x_{i-1})|}{(x - x_i)^2 + (x - x_{i-1})^2}.$$

It can be proved that the values v_i from (27) are such that

$$v_i = \max\{\tau(x) : x \in [x_{i-1}, x_i]\}.$$

As was already mentioned, strategies 1 and 2 described above can slow down the global search. In order to overcome this problem, the following local tuning technique has been introduced in [40].

3. Maximum Local Tuning. Compute estimates

$$m_i = r \cdot \max\{\lambda_i, \gamma_i, \xi\}, \tag{31}$$

where ξ is a small positive number,

$$\lambda_i = \max\{v_{i-1}, v_i, v_{i+1}\}, \quad 3 \le i \le k-1, \tag{32}$$

with v_i as in (27); when $k = 2$ only v_2 should be considered and when $i = 2$ and $i = k$ we consider only v_2, v_3 and v_{k-1}, v_k, respectively.

The value γ_i is calculated as follows:

$$\gamma_i = H^k \frac{(x_i - x_{i-1})}{X^{max}}, \tag{33}$$

$$X^{max} = \max\{(x_i - x_{i-1}), \quad 2 \le i \le k\}. \tag{34}$$

Notice that these adaptive local estimates of local Lipschitz constants are obtained during the search balancing through (31) local and global information obtained in the course of the previous iterations. Indeed, when the interval $[x_{i-1}, x_i]$ is small then the local information managed by λ_i has a decisive rule; in contrast, when the interval $[x_{i-1}, x_i]$ is wide the global information represented by γ_i is used.

It should be stressed that in the case of the a priori given Lipschitz constant, condition (13) ensures that $\psi_i(x)$ from (2) is a minorant for $f(x)$. Since strategies 2 and 3 *estimate* Lipschitz constants, the resulting functions $\psi_i(x)$ can violate the inequality $\psi_i(x) \le f(x)$. This fact can lead to the loss of the global solution during the search. Conditions ensuring convergence to global solutions of optimization methods using global and local estimates of Lipschitz constants for $f'(x)$ will be established in the following pages.

In order to introduce the Maximum-Additive Local Tuning for derivatives let us notice that the Maximum Local Tuning technique described above balances local and global information about $f'(x)$ computing the maximum among values λ_i, γ_i, and ξ (see (31)). It has been recently shown in [48] that for functions $g(x)$ satisfying the Lipschitz condition

$$|g(x) - g(y)| \le L|x - y|, \quad x, y \in D, \tag{35}$$

with a constant L, $0 < L < \infty$, it makes sense to use a maximum-additive convolution of values representing local and global information collected during the search (that, clearly, are collected in a different way w.r.t. (32) and (33) since functions satisfying (35) can be non-differentiable whereas λ_i and γ_i estimate Lipschitz constants for derivatives). Let us illustrate a maximum-additive convolution of (32) and (33) for our problem (12)–(13).

4. Maximum-Additive Local Tuning for derivatives. Compute estimates

$$m_i = r \cdot \max\{v_i, \frac{1}{2}(\lambda_i + \gamma_i), \xi\}, \quad 2 \leq i \leq k, \tag{36}$$

where values v_i, λ_i, and γ_i are from (27), (32), and (33), respectively.

This estimate provides an additive mixture of λ_i and γ_i with the equal usage of local and global information. The presence of v_i in (36) is explained by the fact that for small intervals the estimate γ_i can be very small (see (33)) leading so to the prohibited situation $v_i > 0.5(\lambda_i + \gamma_i)$. If v_i was not in (36), this could lead to the possibility that points y_i' and y_i can be generated outside the interval $[x_{i-1}, x_i]$ and this could produce errors in the work of algorithm. The presence of v_i in (36) avoids this case. Notice also that (36) can be generalized to the case of a weighted usage of local and global estimates depending on a parameter $0 < \rho < 1$ as follows

$$m_i = r \cdot \max\{v_i, \rho\lambda_i + (1 - \rho)\gamma_i, \xi\}.$$

The idea for the last local tuning technique we are going to introduce also comes from algorithms developed to deal with problems satisfying Lipschitz condition (35). Among the first methods proposed to solve this problem we find algorithms of Piyavskij (see [36]) and Strongin (see [27]). The former is based on geometric ideas and constructs for the objective function a piece-wise linear minorant having slopes $\pm L$ defined by (35) whereas the latter uses a stochastic model for its work and adaptively estimates L by the value

$$\mu = \max\{\mu_i : 2 \leq i \leq k\}, \mu_i = |z_i - z_{i-1}|(x_i - x_{i-1})^{-1}.$$

It has been proven in [50] that the method of Strongin has a geometric interpretation. It can be viewed as a procedure constructing over each interval $[x_{i-1}, x_i]$, $2 \leq i \leq k$, an auxiliary piece-wise linear function (that becomes a minorant under certain conditions) with the local slopes $\pm s_i$, where

$$s_i = 0.5(r\mu + \frac{\mu_i^2}{r\mu}), 2 \leq i \leq k. \tag{37}$$

It can be seen from (37) that in s_i we have a mixture of a local information represented by μ_i and the global one represented by μ. Methods using the local tuning technique (37) have been tested in [27] showing a promising performance on a broad class of test problems satisfying (35). Let us now introduce the following Mixed Local Tuning technique for derivatives evolving the idea of the convolution (37) from algorithms working with functions satisfying (35) to methods that can be used to solve our problem (12)–(13).

5. Mixed Local Tuning for derivatives. Compute values

$$m_i = \max\{rv_i, 0.5(r\eta + v_i^2(r\eta)^{-1})\}, \quad \eta = \max\{H^k, \xi\}, \quad 2 \le i \le k, \tag{38}$$

where H^k is from (26). In this local tuning the local information is represented by v_i and the global one is represented by η.

Let us now make a remark regarding the correct construction of the functions $\psi_i(x)$ from (18). As was already mentioned above, if over an interval $[x_{i-1}, x_i]$ the value m_i is underestimated, the points y_i' and y_i can be generated outside the interval $[x_{i-1}, x_i]$ and this can produce errors in the work of the algorithms. It has been proven in [40, 50] that strategies (24), (25), and (31) ensure that the points y_i' and y_i are inside the interval $[x_{i-1}, x_i]$. An analogous result can be proven (see [49]) for the strategies (36) and (38).

Theorem 2 *If β is a finite number and the following condition*

$$v_i < m_i \le \beta < \infty, \tag{39}$$

takes place for the strategies (36) and (38), then for functions $\psi_i(x)$ constructed by these algorithms

$$y_i' \in (x_{i-1}, x_i), \, y_i \in (x_{i-1}, x_i).$$

Moreover, it follows

$$y_i' - x_{i-1} \ge \frac{(\frac{m_i}{v_i} - 1)^2}{4\frac{m_i}{v_i}(\frac{m_i}{v_i} + 1)}(x_i - x_{i-1}), \tag{40}$$

$$x_i - y_i \ge \frac{(\frac{m_i}{v_i} - 1)^2}{4\frac{m_i}{v_i}(\frac{m_i}{v_i} + 1)}(x_i - x_{i-1}). \tag{41}$$

Proof The proof is analogous to the proof of Theorem 4.11 from [50] with the remark that strategies (36) and (38) ensure that the inequality (39) is satisfied automatically since the parameter r is a finite number $r > 1$.

5 General scheme and convergence conditions

We are now ready to describe decision rules of five global optimization algorithms using the strategies illustrated above to estimate Lipschitz constants. The methods have a similar structure that is reported in the following General Scheme incorporating the five algorithms using Derivatives (GSD).

STEP 0. The first two trials are executed at the points $x^1 = a$ and $x^2 = b$. For $k \geq 2$ we choose the point x^{k+1} using the following steps:

STEP 1. Renumber the points $x^1, \ldots, x^k$ and the corresponding function values $z^1, \ldots, z^k$ of the previous iterations by subscripts so that

$$a = x_1 < \cdots < x_k = b, \quad z_i = f(x_i), \quad 1 \leq i \leq k.$$

STEP 2. Calculate the current estimate m_i of the Lipschitz constants of $f'(x)$ over the intervals $[x_{i-1}, x_i]$, $2 \leq i \leq k$, in one of the ways previously described in (24), (25), (31), (36), and (38).

STEP 3. Calculate for each interval $[x_{i-1}, x_i]$, $2 \leq i \leq k$ its characteristic R_i.

STEP 4. Select the interval (x_{t-1}, x_t) corresponding to the minimal characteristic, i.e., such that

$$R_t = \min\{R_i : 2 \leq i \leq k\}.$$

STEP 5. If the stopping rule is not satisfied, i.e.,

$$|x_t - x_{t-1}| > \varepsilon,$$

where ε is the accuracy of the search, then execute the next trial at the point

$$x^{k+1} = p_t = \min\{\psi_t(x) : x \in [x_{t-1}, x_t]\} \tag{42}$$

Using the procedures proposed to estimate M in the previous section and the general scheme described above we obtained the following 5 algorithms:

- **DKC**: The method using the first **D**erivatives and the a priori **K**nown Lipschitz **C**onstant M, see (24).
- **DGE**: The method using the first **D**erivatives and the **G**lobal **E**stimate of the Lipschitz Constant M, see (25).
- **DML**: The method using the first **D**erivatives and the **M**aximum **L**ocal Tuning, see (31).
- **DMAL**: The method using the first **D**erivatives and the **M**aximum-**A**dditive **L**ocal Tuning, see (36).

- **DMXL**: The method using the first **D**erivatives and the **M**i**x**ed **L**ocal Tuning, see (38).

The following global convergence properties have been studied in [40, 49, 50]. Let us consider an infinite trial sequence $\{x^k\}$ generated by an algorithm belonging to the general scheme GSD with the accuracy $\varepsilon = 0$. The following results for each algorithm belonging to GSD hold:

Theorem 3 *Let the point $\overline{x}$, ($\overline{x} \neq a$, $\overline{x} \neq b$) be a limit point of the sequence $\{x^k\}$ generated by an algorithm belonging to the general scheme GSD during the course of minimizing a function $f(x)$, $x \in [a, b]$. If the values m_i satisfy conditions (39) then the point $\overline{x}$ will be a local minimizer of the function $f(x)$.*

Theorem 4 *Let $\overline{x}$, ($\overline{x} \neq a, \overline{x} \neq b$) be a limit point of the sequence $\{x^k\}$ generated by an algorithm belonging to the general scheme GSD during the course of minimizing a function $f(x)$, $x \in [a, b]$. Then, if condition (39) is fulfilled for the intervals containing $\overline{x}$, there exist two subsequences of $\{x^k\}$ converging to $\overline{x}$, one from the left, the other from the right.*

Theorem 5 *Let $\overline{x}$ be a limit point of the sequence $\{x^k\}$ generated by an algorithm belonging to the general scheme GSD and condition (39) be fulfilled. We then have for trial points x^k*

$$f(x^k) \geq f(\overline{x}), \quad k \geq 1. \tag{43}$$

Corollary 1 *If, given the conditions of the theorem, alongside $\overline{x}$ there exists another limit point x' of the sequence $\{x^k\}$ then $f(\overline{x}) = f(x')$.*

Let us denote by M_j the local Lipschitz constant of $f'(x)$ over an interval $[x_{j-1}, x_j]$, by $\overline{X}$ the set of limit points of $\{x^k\}$ and by X^* the set of global minimizers of $f(x)$. The following theorem then takes place.

Theorem 6 *Let x^* be a global minimizer of $f(x)$ and $[x_{j-1}, x_j]$, $j = j(k)$, be an interval containing this point during the course of k-th iteration of GSD. If there exists an iteration number s such that, for all $k \geq s$ for the values m_j, $j = j(k)$, the inequality*

$$M_j \leq m_j \leq \beta \tag{44}$$

takes place for $[x_{j-1}, x_j]$ where β is a finite number and (39) holds for all the other intervals, then the point x^* is a limit point of the sequence $\{x^k\}$.

Corollary 2 *Given the conditions of the theorem, all limit points of $\{x^k\}$ are global minimizers of $f(x)$.*

Corollary 3 *If, given the condition of the theorem, (44) holds for all points* $x^* \in X^*$, *then* $\overline{X} = X^*$.

6 Numerical experiments

In this section, we present results of numerical experiments executed in order to apply the Local Tuning techniques described above and to compute exact (recall that the word exact means: up to the machine precision) derivatives for black-box functions during the work of the algorithms without the necessity to have a code for evaluating $f'(x)$. In order to do so we can use the Infinity Computer and the opportunity offered by Theorem 1.

The five methods using previously discussed smooth support functions and Local Tuning techniques were implemented both in the traditional floating-point arithmetic and in the Infinity Computing framework. Two classes of randomly generated test functions have been used to test the methods. Class 1 contains 100 Shekel test functions $f(x)$ (see [50]) where

$$f(x) = -\sum_{i=1}^{10} \left[k_i^2(10x - a_i)^2 + c_i\right]^{-1}, \quad 0 \le x \le 1, \tag{45}$$

with randomly generated parameters $1 \le k_i \le 3$, $0.1 \le c_i \le 0.3$, $0 \le a_i \le 10$, $1 \le i \le 10$.

Class 2 contains functions $g(x) = -f(x)$, where $f(x)$ is from (45). Functions $g(x)$ were randomly generated and chosen in such way that the global solution x^* from (12) is an internal point of the search region $[a, b]$, i.e., $x^* \ne a \wedge x^* \ne b$. In the traditional floating-point arithmetic, analytical formulae for the first derivative have been used providing so exact values of $f'(x)$. In the Infinity Computing implementation the derivatives have been calculated numerically also giving exact values of $f'(x)$. Both implementations have given identical results in the numerical experiments described below confirming so that the Infinity Computing opens very interesting horizons in numerical global optimization in situations where the objective function is given as a black box and a code for computing derivatives is absent.

Three series of experiments have been executed. For each method the technical parameter ξ from (25) was set to 10^{-8}. As concerns the first two series, the reliability parameter r_1 was obtained starting from the initial value 1.1 and it was increased with step equal to 0.1 until at least the 90% of all test problems were solved, i.e., the tested algorithm has generated a point x^k after k trials such that $|x^k - x^*| \le \varepsilon$ with $\varepsilon = 10^{-4}$ for the first series and

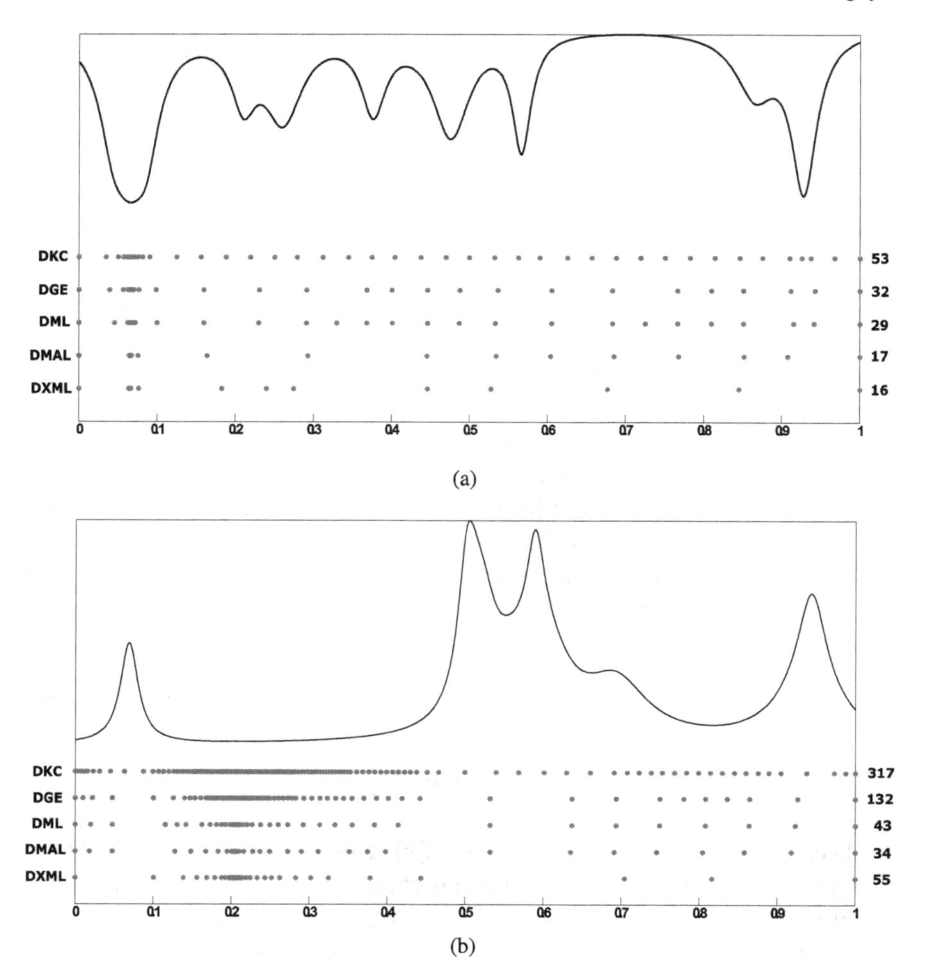

Fig. 3 Graphs of a function from Class 1 (**a**) and from Class 2 (**b**) and trial points generated by the five methods during their work

$\varepsilon = 10^{-6}$ for the second series of experiments. For the remaining unsolved problems the parameter r_2 was used. It was obtained starting from r_1 and it was increased with step equal to 0.1 until the remaining problems were all solved. Figures 3a and 3b show two examples of application of the methods respectively on a function from Class 1 and on a function from Class 2 together with trial points generated by the five methods during their work with these functions.

As the objective functions $f(x)$ are considered to be hard to evaluate, the number of trials was chosen as the comparison criterion. We reported

Table 1 Results of numerical experiments with $\varepsilon = 10^{-4}$ on Class 1

Method	r_1	AVG 1	Success (%)	r_2	AVG 2	Weighted AVG
DKC	–	41.32	100	–	–	41.32
DGE	1.1	36.64	92	1.5	56.50	38.23
DML	1.1	34.42	90	1.8	60.80	37.06
DMAL	1.2	29.92	90	2.4	52.90	32.22
DMXL	1.1	26.78	90	2.8	59.20	**30.02**

in Tables 1, 2, 3 and 4, for each method on both classes, the parameters r_1, the averages of trials (AVG 1) and the percentages of test problems solved using r_1 in the columns 2–4, respectively; the parameters r_2, the averages of trials (AVG 2) to solve the remaining problems and the weighted averages (Weighted AVG) in the last columns 5–7, where

$$\text{Weighted AVG} = \frac{\text{AVG 1} \cdot n_1 + \text{AVG 2} \cdot (100 - n_1)}{100},$$

and n_1 is the number of problems solved using r_1. Best results are shown in all tables in bold.

As can be seen from Tables 1, 2, 3 and 4, the algorithms DML, DMAL, and DMXL using the local tuning strategies show the most promising behavior. Due to this reason, the third series of experiments was carried out with these three methods only. This series of experiments has been done starting from the following observation regarding unsolved problems of the first two series of experiments with the parameter r_1. It has been realized that the main reason for the failure was that algorithms did not have enough information about the local Lipschitz constants during the first iterations and used therefore significantly smaller values of the estimates m_i w.r.t. the real values causing a premature stop of the methods after a very small number of iterations being so an indicator of this abnormal situation.

To overcome this problem, the parameter $r = r(k)$ dependent on the number of trials k has been chosen. The main idea was to choose a value $n \geq 2$ of trials such that for $k \leq n$, the parameter $r_1(k)$ is relatively high in order to obtain a sufficient information on the behavior of the objective function, while for $k > n$ the parameter was stated to the previously tested value, i.e.,

Table 2 Results of numerical experiments with $\varepsilon = 10^{-4}$ on Class 2

Method	r_1	AVG 1	Success (%)	r_2	AVG 2	Weighted AVG
DKC	–	237.36	100	–	–	237.36
DGE	1.1	98.72	97	1.3	94.67	98.60
DML	1.1	38.61	96	1.3	31.00	38.31
DMAL	1.1	31.11	95	1.3	38.40	**31.47**
DMXL	1.3	73.55	95	1.5	74.40	73.59

Table 3 Results of numerical experiments with $\varepsilon = 10^{-6}$ on Class 1

Method	r_1	AVG 1	Success (%)	r_2	AVG 2	Weighted AVG
DKC	–	45.79	100	–	–	45.79
DGE	1.1	40.25	92	1.5	61.13	41.92
DML	1.1	37.39	90	1.8	65.90	40.24
DMAL	1.2	33.37	90	2.4	59.20	35.95
DMXL	1.1	29.82	90	2.8	66.40	**33.48**

Table 4 Results of numerical experiments with $\varepsilon = 10^{-6}$ on Class 2

Method	r_1	AVG 1	Success (%)	r_2	AVG 2	Weighted AVG
DKC	–	377.24	100	–	–	377.24
DGE	1.1	156.18	97	1.3	158.33	156.24
DML	1.1	41.49	96	1.3	35.00	41.23
DMAL	1.1	34.09	95	1.3	42.60	**34.52**
DMXL	1.3	115.06	95	1.5	115.40	115.08

$r_1(k) = 1.1$. For the remaining unsolved problems the parameter r_2 equal to the previously used value of $r(k)$, $k \leq n$, has been taken. Two values, $n = 5$ and $n = 10$, have been applied and the best result was included in Table 5. The choice of the parameter $r_1(k)$ has been done using the values r_2 from Table 1, i.e., $r_1(k) = r_2$, $k \leq n$, for all the cases but DMXL on Class 2

Table 5 Results for the third series of experiments

Method	Class 1			Class 2		
	DML	DMAL	DMXL	DML	DMAL	DMXL
$r_1(k)$ for $k \leq n$	1.8	2.4	2.8	1.3	1.3	1.7
n	10	10	10	5	5	5
$r_1(k)$ for $k > n$	1.1	1.1	1.1	1.1	1.1	1.1
AVG 1	34.91	28.56	27.12	36.55	29.06	60.06
Success	92%	90%	91%	100%	100%	97%
r_2	1.8	2.4	2.8	–	–	1.7
AVG 2	60.00	52.00	58.56	–	–	98.00
Weighted AVG	36.92	30.90	**29.95**	36.55	**29.06**	61.20

where $r_1(k) = r_2 + 0.2$ has been taken. Accuracy $\varepsilon = 10^{-4}$ was used in all the experiments. As can be seen from Table 1 and Table 5, using a high value of the parameter r at the initial iterations improves performance of the algorithms.

References

1. Amodio, P., Iavernaro, F., Mazzia, F., Mukhametzhanov, M.S., Sergeyev, Y.D.: A generalized Taylor method of order three for the solution of initial value problems in standard and infinity floating-point arithmetic. Math. Comput. Simul. **141**, 24–39 (2017)
2. Antoniotti, L., Caldarola, F., d'Atri, G., Pellegrini, M.: New approaches to basic calculus: an experimentation via numerical computation. Lect. Notes Comput. Sci. **11973 LNCS**, 329–342 (2020). https://doi.org/10.1007/978-3-030-39081-5_29
3. Antoniotti, L., Caldarola, F., Maiolo, M.: Infinite numerical computing applied to Hilbert's, Peano's, and Moore's curves. Mediterr. J. Math. **17**(3) (2020)
4. Berz, M.: Automatic differentiation as nonarchimedean analysis. In: Computer Arithmetic and Enclosure Methods, pp. 439–450. Elsevier, Amsterdam (1992)
5. Bischof, C., Bücker, M.: Computing derivatives of computer programs. In: Modern Methods and Algorithms of Quantum Chemistry Proceedings, NIC Series, vol. 3, 2 edn., pp. 315–327. John von Neumann Institute for Computing, Jülich (2000)
6. Breiman, L., Cutler, A.: A deterministic algorithm for global optimization. Math. Program. **58**(1–3), 179–199 (1993)
7. Caldarola, F.: The Sierpinski curve viewed by numerical computations with infinities and infinitesimals. Appl. Math. Comput. **318**, 321–328 (2018)
8. Calude, C.S., Dumitrescu, M.: Infinitesimal probabilities based on grossone. SN Comput. Sci. **1**(36) (2020)
9. Cohen, J.S.: Computer Algebra and Symbolic Computation: Mathematical Methods. A K Peters Ltd, Wellesley, MA (1966)

10. Corliss, G., Faure, C., Griewank, A., Hascoet, L., Naumann, U. (eds.): Automatic Differentiation of Algorithms: From Simulation to Optimization. Springer, New York (2002)
11. D'Alotto, L.: Cellular automata using infinite computations. Appl. Math. Comput. **218**(16), 8077–8082 (2012)
12. D'Alotto, L.: Infinite games on finite graphs using grossone. Soft Comput. **55**, 143–158 (2020)
13. Daponte, P., Grimaldi, D., Molinaro, A., Sergeyev, Y.D.: An algorithm for finding the zero-crossing of time signals with lipschitzean derivatives. Measurement **16**(1), 37–49 (1995)
14. De Leone, R., Fasano, G., Sergeyev, Y.D.: Planar methods and grossone for the conjugate gradient breakdown in nonlinear programming. Comput. Optim. Appl. **71**(1), 73–93 (2018)
15. Falcone, A., Garro, A., Mukhametzhanov, M.S., Sergeyev, Y.D.: Representation of Grossone-based arithmetic in Simulink and applications to scientific computing. Soft Comput. **24**, 17525–17539 (2020)
16. Fiaschi, L., Cococcioni, M.: Non-archimedean game theory: a numerical approach. Appl. Math. Comput. (125356) (2020). https://doi.org/10.1016/j.amc.2020.125356
17. Gaudioso, M., Giallombardo, G., Mukhametzhanov, M.S.: Numerical infinitesimals in a variable metric method for convex nonsmooth optimization. Appl. Math. Comput. **318**, 312–320 (2018)
18. Gergel, V.P.: A global search algorithm using derivatives. In: Yu I. Neymark (Ed.), Systems Dynamics and Optimization, pp. 161–178 (1992)
19. Gergel, V.P., Grishagin, V.A., Israfilov, R.A.: Local tuning in nested scheme of global optimization. Procedia Comput. Sci. **51**, 865–874 (2015)
20. Grishagin, V.A.: On convergence conditions for a class of global search algorithms. In: Numerical Methods of Nonlinear Programming, pp. 82–84. KSU, Kharkov (1979). (In Russian)
21. Grishagin, V.A., Israfilov, R.A.: Global search acceleration in the nested optimization scheme. In: T. Simos, C. Tsitouras (eds.) Proceedings of International Conference on Numerical Analysis and Applied Mathematics (ICNAAM 2015), vol. 1738, p. 400010. AIP Publishing, NY (2016). https://doi.org/10.1063/1.4952198
22. https://www.numericalinfinities.com
23. Iannone, P., Rizza, D., Thoma, A.: Investigating secondary school students' epistemologies through a class activity concerning infinity. In: Bergqvist, E., Österholm, M., Granberg, C., Sumpter, L. (eds.) Proceedings of the 42nd Conference of the International Group for the Psychology of Math. Education, vol. 3, pp. 131–138. PME, Umeå (2018)
24. Iavernaro, F., Mazzia, F., Mukhametzhanov, M.S., Sergeyev, Y.D.: Computation of higher order Lie derivatives on the Infinity Computer. J. Comput. Appl. Math. **383**(113135) (2021)
25. Ingarozza, F., Adamo, M.T., Martino, M., Piscitelli, A.: A grossone-based numerical model for computations with infinity: a case study in an italian high school. Lect. Notes Comput. Sci. LNCS **11973**, 451–462 (2020). https://doi.org/10.1007/978-3-030-39081-5_39
26. Iudin, D.I., Sergeyev, Y.D., Hayakawa, M.: Interpretation of percolation in terms of infinity computations. Appl. Math. Comput. **218**(16), 8099–8111 (2012)
27. Kvasov, D.E., Mukhametzhanov, M.S., Nasso, M.C., Sergeyev, Y.D.: On acceleration of derivative-free univariate Lipschitz global optimization methods. In: Sergeyev,

Y.D., Kvasov, D. (eds), Numerical Computations: Theory and Algorithms. NUMTA 2019. Lecture Notes in Computer Science, vol. 11974, pp. 413–421. Springer, Cham (2020)
28. Kvasov, D.E., Sergeyev, Y.D.: A univariate global search working with a set of Lipschitz constants for the first derivative. Optim. Lett. **3**(2), 303–318 (2009)
29. Kvasov, D.E., Sergeyev, Y.D.: Lipschitz global optimization methods in control problems. Autom. Remote Control **74**(9), 1435–1448 (2013)
30. Lera, D., Sergeyev, Y.D.: Acceleration of univariate global optimization algorithms working with Lipschitz functions and Lipschitz first derivatives. SIAM J. Optim. **23**(1), 508–529 (2013)
31. Lyness, J.N., Moler, C.B.: Numerical differentiation of analytic functions. SIAM J. Numer. Anal. **4**, 202–210 (1967)
32. Margenstern, M.: An application of grossone to the study of a family of tilings of the hyperbolic plane. Appl. Math. Comput. **218**(16), 8005–8018 (2012)
33. Moin, P.: Fundamentals of Engineering Numerical Analysis. Cambridge University Press, Cambridge (2001)
34. Muller, J.M.: Elementary Functions: Algorithms and Implementation. Birkhäuser, Boston (2006)
35. Pintér, J.D.: Global Optimization in Action (Continuous and Lipschitz Optimization: Algorithms, Implementations and Applications). Kluwer Academic Publishers, Dordrecht (1996)
36. Piyavskij, S.A.: An algorithm for finding the absolute extremum of a function. USSR Comput. Math. Math. Phys. **12**(4), 57–67 (1972)
37. Rizza, D.: A study of mathematical determination through Bertrand's Paradox. Philosophia Mathematica **26**(3), 375–395 (2018)
38. Rizza, D.: Numerical methods for infinite decision-making processes. Int. J. Unconv. Comput. **14**(2), 139–158 (2019)
39. Sergeyev, Y.D.: A one-dimensional deterministic global minimization algorithm. Comput. Math. Math. Phys. **35**(5), 705–717 (1995)
40. Sergeyev, Y.D.: Global one-dimensional optimization using smooth auxiliary functions. Math. Program. **81**(1), 127–146 (1998)
41. Sergeyev, Y.D.: On convergence of "divide the best" global optimization algorithms. Optimization **44**(3), 303–325 (1998)
42. Sergeyev, Y.D.: Arithmetic of Infinity. Edizioni Orizzonti Meridionali, CS , 2nd ed. (2013)
43. Sergeyev, Y.D.: Computer system for storing infinite, infinitesimal, and finite quantities and executing arithmetical operations with them. USA patent 7,860,914 (2010)
44. Sergeyev, Y.D.: Higher order numerical differentiation on the Infinity Computer. Optim. Lett. **5**(4), 575–585 (2011)
45. Sergeyev, Y.D.: Numerical infinities and infinitesimals: methodology, applications, and repercussions on two Hilbert problems. EMS Surv. Math. Sci. **4**(2), 219–320 (2017)
46. Sergeyev, Y.D.: Independence of the grossone-based infinity methodology from non-standard analysis and comments upon logical fallacies in some texts asserting the opposite. Found. Sci. **24**(1), 153–170 (2019)
47. Sergeyev, Y.D., Garro, A.: Observability of Turing machines: a refinement of the theory of computation. Informatica **21**(3), 425–454 (2010)
48. Sergeyev, Y.D., Mukhametzhanov, M.S., Kvasov, D.E., Lera, D.: Derivative-free local tuning and local improvement techniques embedded in the univariate global optimization. J. Optim. Theory Appl. **171**(1), 186–208 (2016)

49. Sergeyev, Y.D., Nasso, M.C., Mukhametzhanov, M.S., Kvasov, D.E.: Novel local tuning techniques for speeding up one dimensional algorithms in expensive global optimization using lipschitz derivatives. J. Comput. Appl. Math. **383**(113134) (2021)
50. Strongin, R.G., Sergeyev, Y.D.: Global Optimization with Non-convex Constraints: Sequential and Parallel Algorithms. Kluwer Academic Publishers, Dordrecht (2000)
51. Zhigljavsky, A.: Computing sums of conditionally convergent and divergent series using the concept of grossone. Appl. Math. Comput. **218**(16), 8064–8076 (2012)
52. Žilinskas, A.: On strong homogeneity of two global optimization algorithms based on statistical models of multimodal objective functions. Appl. Math. Comput. **218**(16), 8131–8136 (2012)

Comparing Linear and Spherical Separation Using Grossone-Based Numerical Infinities in Classification Problems

Annabella Astorino and Antonio Fuduli

Abstract We investigate the role played by the linear and spherical separations in binary supervised learning and in Multiple Instance Learning (MIL), in connection with the use of the grossone-based numerical infinities. While in the binary supervised learning the objective is to separate two sets of samples, a binary MIL problem consists in separating two different type of sets (positive and negative), each of them constituted by a finite number of samples. We remind that using the spherical separation in classification problems provides an advantage especially in terms of computational time, since, when the center of the separating sphere is (judiciously) fixed in advance, the corresponding optimization problem reduces to a structured linear program, easily solvable by an ad hoc algorithm. In particular, by embedding the grossone idea, here we analyze the case where the center of the sphere is selected far from both the two sets, obtaining in this way a kind of linear separation. This approach is easily extensible to the margin concept (of the type adopted in the Support Vector Machine technique) and to MIL problems. Some numerical results are reported on classical binary datasets drawn from the literature.

A. Astorino (✉)
Institute for High Performance Computing and Networking (ICAR), National Research Council, Rende, Italy
e-mail: annabella.astorino@icar.cnr.it

A. Fuduli
Department of Mathematics and Computer Science, University of Calabria, Rende, Italy
e-mail: antonio.fuduli@unical.it

Y. D. Sergeyev and R. De Leone (eds.), *Numerical Infinities and Infinitesimals in Optimization*, Emergence, Complexity and Computation 43,
https://doi.org/10.1007/978-3-030-93642-6_10

1 Introduction

Classification problems in mathematical programming concern separation of sample sets by means of an appropriate surface. This field, entered by many researchers in optimization community in the last years, is a part of the more general machine learning area, aimed at providing automated systems able to learn from human experiences.

The objective of pattern classification is to categorize samples into different classes on the basis of their similarities. More formally, given a set of labelled and unlabelled samples, characterized by some features, for each of them we want to express a particular feature, the class label, as a function of the remaining ones. This is done by constructing a prediction function, by means of which we would like to predict the class of each sample. In machine learning literature many approaches [14] have been indeed devised for automatically distinguishing among different samples on the basis of their patterns: approaches of supervised, unsupervised and semi-supervised learning, and more recently approaches of Multiple Instance Learning. In particular, in the supervised case most of the learning models apply the inductive inference concept, where the prediction function, derived only from the labelled input data, is used to predict the label of any future object. A well established supervised technique is the Support Vector Machine (SVM) [33, 58], which has revealed a powerful classification tool in many application areas.

A widely adopted alternative to supervised classification is the unsupervised one, where all the objects are unlabelled: as a consequence, in such case, the prediction function is constructed by clustering the data on the basis of their similarities [22, 28]. In the middle we find the semisupervised techniques [29], that apply the transductive inference concept: the prediction function is derived from the information concerning all the available data (both labelled and unlabelled samples). This function is not aimed at predicting the class label of newly incoming samples, but only at making a decision about the currently available unlabelled objects. Some useful references are [5, 30], the latter being a semisupervised version of the SVM technique.

A more recent classification framework is constituted by the Multiple Instance Learning (MIL) [42], whose main difference with respect to the traditional supervised learning scenario resides in the nature of the learning samples. In fact, each sample is not represented by a fixed-length vector of features but by a bag of feature vectors that are referred to as instances. The classification labels are only provided for the entire training bags whereas the labels of the instances inside them are unknown. The task is to learn a model that predicts the labels of the new incoming bags, possibly together with the labels of the instances inside them. A seminal SVM-type MIL paper is [1], while some recent articles are [11–13, 15, 17, 21, 24, 40, 49].

In this work, strictly connected with [7], starting from the spherical binary supervised classification approach reported in [16], we introduce some spherical separation models for both supervised learning and Multiple Instance Learning. Such models are obtained by embedding the grossone-based numerical methodology [53], which allows to select the center of the sphere far from both the two sets (of samples or of bags), providing a kind of linear separation.

Spherical separation falls into the class of the nonlinear separation surfaces [10, 14], differently, for example, from the well known supervised learning SVM technique [33, 58], where a classifier is constructed by generating a hyperplane far away from the points of the two sets. Also the SVM approach allows to obtain general nonlinear classifiers by adopting kernel transformations. In this case the basic idea is to map the data into a higher dimensional space (the feature space) and to separate the two transformed sets by means of a hyperplane, that corresponds to a nonlinear surface in the original input space. The main advantage of spherical separation is that, once the center of the sphere is heuristically fixed in advance, the optimal radius can be found quite effectively by means of simple sorting algorithms such as those ones reported in [9, 16]. No analogous simplification strategy is apparently available if one adopts the SVM approach. Moreover, another advantage is to work directly in the input space. In fact to keep, whenever possible, the data in the original space seems appealing in order to stay close to the real life modeled processes. Of course kernel methods are characterized by high flexibility, even if sometimes they provide results which are hard to be interpreted in the original input space, differently from the nonlinear classifiers acting directly in such space (see for example [19, 20]).

The chapter is organized in the following way. In the next section we focus on supervised classification, distinguishing between linear and spherical separation, the latter suitable for grossone application (see [7]). In Sect. 3 we discuss the possibility to extend the grossone spherical separation to Multiple Instance Learning, while in Sect. 4 we comment the numerical results published in [7], which confirm the practical applicability of the grossone-based numerical infinities in classification problems. Finally some conclusions are drawn in Sect. 5.

2 Linear and Spherical Separability for Supervised Classification

Let

$$\mathcal{A} = \{a_1, \dots, a_m\}, \quad \text{with } a_i \in \mathbb{R}^n,\ i = 1, \dots, m$$

and

$$\mathcal{B} = \{b_1, \dots, b_k\}, \quad \text{with } b_l \in \mathbb{R}^n,\ l = 1, \dots, k,$$

be the two finite sets of samples (points in $\mathbb{R}^n$). The classical binary classification problem consists in discriminating between $\mathcal{A}$ and $\mathcal{B}$ by means of a separating surface, obtained by minimizing any classification error. Such surface can be a hyperplane (linear separation) or a nonlinear surface, such as a sphere (spherical separation). A seminal paper on linear separation appeared in 1965 by Mangasarian [47], while the first approach for pattern classification based on a minimum volume sphere dates back to 1999 by Tax and Duin [56].

2.1 Linear Separation

The two sets $\mathcal{A}$ and $\mathcal{B}$ are linearly separable if and only if there exists a hyperplane

$$H(w, \gamma) = \{x \in \mathbb{R}^n \mid w^T x = \gamma\}, \text{ with } w \in \mathbb{R}^n \text{ and } \gamma \in \mathbb{R},$$

such that

$$w^T a_i \leq \gamma - 1 \qquad i = 1, \dots, m$$

and

$$w^T b_l \geq \gamma + 1 \qquad l = 1, \dots, k.$$

A geometrical characterization of linear separability is that $\mathcal{A}$ and $\mathcal{B}$ are linearly separable if and only if their convex hulls do not intersect, i.e.

$$\text{conv}(\mathcal{A}) \cap \text{conv}(\mathcal{B}) = \emptyset,$$

as depicted in Fig. 1, where the two cases of linearly separable and inseparable sets are considered.

The problem of finding a separating hyperplane can be formulated as a linear program [23], but several other approaches have been proposed, such as the SVM technique [33, 58], where the idea is to generate a separation hyperplane far away from the objects of both the two sets. This is done by

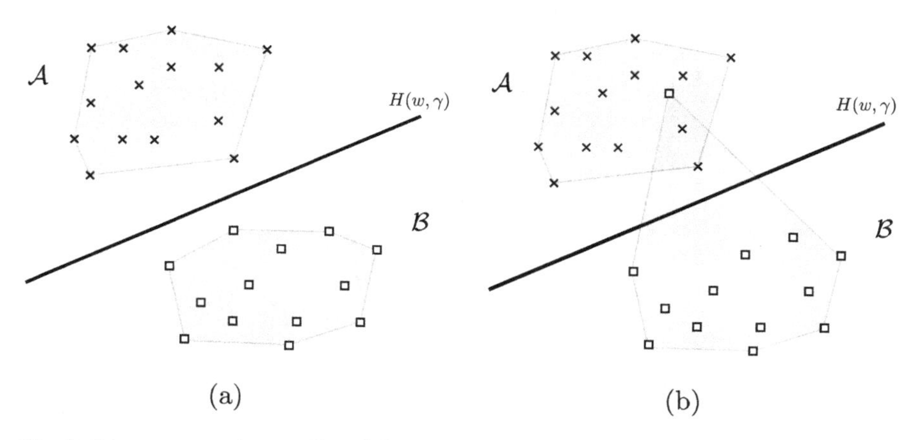

Fig. 1 Linear separation: **a** $\mathcal{A}$ and $\mathcal{B}$ are separable since $\text{conv}(\mathcal{A}) \cap \text{conv}(\mathcal{B}) = \emptyset$; **b** $\mathcal{A}$ and $\mathcal{B}$ are not separable since $\text{conv}(\mathcal{A}) \cap \text{conv}(\mathcal{B}) \neq \emptyset$

maximizing the margin (i.e. the distance between two parallel hyperplanes supporting the sets), representing a measure of the generalization capability, i.e. the ability of the classifier to correctly classify any new sample (see Fig. 2). In particular, from the mathematical point of view, the SVM provides a separating hyperplane $H(w, \gamma)$ by minimizing the following error function:

$$\min_{w,\gamma} \frac{1}{2}\|w\|^2 + C\sum_{i=1}^{m} \max\{0, a_i^T w - \gamma + 1\} + C\sum_{l=1}^{k} \max\{0, -b_l^T w + \gamma + 1\}, \tag{1}$$

where the minimization of first term corresponds to the maximization of the margin, and the last two terms represent the misclassification errors in correspondence to the two point sets $\mathcal{A}$ and $\mathcal{B}$, respectively. The parameter C is a positive constant giving the tradeoff between these two objectives. We conclude this subsection, reminding that the above nonsmooth minimization problem can be easily rewritten as a smooth quadratic programming problem.

2.2 *Spherical Separation*

In the spherical separation the idea is to find a sphere

$$S(x_0, R) = \{x \in \mathbb{R}^n \mid \|x - x_0\|^2 = R^2\},$$

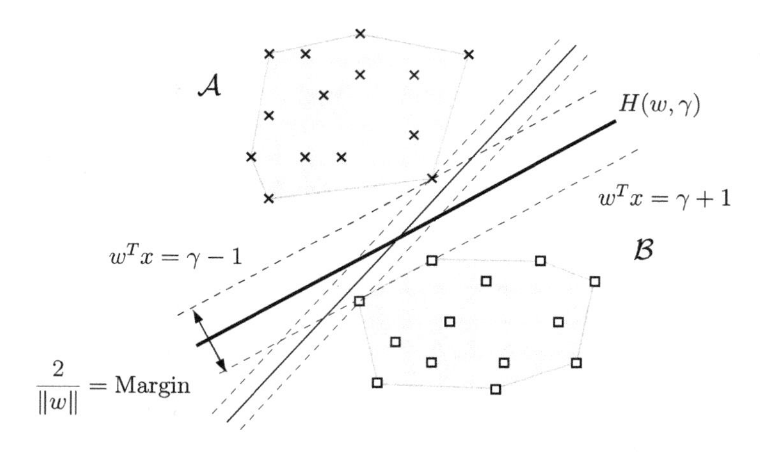

Fig. 2 Among all the separating hyperplanes, the SVM approach selects that one with the largest margin

with center $x_0 \in \mathbb{R}^n$ and radius R, enclosing all points of $\mathcal{A}$ and no points of $\mathcal{B}$.

2.2.1 Spherical Separation Without Margin

The set $\mathcal{A}$ is spherically separable from the set $\mathcal{B}$ if and only if there exists a sphere $S(x_0, R)$ such that

$$\|a_i - x_0\|^2 \leq R^2 \qquad i = 1, \ldots, m$$

and

$$\|b_l - x_0\|^2 \geq R^2 \qquad l = 1, \ldots, k.$$

We observe that, in this case, the role played by the two sets is not symmetric; in fact a necessary (but not sufficient) condition for the existence of a separation sphere is the following (see Fig. 3):

$$\text{conv}(\mathcal{A}) \cap \mathcal{B} = \emptyset.$$

Based on the above spherical separability definition, the classification error associated to any sphere $S(x_0, R)$ is

$$\sum_{i=1}^{m} \max\{0, \|a_i - x_0\|^2 - R^2\} + \sum_{l=1}^{k} \max\{0, R^2 - \|b_l - x_0\|^2\}.$$

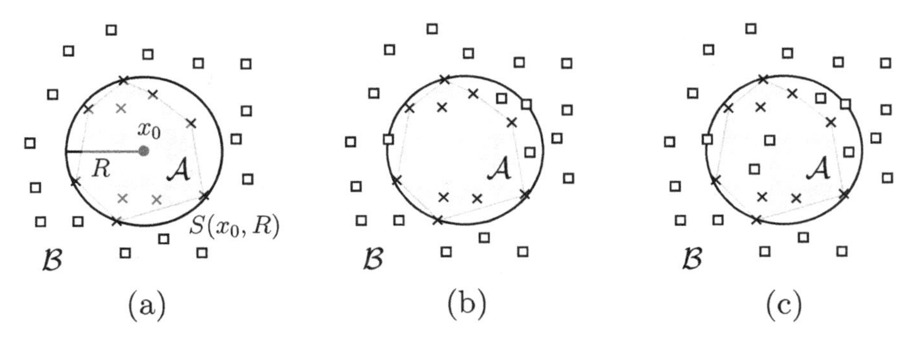

Fig. 3 Spherical separation of $\mathcal{A}$ from $\mathcal{B}$: **a** $\mathcal{A}$ is separable from $\mathcal{B}$ and then $\text{conv}(\mathcal{A}) \cap \mathcal{B} = \emptyset$; **b** $\mathcal{A}$ is not separable from $\mathcal{B}$ even if $\text{conv}(\mathcal{A}) \cap \mathcal{B} = \emptyset$; **c** $\mathcal{A}$ is not separable from $\mathcal{B}$ since $\text{conv}(\mathcal{A}) \cap \mathcal{B} \neq \emptyset$

To take into account the generalization capability, in [16] the authors have proposed to construct a minimal volume separation sphere by solving the following problem:

$$\min_{x_0, z} z + C \sum_{i=1}^{m} \max\{0, \|a_i - x_0\|^2 - z\} + C \sum_{l=1}^{k} \max\{0, z - \|b_l - x_0\|^2\}, \tag{2}$$

with $z \triangleq R^2 \geq 0$ and $C > 0$ being the parameter tuning the tradeoff between the minimization of the volume and the minimization of the classification error.

Some works devoted to spherical separation are [2, 3, 8, 9, 16, 18, 44]. In particular, the approach presented in [16] assumes that the center x_0 of the sphere is fixed (for example, equal to the barycenter of $\mathcal{A}$): in such case it is easy to see that problem (2) reduces to a univariate, convex, nonsmooth optimization problem and it is rewritable as a structured linear program, whose dual can be solved in time $O(p \log p)$, where p is the cardinality of the biggest set between $\mathcal{A}$ and $\mathcal{B}$. In fact the optimal value of the variable z (the square of the radius) is computable by simply comparing the distances, preliminarly sorted, between the center x_0 and each point in the two sets. For further technical details on such approach we refer the reader directly to [16].

2.2.2 Spherical Separation with Margin

Now we consider a margin spherical separation, where we extend the SVM concept of margin to the spherical case with the aim at providing a better quality classifier. In particular the set $\mathcal{A}$ is strictly spherically separable from

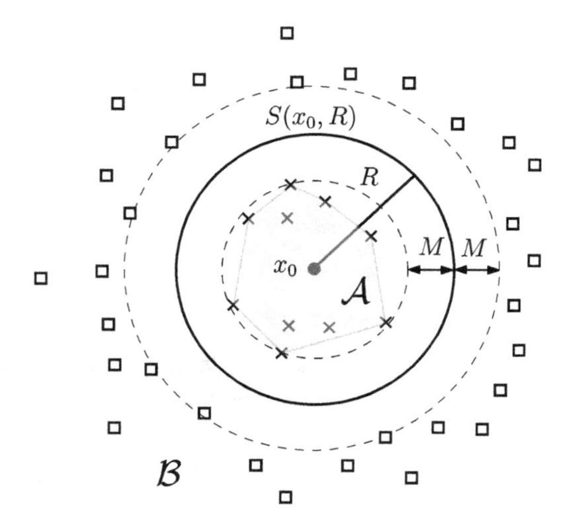

Fig. 4 Strict spherical separation of $\mathcal{A}$ from $\mathcal{B}$

the set $\mathcal{B}$ if there exists a sphere $S(x_0, R)$ such that

$$\|a_i - x_0\|^2 \leq (R - M)^2, \qquad i = 1, \ldots, m$$

and

$$\|b_l - x_0\|^2 \geq (R + M)^2, \qquad l = 1, \ldots, k,$$

for some margin M, $0 < M \leq R$ (see Fig. 4).

Based on the above definition, the classification error becomes

$$\sum_{i=1}^{m} \max\{0, \|a_i - x_0\|^2 - (R - M)^2\} + \sum_{l=1}^{k} \max\{0, (R + M)^2 - \|b_l - x_0\|^2\},$$

which, by setting $z \triangleq R^2 + M^2$ and $q \triangleq 2RM$, can be rewritten as:

$$\sum_{i=1}^{m} \max\{0, q - z + \|a_i - x_0\|^2\} + \sum_{l=1}^{k} \max\{0, q + z - \|b_l - x_0\|^2\}.$$

In [9] the authors have proposed to solve the following optimization problem:

$$\min_{x_0, 0 \leq q \leq z} C \left(\sum_{i=1}^{m} \max\{0, q - z + \|a_i - x_0\|^2\} + \sum_{l=1}^{k} \max\{0, q + z - \|b_l - x_0\|^2\} \right) - q,$$

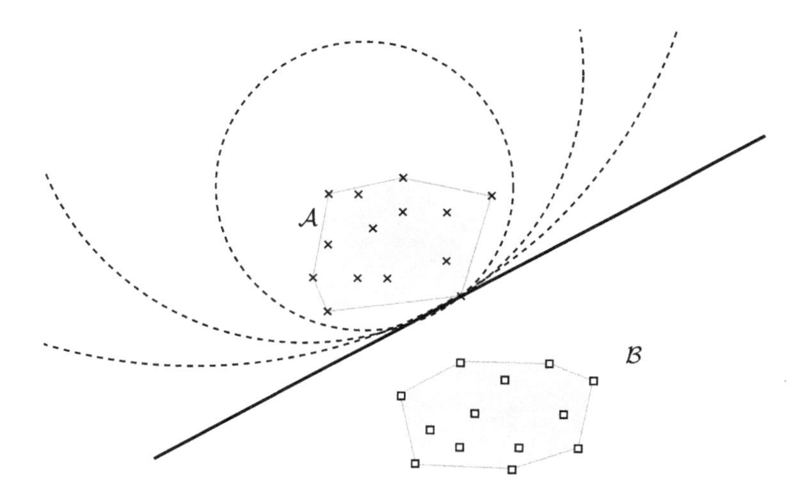

Fig. 5 A hyperplane can be interpreted as a sphere with an infinitely far center

where the objective of margin maximization is represented by the term $-q$, while the tradeoff between classification error and margin is accounted by the positive weighting parameter C.

In case the center x_0 of the sphere is given, the above problem reduces to the minimization of a nonsmooth and convex function [4, 37] in the two variables z and q. Such problem can be easily put in the form of a structured linear program, which is solvable by an extended version of the algorithm presented in [16] (see [9] for the details).

2.3 *Comparing Linear and Spherical Separation in the Grossone Framework*

From the mathematical point of view, both the linear and the spherical separations are characterized by the same number of variables to be determined: in fact a separation hyperplane is identified by the bias and the normal, while a sphere is obtained by computing the center and the radius. In this perspective, a hyperplane can be viewed as a particular sphere where the center is infinitely far (see Fig. 5).

Then a possible choice of the center x_0 is to take a point far from both the sets $\mathcal{A}$ and $\mathcal{B}$, i.e.

$$x_0 = x_0^{\mathcal{A}} + M\left(x_0^{\mathcal{A}} - x_0^{\mathcal{B}}\right), \tag{3}$$

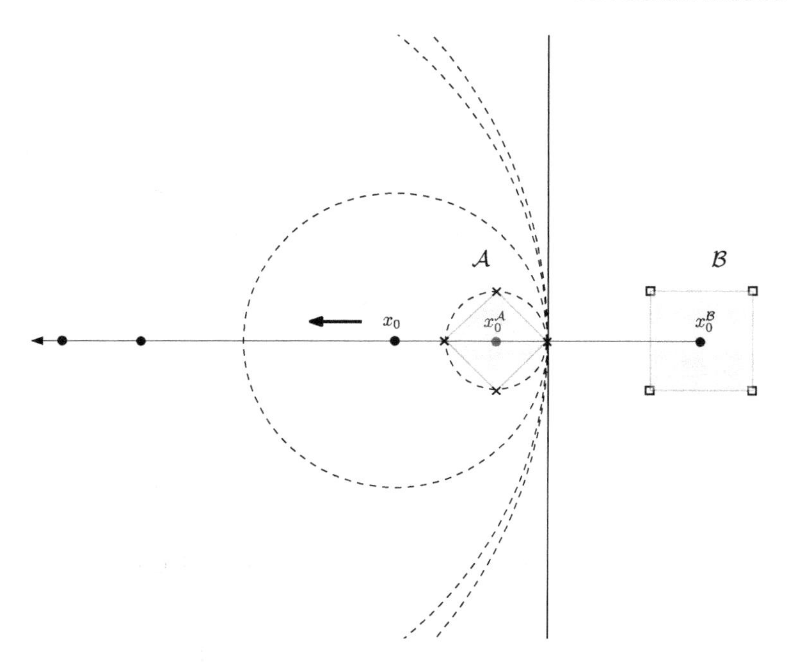

Fig. 6 Spherical separation with a far center

where

$$x_0^{\mathcal{A}} \triangleq \frac{1}{m}\sum_{i=1}^{m} a_i \quad \text{and} \quad x_0^{\mathcal{B}} \triangleq \frac{1}{k}\sum_{l=1}^{k} b_l$$

are the barycenters of $\mathcal{A}$ and $\mathcal{B}$, respectively, while M is a sufficiently large positive parameter, commonly named "big M" (see for example [32]).

Formula (3) corresponds to computing x_0 from $x_0^{\mathcal{A}}$ along the direction $x_0^{\mathcal{A}} - x_0^{\mathcal{B}}$ with stepsize equal to M (see Fig. 6).

Notice that, in general, the "big M" constant is not easy to be managed from the numerical point of view, since indeed it is not evident how to quantify the minimum threshold value such that M could be considered sufficiently big: as a consequence, in the practical cases, the necessity to test many trial values arises. A possible way to overcome this numerical difficulty is to obtain an infinitely far center by exploiting the grossone theory [53], setting M equal to ①, where the symbol ① denotes the new numeral *grossone*.

Differently from [16], where various values of M in formula (3) have been tested in order to obtain a good classification performance, a remarkable advantage in using the grossone resides in avoiding the necessity to repeat several tests with larger and larger values of M.

We conclude the subsection by highlighting that the new grossone-based computational methodology, which is not related to the nonstandard analysis [54], is applied in various fields, such as in optimization [31, 34–36, 41, 55], in numerical differentiation [52], in ordinary differential equations [43] and so on. To the best of our knowledge, it seems that the only machine learning paper involving the grossone idea is [7]. Finally some more theoretical works are in logics and philosophy [45, 46, 50], in probability [27] and in fractals analysis [25, 26].

3 Linear and Spherical Separability for Multiple Instance Learning

Multiple Instance Learning (MIL) [42] is a machine learning paradigm, consisting in classifying sets of samples: the samples are called instances and the sets are called bags. The main peculiarity of a MIL problem stays in the learning phase, where only the labels of the bags are known while the labels of the instances are unknown.

The first MIL paper [38] has appeared in 1997: in such work a drug design problem has been tackled, with the aim at discriminating between active and non-active molecules. A drug molecule is active (i.e. it has the desired drug effect) if one or more of its conformations binds to a particular target site (typically a larger protein molecule): the peculiarity of the problem is that it is not known a priori which conformation makes a molecule active, being available only the label of the overall molecule. In the MIL perspective, each molecule is a bag and the corresponding conformations are the instances.

We focus on binary MIL problems, aimed at discriminating between positive and negative bags, in the presence of only two classes of instances. We adopt the so-called standard MIL assumption (very common in the literature), stating that a bag is positive if and only if it contains at least a positive instance and it is negative otherwise.

Since the considerations reported in Sect. 2.3 for supervised classification can be extended to MIL, in the sequel we first remind the SVM type model for MIL introduced in [1] and, successively, we propose our modification of such model based on the spherical separation.

3.1 The SVM Type Model for MIL

The SVM type model proposed for MIL in [1] provides, in the instance space, a separating hyperplane $H(w, \gamma)$ by solving the following optimization problem:

$$\begin{cases} \min\limits_{w,\gamma,y} \frac{1}{2}\|w\|^2 + C\sum\limits_{i=1}^{m}\sum\limits_{j\in J_i^+} \max\{0, y_j(x_j^T w - \gamma) + 1\} \\ \qquad\qquad + C\sum\limits_{l=1}^{k}\sum\limits_{j\in J_l^-} \max\{0, -x_j^T w + \gamma + 1\} \\ \qquad\qquad \sum\limits_{j\in J_i^+} \frac{y_j + 1}{2} \geq 1 \quad i = 1, \dots, m \\ \qquad\qquad y_j \in \{-1, +1\} \quad j \in J_i^+, \quad i = 1, \dots, m, \end{cases} \tag{4}$$

where m is the number of positive bags indexed by the sets $J_i^+, i = 1, \dots, m$, k is the number of negative bags indexed by the sets $J_l^-, l = 1, \dots, k$, x_j is the jth instance belonging to a bag and y_j is the class label of the instance x_j. The m constraints involved in the above nonlinear mixed integer program impose that at least one instance of each positive bag is labelled positively by $y_j = +1$, i.e. the satisfaction of the standard MIL assumption. A separating MIL hyperplane is depicted in Fig. 7, where the two dashed polygons represent the positive bags and the three continuous polygons are the negative bags.

Notice that in case each bag is a singleton and $y_j = 1$ for any j, problem (4) reduces to the classical SVM problem (1).

3.2 A Grossone MIL Spherical Model

In this subsection, in order to embed the grossone framework into the MIL paradigm, we propose to modify problem (4) by substituting the hyperplane for a sphere. We obtain the following nonlinear mixed integer optimization problem:

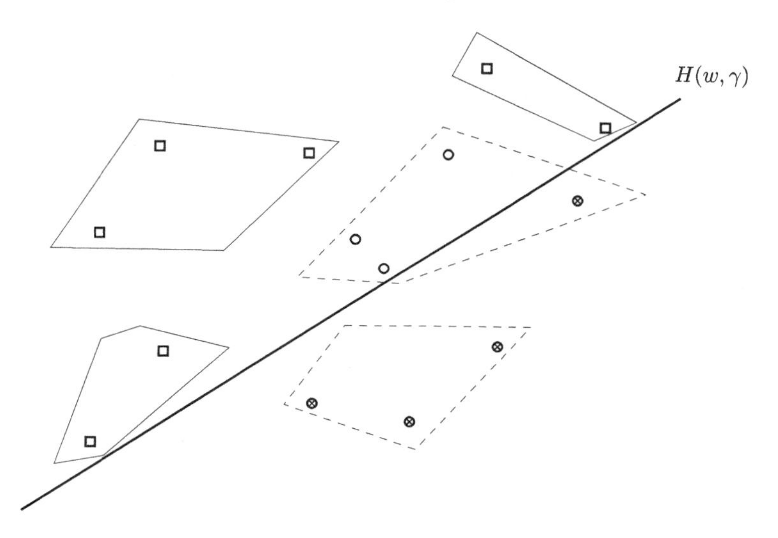

Fig. 7 MIL separating hyperplane: two positive bags (dashed polygons) and three negative bags (continuous polygons). The circles and the squares inside the bags represent the instances

$$\begin{cases} \min\limits_{x_0, R, y} \; R^2 + C \sum\limits_{i=1}^{m} \sum\limits_{j \in J_i^+} \max\{0, \, y_j(\|x_j - x_0\|^2 - R^2)\} \\ \qquad + C \sum\limits_{l=1}^{k} \sum\limits_{j \in J_l^-} \max\{0, \, R^2 - \|x_j - x_0\|^2\} \\ \sum\limits_{j \in J_i^+} \dfrac{y_j + 1}{2} \geq 1 \quad i = 1, \ldots, m \\ y_j \in \{-1, +1\} \quad j \in J_i^+, \quad i = 1, \ldots, m. \end{cases} \tag{5}$$

According to the concept of spherical separation reported in Sect. 2.2.1, the above model takes into account the standard MIL assumption, which, in case of a separating sphere, imposes that a bag is positive if at least one of its instances is inside the sphere and it is negative otherwise (see Fig. 8, where the represented bags are the same as in Fig. 7).

A possible approach to solve heuristically problem (5) could be to use a BDC (Block Coordinate Descent) [57] type algorithm, consisting in the alternation between the computation of the vector y when the couple (x_0, R) is fixed and, vice-versa, the computation of the couple (x_0, R) when y is fixed. In particular, when y is fixed, x_0 could be set by adopting the following formula (analogous to formula (3), with M substituted by ①):

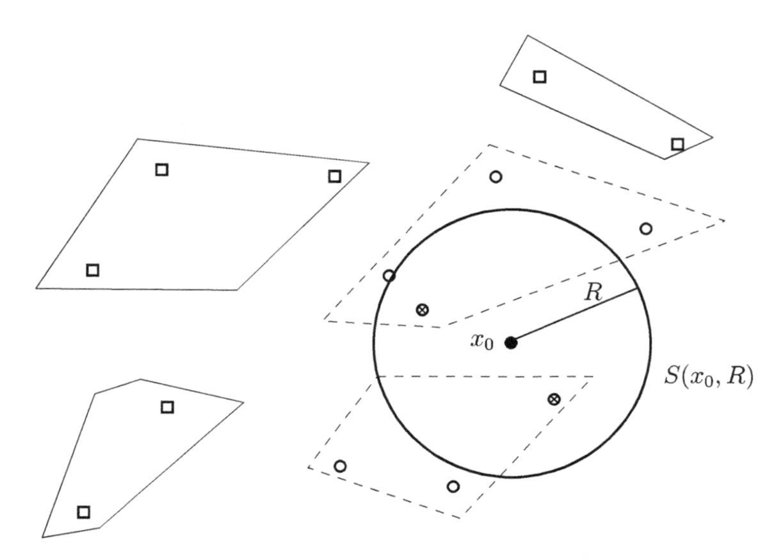

Fig. 8 MIL separating sphere: two positive bags (dashed polygons) and three negative bags (continuous polygons). The circles and the squares inside the bags represent the instances

$$x_0 = x_0^+ + ①(x_0^+ - x_0^-), \tag{6}$$

where x_0^+ and x_0^- are the barycenters of the currently positive and negative instances, respectively. Once y is fixed and x_0 is computed by formula (6), the corresponding optimal radius of the sphere is obtainable by using the ad hoc algorithm presented in [16].

4 Some Numerical Results

In [7] some numerical experiments have been performed to test the grossone idea in the supervised spherical separation without margin. In fact the center of the sphere has been chosen as follows:

$$x_0 = x_0^{\mathcal{A}} + ① \left(x_0^{\mathcal{A}} - x_0^{\mathcal{B}} \right),$$

i.e. by setting $M = ①$ in formula (3).

The code, named in [7] $FC_{①}$, has been implemented in Matlab and it has been tested on thirteen data sets drawn from the literature and listed in Table 1.

Table 1 Data sets

Data set	Dimension	Points
Cancer	9	699
Diagnostic	30	569
Heart	13	297
Pima	9	769
Ionosphere	34	351
Sonar	60	208
Mushrooms	22	8124
Prognosis	32	110
Tic Tac Toe	9	958
Votes	16	435
Galaxy	14	4192
g50c	50	550
g10n	10	550

The first ten test problems have been taken from the UCI Machine Learning Repository [48], a collection of databases, domain theories, and data generators that are used by the machine learning community. Galaxy is the data set used in galaxy discrimination with neural networks [51], while an accurate description of g50c and g10n is reported in [30].

In order to manage the grossone arithmetic operations, the authors have used the Matlab Environment of the new Simulink-based solution of the Infinity Computer [39], where an arithmetic C++ library is integrated within a Matlab environment. In particular, given the two gross-numbers x and y, from such library the following C++ subroutines have been used:

- `TestGrossMatrix(x,y,'-')`, returning the difference between x and y;
- `TestGrossMatrix(x,y,'+')`, returning the sum of x and y;
- `TestGrossMatrix(x,y,'*')`, returning the product of x and y;
- `GROSS_cmp(x,y)`, returning 1 if $x > y$, -1 if $x < y$ and 0 if $x = y$.

Using the Matlab notation, any vector g of n gross-number elements (that in the sequel, for the sake of simplicity, we call gross-vector) has been expressed as a couple `(G, fg)`, with

$$\texttt{G} = [\texttt{g1; g2; \ldots; gn}] \quad \text{and} \quad \texttt{fg} = [\texttt{fg1 fg2} \ldots \texttt{fgn}],$$

where `gj`, $j = 1, \ldots, n$, is an array of appropriate dimension representing a gross-number. For each row of `gj`, the first element contains a gross-digit, while the second one contains the corresponding gross-power. The scalar `fgj`, $j = 1, \ldots, n$, is necessary to provide the position in `G` of the last component of `gj`.

To manage the gross-vectors, in [7] the following new Matlab subroutines have been implemented:

- `realToGrossone(r)`, returning a grossone representation `(G, fg)` of a real vector `r`;
- `extract(G,fg,i)`, returning the `i`th gross-number in the gross-vector `(G, fg)`;
- `normGrossone(G,fg)`, computing the squared Euclidean norm of the gross-vector `(G,fg)`;
- `scalProdG(G1,fg1,G2,fg2)`, computing the scalar product between the two gross-vectors `(G1,fg1)` and `(G2,fg2)`;
- `BubbleSortGrossone(G,fg,sign)`, sorting the gross-vector `(G, fg)` in the ascending order if `sign` $= 1$ and in the descending order if `sign` $= -1$.

For each data set, in order to compute the best value of the parameter C, a bilevel cross-validation strategy [6] has been adopted, by varying C in the grid $\{10^{-1}, 10^{0}, 10^{1}, 10^{2}\}$: such choice of the grid has been suggested by the necessity to obtain a nonzero optimal value of z, which in turn provides the optimal value of the radius R, as shown in [16].

In Table 2 we report the results, provided by Algorithm $\mathrm{FC}_{①}$ and published in [7], expressed in terms of average testing correctness. Such results have been compared by the authors with those ones relative to the two following fixed-center classical variants, obtained by setting

$$x_0 = x_0^{\mathcal{A}} \qquad \text{(Algorithm } \mathrm{FC}_{\mathcal{A}}\text{)}$$

and

$$x_0 = x_0^{\mathcal{A}} + x_0^{\mathcal{B}} \qquad \text{(Algorithm } \mathrm{FC}_{\mathcal{AB}}\text{)},$$

respectively, and with the results obtained by a variant of the standard linear SVM (Algorithm SVM_0), where, in order to have a fair comparison, the margin term has been dropped by setting, in the `fitcsvm` Matlab subroutine, the penalty parameter `BoxConstraint` equal to 10^6. We recall in fact the spherical approach implemented in [7] does not involve any margin concept. In Table 2, for each data set, the best result is underlined.

In comparison with $\mathrm{FC}_{\mathcal{A}}$ and $\mathrm{FC}_{\mathcal{AB}}$, the choice of the infinitely far center appears to be the best one: in fact Algorithm $\mathrm{FC}_{①}$ outperforms the other two

Table 2 Numerical results

Data set	$FC_{\mathcal{A}}$	$FC_{\mathcal{AB}}$	$FC_{①}$	SVM_0
Cancer	97.00	95.71	97.57	71.86
Diagnostic	83.86	53.33	89.65	92.11
Heart	74.33	55.00	87.33	68.67
Pima	69.35	66.23	61.43	61.82
Ionosphere	51.14	40.75	78.86	69.43
Sonar	59.05	52.86	65.71	75.24
Mushrooms	76.44	64.50	78.19	49.59
Prognosis	56.00	45.00	68.00	53.00
Tic Tac Toe	71.79	70.42	57.79	50.11
Votes	82.79	53.35	86.74	76.51
Galaxy	80.24	51.36	89.19	54.32
g50c	67.62	50.26	90.58	86.56
g10n	53.58	45.02	77.66	90.24

approaches on all the data sets except Pima and Tic Tac Toe, where the best performance is got by fixing x_0 as the barycenter of $\mathcal{A}$. We note also that choosing x_0 as the barycenter of all the points is not a good strategy, since the corresponding results are very poor on all the test problems, but Cancer and Tic Tac Toe, where the testing correctnesses appear comparable.

Also with respect to SVM_0, Algorithm $FC_{①}$ is characterized by a good performance, except on Diagnostic, Sonar and g10n, while on Pima both the approaches behave almost the same. These results were expected because, even if taking the radius infinitely far makes the spherical separability tend to the linear separability, the two approaches differ substantially. We recall in fact that, if two sets are linearly separable, they are also spherical separable (even taking a very large radius), but the vice-versa is not true.

5 Conclusions

In this work we have examined the main differences between linear and spherical separation in the light of the grossone theory. In particular, we have recalled the main observations reported in [7] for supervised classification, extending them to the cases of the supervised spherical separation with margin and of the Multiple Instance Learning.

We have focused on the possibility to construct binary spherical classifiers characterized by an infinitely far center. As shown by the preliminary numerical results reported in [7], adopting the grossone theory allows to obtain a good performance in terms of average testing correctness, managing very easily the numerical computations, which do not require any tuning of the "big M" parameter.

Future research could consist in extending such approach to the kernel trick, which is well suitable in the fixed-center spherical separation, as shown in [16], and to practically implement the grossone idea for solving MIL problems.

References

1. Andrews, S., Tsochantaridis, I., Hofmann, T.: Support vector machines for multiple-instance learning. In: Becker, S., Thrun, S., Obermayer, K. (eds.) Advances in Neural Information Processing Systems, pp. 561–568. MIT Press, Cambridge (2003)
2. Astorino, A., Bomze, I., Brito, P. Gaudioso, M.: Two spherical separation procedures via non-smooth convex optimization. In: De Simone, V., Di Serafino, D., Toraldo, G. (eds.) Recent Advances in Nonlinear Optimization and Equilibrium Problems: A Tribute to Marco D'Apuzzo, Quaderni di Matematica, Dipartimento di Matematica della Seconda Universitá di Napoli, vol. 27, pp. 1–16. Aracne (2012)
3. Astorino, A., Bomze, I., Fuduli, A., Gaudioso, M.: Robust spherical separation. Optimization **66**(6), 925–938 (2017)
4. Astorino, A., Frangioni, A., Fuduli, A., Gorgone, E.: A nonmonotone proximal bundle method with (potentially) continuous step decisions. SIAM J. Optim. **23**(3), 1784–1809 (2013)
5. Astorino, A., Fuduli, A.: Nonsmooth optimization techniques for semisupervised classification. IEEE Trans. Pattern Anal. Mach. Intell. **29**(12), 2135–2142 (2007)
6. Astorino, A., Fuduli, A.: The proximal trajectory algorithm in SVM cross validation. IEEE Trans. Neural Netw. Learn. Syst. **27**(5), 966–977 (2016)
7. Astorino, A., Fuduli, A.: Spherical separation with infinitely far center. Soft Comput. **24**(23), 17751–17759 (2020)
8. Astorino, A., Fuduli, A., Gaudioso, M.: DC models for spherical separation. J. Global Optim. **48**(4), 657–669 (2010)
9. Astorino, A., Fuduli, A., Gaudioso, M.: Margin maximization in spherical separation. Comput. Optim. Appl. **53**(2), 301–322 (2012)
10. Astorino, A., Fuduli, A., Gaudioso, M.: Nonlinear programming for classification problems in machine learning. In: AIP Conference Proceedings, vol. 1776 (2016)
11. Astorino, A., Fuduli, A., Gaudioso, M.: A Lagrangian relaxation approach for binary multiple instance classification. IEEE Trans. Neural Netw. Learn. Syst. **30**(9), 2662–2671 (2019)
12. Astorino, A., Fuduli, A., Gaudioso, M., Vocaturo, E.: Multiple instance learning algorithm for medical image classification. In: CEUR Workshop Proceedings, vol. 2400 (2019)

13. Astorino, A., Fuduli, A., Giallombardo, G., Miglionico, G.: SVM-based multiple instance classification via DC optimization. Algorithms **12**(12) (2019)
14. Astorino, A., Fuduli, A., Gorgone, E.: Non-smoothness in classification problems. Optim. Methods Softw. **23**(5), 675–688 (2008)
15. Astorino, A., Fuduli, A., Veltri, P., Vocaturo, E.: Melanoma detection by means of multiple instance learning. Interdiscip. Sci.: Comput. Life Sci. **12**(1), 24–31 (2020)
16. Astorino, A., Gaudioso, M.: A fixed-center spherical separation algorithm with kernel transformations for classification problems. Comput. Manag. Sci. **6**(3), 357–372 (2009)
17. Astorino, A., Gaudioso, M., Fuduli, A., Vocaturo, E.: A multiple instance learning algorithm for color images classification. In: ACM International Conference Proceeding Series, pp. 262–266 (2018). www.scopus.com. Cited By :15
18. Astorino, A., Gaudioso, M., Khalaf, W.: Edge detection by spherical separation. Comput. Manag. Sci. **11**(4), 517–530 (2014)
19. Astorino, A., Gaudioso, M., Seeger, A.: Conic separation of finite sets. I. The homogeneous case. J. Convex Anal. **21** (2014)
20. Astorino, A., Gaudioso, M., Seeger, A.: Conic separation of finite sets II. The non-homogeneous case. J. Convex Anal. **21**(3), 819–831 (2014)
21. Avolio, M., Fuduli, A.: A semiproximal support vector machine approach for binary multiple instance learning. IEEE Trans. Neural Netw. Learn. Syst. **32**(8), 3566–3577 (2021)
22. Bagirov, A., Karmitsa, N., Taheri, S.: Partitional Clustering via Nonsmooth Optimization. Springer, Berlin (2020)
23. Bennett, K.P., Mangasarian, O.L.: Robust linear programming discrimination of two linearly inseparable sets. Optim. Methods Softw. **1**, 23–34 (1992)
24. Bergeron, C., Moore, G., Zaretzki, J., Breneman, C., Bennett, K.: Fast bundle algorithm for multiple instance learning. IEEE Trans. Pattern Anal. Mach. Intell. **34**(6), 1068–1079 (2012)
25. Caldarola, F.: The exact measures of the Sierpinski d-dimensional tetrahedron in connection with a diophantine nonlinear system. Commun. Nonlinear Sci. Numer. Simul. **63**, 228–238 (2018)
26. Caldarola, F.: The Sierpinski curve viewed by numerical computations with infinities and infinitesimals. Appl. Math. Comput. **318**, 321–328 (2018)
27. Calude, C.S., Dumitrescu, M.: Infinitesimal probabilities based on grossone. SN Comput. Sci. **1**, article number: 36 (2020)
28. Celebi, M.E. (ed.): Partitional Clustering Algorithms. Springer International Publishing, Berlin (2015)
29. Chapelle, O., Schölkopf, B., Zien, A. (eds.): Semi-supervised learning. MIT Press, Cambridge (2006)
30. Chapelle, O., Zien, A.: Semi-supervised classification by low density separation. In: Proceedings of the Tenth International Workshop on Artificial Intelligence and Statistics, pp. 57–64 (2005)
31. Cococcioni, M., Cudazzo, A., Pappalardo, M., Sergeyev, Y.D.: Solving the lexicographic multi-objective mixed-integer linear programming problem using branch-and-bound and grossone methodology. Commun. Nonlinear Sci. Numer. Simul. **84**, 105177 (2020)
32. Cococcioni, M., Fiaschi, L.: The Big-M method with the numerical infinite M. Optim. Lett, **15**, 2455–2468 (2021)

33. Cristianini, N., Shawe-Taylor, J.: An Introduction to Support Vector Machines and Other Kernel-based Learning Methods. Cambridge University Press, Cambridge (2000)
34. De Cosmis, S., De Leone, R.: The use of grossone in mathematical programming and operations research. Appl. Math. Comput. **218**(16), 8029–8038 (2012)
35. De Leone, R.: Nonlinear programming and grossone: quadratic programming and the role of constraint qualifications. Appl. Math. Comput. **318**, 290–297 (2018)
36. De Leone, R., Fasano, G., Sergeyev, Y.D.: Planar methods and grossone for the conjugate gradient breakdown in nonlinear programming. Comput. Optim. Appl. **71**(1), 73–93 (2018)
37. Demyanov, A., Fuduli, A., Miglionico, G.: A bundle modification strategy for convex minimization. Eur. J. Oper. Res. **180**, 38–47 (2007)
38. Dietterich, T.G., Lathrop, R.H., Lozano-Pérez, T.: Solving the multiple instance problem with axis-parallel rectangles. Artif. Intell. **89**(1–2), 31–71 (1997)
39. Falcone, A., Garro, A., Mukhametzhanov, M.S., Sergeyev, Y.D.: A simulink-based infinity computer simulator and some applications. In: Sergeyev, Y.D., Kvasov, D.E. (eds.) Numerical Computations: Theory and Algorithms, pp. 362–369. Springer International Publishing, Cham (2020)
40. Gaudioso, M., Giallombardo, G., Miglionico, G., Vocaturo, E.: Classification in the multiple instance learning framework via spherical separation. Soft Comput. **24**, 5071–5077 (2020). https://doi.org/10.1007/s00500-019-04255-1
41. Gaudioso, M., Giallombardo, G., Mukhametzhanov, M.S.: Numerical infinitesimals in a variable metric method for convex nonsmooth optimization. Appl. Math. Comput. **318**, 312–320 (2018)
42. Herrera, F., Ventura, S., Bello, R., Cornelis, C., Zafra, A., Sánchez-Tarragó, D., Vluymans, S.: Multiple Instance Learning: Foundations and Algorithms. Springer International Publishing, Berlin (2016)
43. Iavernaro, F., Mazzia, F.: Solving ordinary differential equations by generalized Adams methods: properties and implementation techniques. Appl. Numer. Math. **28**(2–4), 107–126 (1998)
44. Le Thi, H.A., Minh, L.H., Pham Dinh, T., Ngai, V.H.: Binary classification via spherical separator by DC programming and DCA. J. Global Optim. **56**, 1393–1407 (2013)
45. Lolli, G.: Infinitesimals and infinites in the history of mathematics: a brief survey. App. Math. Comput. **218**(16), 7979–7988 (2012)
46. Lolli, G.: Metamathematical investigations on the theory of grossone. Appl. Math. Comput. **255**, 3–14 (2015)
47. Mangasarian, O.L.: Linear and nonlinear separation of patterns by linear programming. Oper. Res. **13**(3), 444–452 (1965)
48. Murphy, P.M., Aha, D.W.: UCI repository of machine learning databases (1992). www.ics.uci.edu/~mlearn/MLRepository.html
49. Plastria, F., Carrizosa, E., Gordillo, J.: Multi-instance classification through spherical separation and VNS. Comput. & Oper. Res. **52**, 326–333 (2014)
50. Rizza, D.: A study of mathematical determination through Bertrand's Paradox. Philos. Math. **26**(3), 375–395 (2018)
51. Odewahn, S., Stockwell, E., Pennington, R., Humphreys, R., Zumach, W.: Automated star/galaxy discrimination with neural networks. Astron. J. **103**(1), 318–331 (1992)
52. Sergeyev, Y.D.: Higher order numerical differentiation on the infinity computer. Optim. Lett. **5**(4), 575–585 (2011)

53. Sergeyev, Y.D.: Numerical infinities and infinitesimals: methodology, applications, and repercussions on two Hilbert problems. EMS Surv. Math. Sci. **4**(2), 219–320 (2017)
54. Sergeyev, Y.D.: Independence of the grossone-based infinity methodology from non-standard analysis and comments upon logical fallacies in some texts asserting the opposite. Found. Sci. **24**(1), 153–170 (2019)
55. Sergeyev, Y.D., Kvasov, D.E., Mukhametzhanov, M.S.: On strong homogeneity of a class of global optimization algorithms working with infinite and infinitesimal scales. Comm. Nonlinear Sci. Num. Sim. **59**, 319–330 (2018)
56. Tax, D.M.J., Duin, R.P.W.: Data domain description using support vectors. In: ESANN'1999 proceedings Bruges, pp. 251–256. Belgium (1999)
57. Tseng, P.: Convergence of a block coordinate descent method for nondifferentiable minimization. J. Optim. Theory App. **109**(3), 475–494 (2001)
58. Vapnik, V.: The Nature of the Statistical Learning Theory. Springer, New York (1995)

Computing Optimal Decision Strategies Using the Infinity Computer: The Case of Non-Archimedean Zero-Sum Games

Marco Cococcioni, Lorenzo Fiaschi, and Luca Lambertini

Abstract As is well known, zero-sum games are appropriate instruments for the analysis of several issues across areas including economics, international relations and engineering, among others. In particular, the Nash equilibria of any two-player finite zero-sum game in mixed-strategies can be found solving a proper linear programming problem. This chapter investigates and solves non-Archimedean zero-sum games, i.e., games satisfying the zero-sum property allowing the payoffs to be infinite, finite and infinitesimal. Since any zero-sum game is coupled with a linear programming problem, the search for Nash equilibria of non-Archimedean games requires the optimization of a non-Archimedean linear programming problem whose peculiarity is to have the constraints matrix populated by both infinite and infinitesimal numbers. This fact leads to the implementation of a novel non-Archimedean version of the Simplex algorithm called Gross-Matrix-Simplex. Four numerical experiments served as test cases to verify the effectiveness and correctness of the new algorithm. Moreover, these studies helped in stressing the difference between numerical and symbolic calculations: indeed, the solution output by the Gross-Matrix Simplex is just an approximation of the true Nash equilibrium, but it still satisfies some properties which resemble the idea of a non-Archimedean ε-Nash equilibrium. On the contrary, symbolic tools seem

M. Cococcioni (✉) · L. Fiaschi
Department of Information Engineering, University of Pisa,
Largo Lucio Lazzarino 1, Pisa, Italy
e-mail: marco.cococcioni@unipi.it

L. Fiaschi
e-mail: lorenzo.fiaschi@phd.unipi.it

L. Lambertini
Department of Economics, University of Bologna, Strada Maggiore 45, Bologna, Italy
e-mail: luca.lambertini@unibo.it

Y. D. Sergeyev and R. De Leone (eds.), *Numerical Infinities and Infinitesimals in Optimization*, Emergence, Complexity and Computation 43,
https://doi.org/10.1007/978-3-030-93642-6_11

to be able to compute the "exact" solution, a fact which happens only on very simple benchmarks and at the price of its intelligibility. In the general case, nevertheless, they stuck as soon as the problem becomes a little more challenging, ending up to be of little help in practice, such as in real time computations. Some possible applications related to such non-Archimedean zero-sum games are also discussed.

1 Introduction

Non-Archimedean zero-sum games are a contact point between three distinct branches of Mathematics, namely Non-Archimedean Analysis (NAA), Linear Programming (LP) and Game Theory (GT). The former deals with algebraic structures lacking of the Archimedes' property [4, 39], which means they contain elements that are not finitely comparable, i.e., their elements are infinitely large or infinitely small with respect to each other. Two examples are Sergeyev's Grossone Methodology (GM) [42] (along with his patented Infinity Computer [21, 22, 40]) and Robinson's hyperreal numbers [38]: both allow one to use numbers which can be infinite, infinitesimal, rather than just finite as one may be used to work with.

LP is the well know and deeply studied branch of optimization theory firstly introduced by Kantorovič [28] and Dantzig [13]. It consists in the optimization of linear functions, subject to linear constraints (both equalities and inequalities) [47]. Moreover, LP has a very well know and strict relation with GT, i.e., the discipline which models the behaviour of rational agents within competitive environments, commonly called games [31]. Indicating with $\mathcal{ZG}$ the set of two-person finite zero-sum games in mixed-strategies [33], then any game in $\mathcal{ZG}$ can be always transformed in an LP problem [1] whose optima are the Nash equilibria of the original game. In particular, such optimization task has the game's payoffs matrix as the constraint matrix of the problem. The word "finite" means that the games model competitive environments where each player is endowed with a finite number of possible strategies, the label "zero-sum" refers to the fact that gains in the game are always exactly balanced by other players' losses, while "mixed-strategies" means that the players adopt each strategy they have according to a probability distribution.

So far, the GM has been used both in LP and in GT, but separately. The application of the GM to LP [7, 8, 16] gave birth to the very first generalization of the Simplex algorithm to non-Archimedean quantities [6], namely the Gross-Simplex (G-Simplex in short). It has been the starting point for the developing of the Gross-Matrix Simplex used in this chapter [10]. Another fruitful result from the interaction between GM and LP is an enhancement of the Big-M method [3, 13, 46], namely the Infinitely-Big-M method (I-Big-M) [9].

On the other hand, GT mainly leveraged GM to model both non-Archimedean payoffs (infinite, finite or infinitesimal) and non-Archimedean probabilities (finite or infinitesimal) in the Prisoner's Dilemma [23–25]. Other applications of GM to GT involve games on graphs [11, 12] and infinite decision-making processes [37]. More broadly, GM has also been successfully applied to more general optimization problems ([18, 19, 26, 44]), to machine learning [2, 5] and to many other fields.

All in all, this chapter introduces a new algorithm, namely Gross-Matrix-Simplex (in brief GM-S), able to find Nash equilibria of games in $\mathcal{NZG}$, i.e., the non-Archimedean extension of $\mathcal{ZG}$. In particular, it is able to cope with non-Archimedean LP problems also having the constraint matrix and the unknowns vector filled with non-Archimedean numbers. The efficacy of the algorithm is tested on several experiments. Since GM-S outputs an arbitrarily precise approximation of a true Nash equilibrium of a game, a relevant part of this chapter is dedicated to study such approximation, to characterize it and to show its anti-exploitation properties, which resemble those of standard ε-Nash equilibria. Before continuing, it is important to remark that GM is independent from non-standard analysis, as shown in [43].

2 Zero-Sum Games

Any game $\mathcal{G}$ is typically represented by the triple $\{N, S, u\}$, where $N = \{1, \ldots, n\}$ indicates the set of players which take part to the game. Each of them has at his/her disposal a set of strategies to adopt, namely S_i. The set S is defined as the Cartesian product of all the players sets of strategies, i.e., $S = \prod_{i=1}^{n} S_i$. Finally, $u : S \to \mathbb{R}^n$ is the game utility function, where the entry $u_i : S \to \mathbb{R}$ is the i-th player utility function, which assumes his/her income provided the strategies implemented by all the participants.

The game-theoretic model which will be generalized to the case of non-Archimedean quantities is the one of finite two-person zero-sum game in mixed-strategies. Formally, a zero-sum game [48] is any game satisfying the following property:

$$\sum_{i=1}^{n} u_i(s) = 0 \;\; \forall s \in S,$$

which means that the total utility experienced by the players altogether is always zero (whence the zero-sum label). On the other hand, a game is *finite* if the competitive scenario involves agents each having a finite number of strategies, i.e.,

$$|S_i| \in \mathbb{N} \ \ \forall i \in N.$$

Whenever the number of players is finite too, these games take the name of *matrix games*, since they are fully represented by the n-dimensional matrix $A = [u(s_k)]_k$, where $k = (k_1, \ldots, k_n)$, $s_k = (s_{k_1}, \ldots, s_{k_n})$, $k_i \in \{1, \ldots, |S_i|\}$ and $s_{k_i} \in S_i, i = 1, \ldots, n$. An interesting corner case consists of those games involving just two players, i.e., $n = 2$. Their matrix A simplifies a lot, ending to have the following structure:

$$A = \begin{bmatrix} u(s_1^1, s_1^2) & \cdots & u(s_1^1, s_{|S_2|}^2) \\ \vdots & \ddots & \vdots \\ u(s_{|S_1|}^1, s_1^2) & \cdots & u(s_{|S_1|}^1, s_{|S_2|}^2) \end{bmatrix},$$

where s_j^i means the j-th strategy of player i, i.e., $s_j \in S_i, i = 1, 2$. The label *mixed-strategies* refers to a game $G' = \{N, \Delta, h\}$ for which there exists a game $G = \{N, S, u\}$ and for which the following equalities are satisfied:

$$\Delta_{S_i} = \left\{ \delta_i : S_i \to [0, 1] : \sum_{s \in S_i} \delta_i(s) = 1 \right\}, \quad \Delta = \prod_{i=1}^{n} \Delta_{S_i}$$

$$\delta : S \to [0, 1], \quad \delta(s) = \prod_{i=1}^{n} \delta_i(s_i), \quad \delta_i \in \Delta_{S_i}, \quad s_i \in S_i$$

$$h_i(\delta) = \sum_{s \in S} u_i(s)\delta(s), \quad h = (h_1, \ldots, h_n).$$

As a result, $\mathcal{G}'$ describes the very same environment of $\mathcal{G}$, but this time each player can adopt a behavior δ_i which is a probability distribution over the strategies, rather than choosing a single one (whence the *mixed* nature of their strategies).

The choice of two-player finite zero-sum games in mixed-strategies is threefold:

- Finiteness and mixed-strategies guarantee the existence of at least one Nash equilibrium, as a corollary of the Nikaido-Isoda theorem [32];

- The presence of two players and the zero-sum property jointly imply that every Nash equilibrium is at the intersection of minimax (or maximin) strategies;
- Together, these four properties and their consequences allow one to reformulate the search for Nash equilibria as an LP problem [1].

Let one give a better look to the last dot. The Nash equilibria of games in $\mathcal{ZG}$ are all and only the mixed-strategies x and y which satisfy

$$\min_{x \in \Delta_{S_1}} \max_{y \in \Delta_{S_2}} x^T A y = \max_{y \in \Delta_{S_2}} \min_{x \in \Delta_{S_1}} x^T A y \,, \tag{1}$$

where A is the game matrix. The primal LP reformulation of (1) is the one in (2), while the dual is (3).

$$\begin{aligned} &\max \lambda \\ &\text{s.t. } \lambda \mathbf{1}^T \leq x^T A \\ &\quad x^T \mathbf{1} = 1 \\ &\quad x \geq 0 \end{aligned} \tag{2}$$

$$\begin{aligned} &\min \mu \\ &\text{s.t. } Ay \leq \mu \mathbf{1} \\ &\quad \mathbf{1}^T y = 1 \\ &\quad y \geq 0 \end{aligned} \tag{3}$$

where $\mathbf{1}$ is the vector, of proper dimension, filled by ones, and $\lambda, \mu \in \mathbb{R}$. Both the cases allow a rewriting into the canonical form reported below:

$$\begin{aligned} \min \quad & c^T x \\ \text{s.t.} \quad & Ax = b \\ & x^T \mathbf{1} = 1 \\ & x \geq 0 \end{aligned}$$

Nash and ε-Nash Equilibria in Numerical Algorithms In most of the cases, numerical routines implemented to find Nash equilibria, e.g. Simplex algorithm for games in $\mathcal{ZG}$, output just an approximation of them. This is due to the rounding errors induced by the finiteness of the machine on the computations, which may be themselves ill-conditioned. Such errors propagate during the algorithm execution and persist in its output, playing a crucial role from a game theoretical perspective since the result may not form a Nash equilibrium anymore. This implies that unilateral deviation from that strategy profile may lead to benefits for the deviator. In finite games however, such a benefit is bounded, testifying that the approximated strategies form a ε-Nash equilibrium [35], a relaxed version of the Nash equilibrium.

Originally introduced in [36], ε-Nash equilibria are one of the key ingredients of Proposition 1, which links the deviator benefit bound to the quality of the optimal strategy approximation.

Definition 1 (*Nash equilibrium*) Let $\mathcal{G}$ be a two-person zero-sum game and $H(x, y)$ be the game value function. Then, (x^*, y^*) is said to be a Nash equilibrium if and only if $H(x^*, y) \leq H(x^*, y^*) \leq H(x, y^*)$ $\forall x \in \Delta_{S_1}$, $y \in \Delta_{S_2}$.

Definition 2 (*ε-Nash equilibrium*) Let $\mathcal{G}$ be a two-person zero-sum game and $H(x, y)$ be the game value function. Then, (x^*, y^*) is said to be an ε-Nash equilibrium, indicated by $(x_\varepsilon, y_\varepsilon)$, if and only if $H(x^*, y) - \varepsilon \leq H(x^*, y^*) \leq H(x, y^*) + \varepsilon$ $\forall x \in \Delta_{S_1}, y \in \Delta_{S_2}, \varepsilon \geq 0$.

Lemma 1 *Let $\mathcal{G}$ be a finite two-person zero-sum game and $H(x, y) = x^T Ay$ be the game value function. If (x^*, y^*) is a Nash equilibrium, then $\min_i (Ay^*)_i = H(x^*, y^*) = \max_j (x^{*T} A)_j$.*

Proof The proof comes straightforwardly from Definition 1; indeed, if there exists $\overline{j}$ such that $H(x^*, y^*) < (x^{*T} A)_{\overline{j}}$, then (x^*, y^*) is no more a Nash equilibrium. To prove it, let $\overline{y}$ be the deterministic strategy where $\overline{y}_j = 1$ if $j = \overline{j}$ and zero otherwise. Then, it happens that $H(x^*, y^*) = x^{*T} Ay^* < x^{*T} A\overline{y} = H(x^*, \overline{y}) \leq \max_j (x^{*T} A)_j$, which is in contrast with the assumption that $H(x^*, y^*)$ is a Nash equilibrium.

Lemma 2 *Let $x \in \mathbb{R}^n$ and $A \in \mathbb{R}^{n \times m}$. Then, $\max_j (x^T A)_j \leq ||x|| \, ||A_x||$, where $||A_x|| = \sup_{u \in \mathcal{S}^n} ||u^T A||$.*

Proof The following two inequalities are true and prove the lemma: $\max_j (x^T A)_j \leq ||x^T A|| \leq ||x|| \, ||A_x||$. The first one holds by construction, while the second follows from the Cauchy-Schwarz inequality.

Hereinafter, always assume $x \in \Delta_{S_1}$ and $y \in \Delta_{S_2}$. This shall increase the conciseness and improve the readability of the text.

Definition 3 (*δ-approximating strategy*) Let x be an approximation of the strategy x^*. Then, x is said to δ-approximate x^*, indicated by x_δ, if and only if $||x^* - x_\delta|| \leq \frac{\delta}{2}$, $\delta \in \mathbb{R}^+$.

Proposition 1 *Let $\mathcal{G}$ be a finite two-person zero-sum game and $H(x, y) = x^T Ay$ be the game value function. Let also (x_δ, y_δ) be the δ-approximation of the game Nash equilibrium (x^*, y^*), i.e., $||x^* - x_\delta|| \le \frac{\delta}{2}$ and $||y^* - y_\delta|| \le \frac{\delta}{2}$. Then, (x_δ, y_δ) is an ε-Nash equilibrium with $\varepsilon = \delta||A||$.*

Proof Let $\Delta_x = x^* - x_\delta$ and $\Delta_y = y^* - y_\delta$. By construction and applying Lemma 2 to the last inequality, one has

$$\begin{aligned}
x_\delta^T Ay_\delta &\le \max_y x_\delta^T Ay = \max_j (x_\delta^T A)_j = \max_j ((x_\delta + \Delta_x - \Delta_x)^T A)_j = \\
&= \max_j ((x^* - \Delta_x)^T A)_j = \max_j (x^{*T} A - \Delta_x^T A)_j \le \max_j (x^{*T} A)_j - \min_j (\Delta_x^T A)_j \le \\
&\le \max_j (x^{*T} A)_j + \frac{\delta}{2}||A||,
\end{aligned} \tag{4}$$

where $||A|| = \max(||A||_x, ||A||_y)$. By means of analogous manipulations, it holds true that

$$\min_i (Ay^*)_i - \frac{\delta}{2}||A|| \le \min_i (Ay_\delta)_i \le x_\delta^T Ay_\delta.$$

Leveraging Lemma 1, one can rewrite it as

$$H(x^*, y^*) - \frac{1}{2}\varepsilon \le \min_x H(x, y_\delta) \le H(x_\delta, y_\delta) \le \max_y H(x_\delta, y) \le H(x^*, y^*) + \frac{1}{2}\varepsilon, \tag{5}$$

having imposed $\varepsilon = \delta||A||$. The proof completes using the first and last inequalities of (5), which give

$$H(x^*, y^*) - \frac{1}{2}\varepsilon \le \min_x H(x, y_\delta) \Longrightarrow H(x^*, y^*) \le \min_x H(x, y_\delta) + \frac{1}{2}\varepsilon$$

$$\max_y H(x_\delta, y) \le H(x^*, y^*) + \frac{1}{2}\varepsilon \Longrightarrow \max_y H(x_\delta, y) - \frac{1}{2}\varepsilon \le H(x^*, y^*)$$

and plugging them back into (5), respectively at the end and at the beginning of the inequalities chain, obtaining:

$$\max_y H(x_\delta, y) - \varepsilon \le H(x_\delta, y_\delta) \le \min_x H(x, y_\delta) + \varepsilon.$$

The interpretation of Proposition 1 is the following: in real-world standard applications, one can arbitrarily approximate the game value $H(x^*, y^*)$, i.e., up to a certain user-defined tolerance ε by opportunely setting δ in the optimizing algorithm (of course, both ε and δ must be greater than the machine precision). Indeed, δ measures the closeness of (x_δ, y_δ) to the truly optimal Nash-equilibrium and at the same time it bounds the value of ε. Therefore, when δ approaches zero also ε does, forcing the deviator's benefit towards values sufficiently small to be considered negligible for practical purposes.

3 Non-Archimedean Zero-Sum Games and the Gross-Matrix-Simplex Algorithm

A non-Archimedean zero-sum game is a zero-sum game where the payoffs can assume also infinite and infinitesimal values. A very intuitive method to represent them within a machine is leveraging Grossone Methodology [42], where each payoff $\tilde{z}$, which can be made by different infinite, finite and infinitesimal components, is represented as follows:

$$\tilde{z} = \sum_{i \in \mathbb{P}} z_i ①^i, \tag{6}$$

where $\mathbb{P} \subset \mathbb{Z}$ is a finite set of G-powers, $z_i \in \mathbb{R}\ \forall i \in \mathbb{P}$. As an example, consider the payoff $4① + 3 - 2①^{-1}$. One can interpret it as a reward constituted by three goods ordered by priority, approach which was originally proposed in [41]. In fact, it represents four units of the most important ware, three of the second one and a debt of two units of the third one. This is possible because the value of each good (a finite number) contributes to the payoff weighted by a scalar which is infinitely incommensurable with all the others. In the current case, the first and most important commodity has an infinite impact on the value of the payoff (indeed, it is multiplied by the infinite scalar ①), the second commodity has a finite effect (it is multiplied by $1 = ①^0$), while the third commodity an infinitesimal one.

Section 2 showed that one can find at least one Nash equilibrium for each game in $\mathcal{ZG}$ solving the LP problem coupled with it. In case games in $\mathcal{NZG}$ are considered, the LP problems turns into a non-Archimedean one (in brief, NALP). In literature, there already exists a Simplex-like algorithm able to deal with certain NALP problems, namely the Gross-Simplex algorithm [6]. It consists of an improved version of the Dantzig's algorithm enhanced by means of GM in order to cope with non-Archimedean cost functions. Nevertheless, its implementation is useless in the current context since the NALP problem associated to games in $\mathcal{NZG}$ have the constraints matrix A which is non-Archimedean.

This fact motivates the implementation of a further generalization of the Simplex algorithm, namely the Gross-Matrix-Simplex, which is able to solve such class of NALP and, as a consequence, to output strategy profiles which

Algorithm 1: The Gross-Matrix-Simplex algorithm

Step 0. The user has to provide the initial set B of basic indices.

Step 1. Compute $\tilde{x}_{\mathrm{B}}$ as: $\tilde{x}_{\mathrm{B}} = \tilde{A}_{\mathrm{B}}^{-1} b$ (where $\tilde{A}_{\mathrm{B}}$ is the sub-matrix obtained by $\tilde{A}$ by considering the columns indexed by B, while $\tilde{A}_{\mathrm{B}}^{-1}$ is the inverse of $\tilde{A}_{\mathrm{B}}$, i.e., the non-Archimedean matrix which satisfies the equality $\tilde{A}_{\mathrm{B}}^{-1}\tilde{A}_{\mathrm{B}} = \tilde{A}_{\mathrm{B}}\tilde{A}_{\mathrm{B}}^{-1} = I$).

Step 2. Compute $\tilde{y}$ as: $\tilde{y} = c_{\mathrm{B}}^T \tilde{A}_{\mathrm{B}}^{-1}$ (where $\tilde{y}$ is a G-vector obtained by linearly combining the
purely finite vector c_{B}^T by the G-scalar elements in the rows of $\tilde{A}_{\mathrm{B}}^{-1}$).

Step 3. Compute $\tilde{s}$ as: $\tilde{s} = c_{\mathrm{N}}^T - \tilde{y}\tilde{A}_{\mathrm{N}}$ (where N is the complementary set of B) and then select the maximum (gradient rule). When this maximum is negative (being it infinite, finite or infinitesimal), then the current solution is optimal and the algorithm stops. Otherwise, the position of the maximum in the G-vector $\tilde{s}$ is the index k of the entering variable $\mathrm{N}(k)$.

Step 4. Compute $\tilde{d}$ as: $\tilde{d} = \tilde{A}_{\mathrm{B}}^{-1}\tilde{A}_{\mathrm{N}(k)}$, where $\tilde{A}_{\mathrm{N}(k)}$ is the G-vector corresponding to the $\mathrm{N}(k)$-th column of $\tilde{A}$.

Step 5. Find the largest G-scalar $\tilde{t} > 0$ such that $\tilde{x}_{\mathrm{B}} - \tilde{t}\tilde{d} \geqslant 0$. If there is not such a $\tilde{t}$, then the problem is unbounded (STOP); otherwise, at least one component of $\tilde{x}_{\mathrm{B}} - \tilde{t}\tilde{d}$, say h,
equals to zero and the corresponding variable is the leaving variable.

Step 6. Update sets B and N by swapping $\mathrm{B}(h)$ and $\mathrm{N}(k)$, then return to Step 1.

are themselves non-Archimedean. The problem to solve is the following:

$$\begin{aligned} \min \ & c^T \tilde{x} \\ \text{s.t.} \ & \tilde{A}\tilde{x} = b \\ & \tilde{x} \geq 0 \end{aligned} \tag{7}$$

where $\tilde{A}$ is the G-matrix containing the payoffs, c and b are the (purely finite) objective and constant-terms vectors.

Algorithm 1 reports the pseudocode of the Gross-Matrix-Simplex. A posteriori, one could have achieved a better coherency if the Gross-Simplex algorithm would have been named Gross-Cost-Simplex, since it is only able to handle non-Archimedean cost functions $\tilde{c}$. The most interesting instruction Algorithm 1 is the computation of the inverse of a non-Archimedean matrix, which is achieved by means of the Gaussian-Jordan elimination algorithm tailored to cope with non-Archimedean quantities. A very appealing way to improve the algorithm is to substitute the explicit computation of the inverse with an incremental LU factorization. This should lead to better numerical stability, as it happens in the original Simplex algorithm. A dedicated work about it seems perfectly reasonable and absolutely promising

The next subsection investigates a property similar to Proposition 1 for games in $\mathcal{NZG}$, i.e., the reliability of numerical solvers for Nash equilibria in case of non-Archimedean payoffs. Then, the discussion will move to show the effectiveness of GM-S to solve NALP problems, i.e., to numerically find Nash equilibria of games in $\mathcal{NZG}$. To do this, several test cases are presented, solved and discussed.

$\mathcal{E}$-Nash Optimality in Non-Archimedean Contexts

Similarly to standard routines, algorithms working with non-Archimedean numbers suffer from rounding errors the G-digits: for instance, the machine represents the G-number $\frac{1}{3}①$ as $0.3333①$ (where the number of 3s is architecture-dependent). However, they are afflicted by one further source of inaccuracy: the finite space for G-scalars components, e.g., a computer cannot exactly represent the G-scalar $\frac{1}{①+1} = \frac{1}{①} - \frac{1}{①^2} + \frac{1}{①^3} + \ldots = \sum_{i \in \mathbb{N}} (-1)^{i-1} ①^{-i}$ because it would need an infinite memory. Therefore, it is interesting to investigate whether the approximated non-Archimedean Nash equilibria found by numerical routines are somehow related to a non-Archimedean version of ε-Nash equilibria, as it happens in the standard case (see Sect. 2). This time, the number of G-scalar components the machine is able to store, say k, shall play a role as well as the rounding G-digits tolerance δ. Proposition 2 formally extends the result in Proposition 1 to the non-Archimedean scenario. As before, the definitions of (δ, k)-approximating strategy and (ε, k)-Nash equilibrium are needed, i.e., the non-Archimedean extension of Definition 3 and Definition 2, respectively.

Roughly speaking, the former states that a non-Archimedean strategy $\tilde{x}$ (δ, k)-approximates $\tilde{x}^*$ if the G-digits of the first k components of $||\tilde{x}^* - \tilde{x}||$ (which is a G-scalar) are bounded by a value proportional to δ. Notice that (δ, k)-approximating strategies are exactly the type of solutions GM-S is designed to seek for and to output. The notion (ε, k)-Nash equilibrium works in a very similar way, and Proposition 2 shall prove that they are the type of Nash equilibrium output by the GM-S algorithm. For the sake of simplicity, hereinafter assume that $||\tilde{A}||$ is a finite number. This fact does not affect generality since the following scaling is always possible whenever $\tilde{A}$ does not satisfy it: $\tilde{A} \leftarrow \tilde{A}(||\tilde{A}||)^{-1}$.

Definition 4 (*(δ, k)-approximating strategy*) Let $\tilde{x}$ be an approximation of the non-Archimedean strategy $\tilde{x}^*$. Then, $\tilde{x}$ is said to be a (δ, k)-approximation of $\tilde{x}^*$, indicated by $\tilde{x}^k_\delta$, if and only if $||\tilde{x}^* - \tilde{x}|| = ||\tilde{\Delta}_x|| = \sum_{i \in \mathbb{Z}_0^-} \Delta_i ①^i$ is such that $|\Delta_i| < \frac{\delta}{2} \in \mathbb{R}^+ \; \forall i = 0, \ldots, 1-k, k \in \mathbb{N}$.

Definition 5 (*(ε, k)-Nash equilibrium*) Let $\mathcal{G}$ be a finite non-Archimedean two-person zero-sum game and $H(\tilde{x}, \tilde{y})$ be the game non-Archimedean value function. Also suppose that $H(\tilde{x}, \tilde{y})$ assumes at most finite values (if not shrink it so). Then, the feasible strategy profile $(\tilde{x}^*, \tilde{y}^*)$ is said to be an (ε, k)-Nash equilibrium, indicated by $(\tilde{x}^*, \tilde{y}^*)^k_\varepsilon$, if and only if $\max_{\tilde{y}} H(\tilde{x}^*, \tilde{y}) - \tilde{\varepsilon} \leq H(\tilde{x}^*, \tilde{y}^*) \leq \min_{\tilde{x}} H(\tilde{x}, \tilde{y}^*) + \tilde{\varepsilon}$, where $\tilde{\varepsilon} = \sum_{i \in \mathbb{Z}_0^-} \varepsilon_i ①^i \geq 0, \varepsilon \in \mathbb{R}_0^+$ and $|\varepsilon_i| \leq \varepsilon \; \forall i = 0, \ldots, 1-k, k \in \mathbb{N}$.

Finally, Proposition 2 shows that the strategy profile output by GM-S is an (ε, k)-Nash equilibrium. This come from the fact that GM-S is designed to output (δ, k)-approximating strategies, while Proposition 2 links (δ, k)-approximations of Nash equilibria strategies to the property of being (ε, k)-Nash equilibria.

Proposition 2 *Let $\mathcal{G}$ be a finite non-Archimedean two-person zero-sum game and $\tilde{A}$ is the non-Archimedean payoff matrix. If all the entries of $\tilde{A}$ have the form of* (6), *then any (δ, k)-approximation $(\tilde{x}^k_\delta, \tilde{y}^k_\delta)$ of a Nash equilibrium $(\tilde{x}^*, \tilde{y}^*)$ is an (ε, k)-Nash equilibrium.*

Proof The result of Eq. (4) holds true until the last inequality even in a non-Archimedean context, i.e.,

$$\tilde{x}_\delta^T \tilde{A} \tilde{y}_\delta \leq \max_j (\tilde{x}^{*T} \tilde{A})_j - \min_j (\tilde{\Delta}_x^T \tilde{A})_j.$$

To continue with the inequality chain, let one introduce two new G-scalars $\tilde{\nu}^k_x$ and $\tilde{\nu}^{-k}_x$ such that $\tilde{\nu}^k_x$ is the first k components of $||\tilde{x}^* - \tilde{x}^k_\delta|| = ||\tilde{\Delta}_x|| = \sum_{i \in \mathbb{Z}_0^-} \Delta_i ①^i$, i.e., $\tilde{\nu}^k_x = \sum_{i=0}^{1-k} \Delta_i ①^i$, while $\tilde{\nu}^{-k}_x$ is the remaining components of $||\tilde{\Delta}_x||$, i.e., $\tilde{\nu}^{-k}_x = ||\tilde{\Delta}_x|| - \tilde{\nu}^k_x$. Then, Leveraging Lemma 2 (which holds even in a non-Archimedean context) it holds true that $-\min_j (\tilde{\Delta}_x^T \tilde{A})_j \leq ||\tilde{\Delta}_x||\,||A|| = ||\tilde{\nu}^k_x||\,||A|| + ||\tilde{\nu}^{-k}_x||\,||A|| \leq \frac{\tilde{\delta}}{2}||\tilde{A}|| + ||\tilde{\nu}^{-k}_x||\,||A|| = \frac{1}{2}\tilde{\varepsilon}^k_x + \frac{1}{2}\tilde{\varepsilon}^{-k}_x = \frac{1}{2}\tilde{\varepsilon}_x$, where $\tilde{\delta} = \sum_{j=0}^{1-k} \delta ①^j, \tilde{\varepsilon}^k_x = \delta||\tilde{A}||, \tilde{\varepsilon}^{-k}_x = 2||\tilde{\nu}^{-k}_x||\,||A||$. Finally, define $\tilde{\varepsilon}_y$ accordingly and set $\tilde{\varepsilon} = \max(\tilde{\varepsilon}_x, \tilde{\varepsilon}_y)$, i.e., $\tilde{\varepsilon} = \tilde{\varepsilon}^k + \tilde{\varepsilon}^{-k}$, where $\tilde{\varepsilon}^k = \tilde{\varepsilon}^k_x = \tilde{\varepsilon}^k_y$ and $\tilde{\varepsilon}^{-k} = \max(\tilde{\varepsilon}^{-k}_x, \tilde{\varepsilon}^{-k}_y)$. Reasoning similarly to Proposition 1, one obtains:

$$\max_{\tilde{y}} H(\tilde{x}^k_\delta, \tilde{y}) - \tilde{\varepsilon} \leq H(\tilde{x}^k_\delta, \tilde{y}^k_\delta) \leq \min_{\tilde{x}} H(\tilde{x}, \tilde{y}^k_\delta) + \tilde{\varepsilon}.$$

By definition, one can rewrite $\tilde{\varepsilon}^k$ as follows:

$$\tilde{\varepsilon}^k = \sum_{h=0}^{1-k} \varepsilon_h ①^h = \tilde{\delta}||\tilde{A}|| = \left(\sum_{i=0}^{1-k} \delta ①^i\right)\left(\sum_{j\in\mathbb{P}} a_j ①^j\right) = \delta \sum_{i=0}^{1-k}\sum_{j\in\mathbb{P}} a_j ①^{i+j},$$

$$\varepsilon_h = \delta \sum_{(i,j)\in\mathcal{H}_h} a_j ①^{i+j} \quad \forall h = 0, \dots, 1-k,$$

where $\mathcal{H}_h = \{(i, j) \in \{0, \dots, 1-k\} \times \mathbb{P} \mid i + j = h\}$. Of course, $\mathcal{H}_h = \emptyset \Longrightarrow \varepsilon_h = 0$ by construction. In addition, it holds true that

$$|\varepsilon_h| = \delta\Big|\sum_{(i,j)\in\mathcal{H}_h} a_j ①^{i+j}\Big| \le \delta \sum_{(i,j)\in\mathcal{H}_h} |a_j| ①^{i+j} \le \delta \max_h |\mathcal{H}_h| \max_{j\in\mathbb{P}} |a_j| \le \delta k \max_{j\in\mathbb{P}} |a_j| = \varepsilon,$$

which means that the absolute value of all the G-digits of $\tilde{\varepsilon}^k$ are bounded by a positive constant depending on δ, similarly to Definition 5. The proof completes noticing that, by construction, $\tilde{\varepsilon}^k$ and $\tilde{\varepsilon}^{-k}$ are completely separated from components perspective, i.e., the smallest G-power in $\tilde{\varepsilon}^k$ is greater than the biggest one in $\tilde{\varepsilon}^{-k}$ (which comes from the particular choice of $\tilde{\nu}_x^k$ and $\tilde{\nu}_x^{-k}$). This implies that the first k components of $\tilde{\varepsilon}$ form exactly $\tilde{\varepsilon}^k$, and therefore $(\tilde{x}_\delta^k, \tilde{y}_\delta^k)$ satisfies Definition 5, i.e., it is an (ε, k)-Nash equilibrium.

In fact, Proposition 2 states a finer the approximation of the optimal strategy profile corresponds with a higher negligiblity of the first k components of the deviation benefit. Notice that nothing can be said about the term $\tilde{\varepsilon}^{-k}$ except that its magnitude is not greater than $①^{-k}$, which is reasonable since the machine representation truncates G-scalars up to the (k-1)-th component, i.e., the computations are not aware of the components with magnitude smaller or equal to $①^{-k}$.

4 Numerical Illustrations

This section aims at assess the effectiveness of GM-S algorithm to find Nash equilibria of games in $\mathcal{NZG}$. Furthermore, Sect. 4.1 spends some lines to verify that GM-S outputs are both (δ, k)-approximations and (ε, k)-Nash equilibria. The majority of the experiments refer to non-Archimedean variations of the very well known zero-sum game called rock-paper-scissors. The choice of this model comes from the fact it is simple enough to make its non-Archimedean modifications easy to understand and predictable in terms of Nash equilibria. Table 1 reports its canonical form, while its mathematical

Table 1 Rock-Paper-Scissors canonical form

$\wp_1/\wp_2$	R	P	S
R	0	1	−1
P	−1	0	1
S	1	−1	0

Algorithm 2: Procedure to double-check the results provided by the GM-S algorithm

step 0. Let (x^*, y^*) be a Nash equilibrium for a given n-dimensional game obtained somehow
(in our case by GM-S).
Step 1. Indicate with $\mathcal{A}_x$ and $\mathcal{A}_y$ the index set of active strategies in x^* and y^*, respectively. This means that $i \in \mathcal{A}_z \Leftrightarrow z_i > 0, i \in \{1, \ldots, n\}, z = x, y$
Step 2. Define the active matrix B as the payoff matrix reduced to the row indexes in $\mathcal{A}_x$ and the column indexes in $\mathcal{A}_y$, i.e., $B = A_{\mathcal{A}_x}^{\mathcal{A}_y}$.
Step 3. Define C and D such that $C_i = B^i - B^{i+1}\ \forall i = 1, \ldots, |\mathcal{A}_y| - 1$ and $C_{|\mathcal{A}_y|} = \mathbf{1}$, while for D holds true that $D_i = B_i - B_{i+1}\ \forall i = 1, \ldots, |\mathcal{A}_x| - 1$ and $D_{|\mathcal{A}_x|} = \mathbf{1}$.
Step 4. Verify that $Cx^* = e$ and $Dy^* = e$ hold true, where we have indicated with e the last (under the natural ordering) vector of the canonical base with proper dimension, i.e., $e = (0, \ldots, 0, 1)^T$.

formulation is reported in (8), where A indicates the payoffs matrix whose content coincides with Table 1.

$$\min_{x \in \Delta_1} \max_{y \in \Delta_2} x^T A y \tag{8}$$

$$\Delta_1 = \Delta_2 = \left\{ (\rho_1, \rho_2, \rho_3) \in \mathbb{R}^3 \,\middle|\, \sum_{i=1}^{3} \rho_i = 1,\ \rho_i \geq 0 \forall i \right\}$$

In this game there exists only one Nash equilibrium, which is shown in Eq. (9):

$$x^* = y^* = \left[\frac{1}{3}, \frac{1}{3}, \frac{1}{3}\right] \tag{9}$$

It is right to say that all the experiments outcomes have been double-checked verifying that they are *basic* Nash equilibria [45]. The procedure to do it is reported in Algorithm 2.

4.1 Experiment 1: Infinitesimally Perturbed Rock-paper-scissors

The first experiment involves the very same problem of Table 1 with one single payoff infinitesimally perturbed, namely the $\tilde{A}_{2,1}$ (see Eq. (10)). The main idea is to check the GM-S algorithm sensibility to infinitesimal changes in the matrix game. To improve the readability, the infinitesimal components shall be colored in blue. Since the original problem admits a unique Nash equilibrium, it is reasonable to expect that the new game's one is infinitely close to the former. Table 2 reports GM-S iterations executed during the optimization, while Eq. (11) shows the new Nash equilibrium, which confirms the intuition of infinity closeness. However, GM-S algorithm is able to tell *exactly how much* the two are similar (up to the machine precision, of course).

$$\tilde{A} = \begin{bmatrix} 0 & 1 & -1 \\ -1-①^{-1} & 0 & 1 \\ 1 & -1 & 0 \end{bmatrix} \tag{10}$$

$$\tilde{x}_\delta^k = \begin{bmatrix} \frac{1}{3} \\ \frac{1}{3} - \frac{1}{9}①^{-1} + \frac{1}{27}①^{-2} \\ \frac{1}{3} + \frac{1}{9}①^{-1} - \frac{1}{27}①^{-2} \end{bmatrix}, \qquad \tilde{y}_\delta^k = \begin{bmatrix} \frac{1}{3} - \frac{1}{9}①^{-1} + \frac{1}{27}①^{-2} \\ \frac{1}{3} \\ \frac{1}{3} + \frac{1}{9}①^{-1} - \frac{1}{27}①^{-2} \end{bmatrix} \tag{11}$$

The reason why in Table 2 the cost function starts from an infinite value and the length of vector $\tilde{x}$ is larger than expected is due to the use of I-Big-M method [9] as wrapper for the GM-S solver. It provides an easy way to retrieve a feasible starting basis adding a set of artificial variables to the problem and infinitely penalizing them by the term ① in the cost function (as pioneered in [16, 17]). The fact that at the end of the optimization the cost function has a finite value means that all the artificial variables have exited the basis and the solution is a feasible one.

Finally, let one verify that the $(\delta, 3)$-approximating strategy profile $(\tilde{x}_\delta^k, \tilde{y}_\delta^k)$ is really an $(\varepsilon, 3)$-Nash equilibrium. Theoretically, the strategy pair $(\tilde{x}_\delta^k, \tilde{y}_\delta^k)$ approximates the Nash equilibrium $(\tilde{x}^*, \tilde{y}^*)$ defined as

$$\tilde{x}^* = \left[\frac{1}{3}, \frac{①}{3①+1}, \frac{3①+2}{3(3①+1)}\right]^T, \qquad \tilde{y}^* = \left[\frac{①}{3①+1}, \frac{1}{3}, \frac{3①+2}{3(3①+1)}\right]^T.$$

Expanding the second entry of $\tilde{x}^*$ one gets $\frac{①}{3①+1} = \sum_{i \in \mathbb{Z}_0^-} \frac{1}{3(-3)^{-i}}①^i = \frac{1}{3} - \frac{1}{9}①^{-1} +$

Table 2 GM-S iterations for Game 1 of Eq. (10): infinitesimally perturbed rock-paper-scissors game

It.	Base	$\tilde{x}$	$c^T\tilde{x}$
1	{4, 5, 6}	$[0, 0, 0, \frac{1}{3}, \frac{1}{3}, \frac{1}{3}, 0, 0, 0]$	$-\text{①}$
2	{2, 5, 6}	$\left[0, \frac{1}{4} - \frac{1}{8}\text{①}^{-1} + \frac{1}{16}\text{①}^{-2}, 0, 0, \frac{1}{4} + \frac{1}{8}\text{①}^{-1} - \frac{1}{16}\text{①}^{-2}, \frac{1}{2}, 0, 0, 0\right]$	$-\frac{3}{4}\text{①} - \frac{3}{8} + \frac{3}{16}\text{①}^{-1} - \frac{1}{16}\text{①}^{-2}$
3	{2, 3, 6}	$\left[0, \frac{1}{3} - \frac{1}{9}\text{①}^{-1} + \frac{1}{27}\text{①}^{-2}, \frac{1}{6} + \frac{1}{9}\text{①}^{-1} - \frac{1}{27}\text{①}^{-2}, 0, 0, \frac{1}{2}, 0, 0, 0\right]$	$-\frac{1}{2}\text{①} - \frac{1}{2}$
4	{2, 3, 1}	$\left[\frac{1}{3}, \frac{1}{3} - \frac{1}{9}\text{①}^{-1} + \frac{1}{27}\text{①}^{-2}, \frac{1}{3} + \frac{1}{9}\text{①}^{-1} - \frac{1}{27}\text{①}^{-2}, 0, 0, 0, 0, 0, 0\right]$	-1

$\frac{1}{27}①^{-2} - \frac{1}{81}①^{-3} + \ldots$, which becomes exactly the second entry of $\tilde{x}_\delta^k$ when it is truncated up to the second infinitesimal component ($k = 3$). Repeating the same operation with all the other entries, the fact that $\tilde{x}_\delta^k$ and $\tilde{y}_\delta^k$ are $(\delta, 3)$-approximations of $\tilde{x}^*$ and $\tilde{y}^*$ is verified. Then, Proposition 2 states that if $(\tilde{x}_\delta^k, \tilde{y}_\delta^k)$ is a $(\varepsilon, 3)$-Nash equilibrium, then it must hold true that $|H(\tilde{x}_\delta^k, \tilde{y}_\delta^k) - \min_{\tilde{x}}(\tilde{x}, \tilde{y}_\delta^k)| \leq \tilde{\varepsilon} = \tilde{\varepsilon}^k + \tilde{\varepsilon}^{-k} = \tilde{\varepsilon}^{-k} < M①^{-k}$, with $M \in \mathbb{R}^+$ sufficiently big (the last equality comes from the fact that $\delta = 0$). The game value associated to (11) is $H(\tilde{x}_\delta^k, \tilde{y}_\delta^k) = \sum_{i=-1}^{-5} \frac{1}{3(-3)^{-i}}①^i$, while the row player optimal strategy as answer to $\tilde{y}_\delta^k$ is$\overline{x} = [0, 1, 0]^T$, which leads to $H(\overline{x}, \tilde{y}_\delta^k) = -\frac{1}{9}①^{-1} + \frac{1}{27}①^{-2} - \frac{1}{27}①^{-3}$. Therefore, the deviation benefit is $|H(\tilde{x}_\delta^k, \tilde{y}_\delta^k) - H(\overline{x}, \tilde{y}_\delta^k)| = \frac{2}{81}①^{-3} + \frac{1}{243}①^{-4} - \frac{1}{729}①^{-5} = \varepsilon^{-k} < M①^{-3}$ for $M > \frac{2}{81}$. This means that $(\tilde{x}_\delta^k, \tilde{y}_\delta^k)$ is a $(\varepsilon, 3)$-Nash equilibrium since $\tilde{\varepsilon}^k$ is negligible and $\tilde{\varepsilon}^{-k}$ has magnitude not greater than $①^{-3}$ (provided that the same verification is done for column player as deviator, study which is omitted for brevity). In addition, notice how close the approximation of $H(\tilde{x}_\delta^k, \tilde{y}_\delta^k)$ to $H(\tilde{x}^*, \tilde{y}^*)$ is. Since $H(\tilde{x}^*, \tilde{y}^*) = -\frac{1}{3(3①+1)} = \sum_{i \in \mathbb{Z}^-} \frac{1}{3(-3)^{-i}}①^i = -\frac{1}{9}①^{-1} + \frac{1}{27}①^{-2} - \frac{1}{81}①^{-3} + \ldots$, we have $|H(\tilde{x}^*, \tilde{y}^*) - H(\tilde{x}_\delta^k, \tilde{y}_\delta^k)| = |\sum_{\substack{i \in \mathbb{Z} \\ i < -5}} \frac{1}{3(-3)^{-i}}①^i|$: the game value is well approximated up to the fifth order of infinitesimal.

4.2 Experiment 2: A Purely Finite 4-by-3 Game

In games where more than one Nash equilibrium exists, even an infinitesimal perturbation can significantly affect the optimal strategy choice. For instance, consider the game in Eq. (12) which is built adding one strategy for the row player to the rock-paper-scissors of Table 8. The crucial aspect is that the new game has infinitely many Nash equilibria, without any further criterion by means of which to choose among them. Therefore, the choice of which mixed-strategy to play depends only on the actual implementation of the decision making algorithm, e.g., the Simplex algorithm.

Table 3 GM-S iterations for the purely finite Game 2 of Eq. (12)

It.	Base	$\tilde{x}$	$c^T\tilde{x}$
1	{5, 6, 7}	$\left[0, 0, 0, 0, \frac{1}{3}, \frac{1}{3}, \frac{1}{3}, 0, 0, 0\right]$	$-①$
2	{5, 6, 1}	$\left[\frac{1}{4}, 0, 0, 0, \frac{1}{4}, \frac{1}{2}, 0, 0, 0, 0\right]$	$-\frac{3}{4}① - \frac{1}{4}$
3	{2, 6, 1}	$\left[\frac{1}{3}, \frac{1}{6}, 0, 0, 0, \frac{1}{2}, 0, 0, 0, 0\right]$	$-\frac{1}{2}① - \frac{1}{2}$
4	{2, 3, 1}	$\left[\frac{1}{3}, \frac{1}{3}, \frac{1}{3}, 0, 0, 0, 0, 0, 0, 0\right]$	-1

$$A = \begin{bmatrix} 0 & 1 & -1 \\ -1 & 0 & 1 \\ 1 & -1 & 0 \\ \frac{1}{2} & -1 & \frac{1}{2} \end{bmatrix} \tag{12}$$

Table 3 reports the iterations of GM-S algorithm running on the degenerate problem (12). The routine suggests that the row-player adopts the strategy $x_\delta^k = [\frac{1}{3}, \frac{1}{3}, \frac{1}{3}, 0]^T$, which is essentially the same as in the original game (9), i.e., $x^* = [\frac{1}{3}, \frac{1}{3}, \frac{1}{3}]^T$. To improve the readability, the optimal strategies have been reported as fractions. Next subsection shall show that an infinitesimal perturbation of these payoffs causes a finite change on the equilibrium rather than an infinitesimal one.

4.3 Experiment 3: Infinitesimally Perturbed 4-by-3 Game

Assume now that a secondary information to better discriminate among the infinite Nash equilibria in (12) is available. It consists of a second payoff matrix A_s whose importance is subordinated to the one of Eq. (12):

$$A_s = \begin{bmatrix} 0 & 2 & -2 \\ -2 & 0 & 2 \\ 2 & -2 & 0 \\ -1 & -2 & 1 \end{bmatrix}$$

As discussed at the beginning of Sect. 3 and shown in Eq. (6), the two sources of information can be eventually merged without losing the priority information by means of GM. In particular, a new non-Archimedean zero-sum game is built summing the payoffs entry-wise and scaling down the ones having lesser priority by the ①$^{-1}$, i.e., the new payoff matrix $\tilde{A}$ is obtained as

$$\tilde{A} = A + A_s ①^{-1},$$

which in fact means

$$\tilde{A} = \begin{bmatrix} 0 & 1 + 2①^{-1} & -1 - 2①^{-1} \\ -1 - 2①^{-1} & 0 & 1 + 2①^{-1} \\ 1 + 2①^{-1} & -1 - 2①^{-1} & 0 \\ \frac{1}{2} - ①^{-1} & -1 - 2①^{-1} & \frac{1}{2} + ①^{-1} \end{bmatrix}. \quad (13)$$

The overall game is now a non-Archimedean one, falling in the set $\mathcal{NZG}$, since $\tilde{A}$ is non-Archimedean.

In the face of the infinitesimal perturbation on the payoffs matrix induced by the secondary information, the result is all but negligible. As opposed to the problem in Table 1, there exist infinitely many Nash equilibria in (12). This means that the presence of a secondary payoffs in A_s informs the optimization algorithm of an additional discriminating rule to better single out the most preferable Nash equilibria. More importantly, the equilibrium found in Table 3 may not be optimal also with respect to the secondary information, while all those the strategy profiles which are may notably differ from it. This is the case of problem in (13), as testified by Tables 4 and 5 which report the GM-S algorithm iterations computed to solve it for the row and the column-player, respectively.

The new equilibrium is:

$$\tilde{x}^*_\delta = \begin{bmatrix} \frac{2}{5} + \frac{4}{75}①^{-1} - \frac{16}{25}①^{-2} \\ \frac{1}{5} - \frac{28}{75}①^{-1} + \frac{56}{125}①^{-2} \\ 0 \\ \frac{2}{5} + \frac{8}{25}①^{-1} - \frac{28}{125}①^{-2} \end{bmatrix}, \qquad \tilde{y}^*_\delta = \begin{bmatrix} \frac{1}{3} + \frac{4}{15}①^{-1} - \frac{8}{25}①^{-2} \\ \frac{1}{3} - \frac{4}{15}①^{-1} + \frac{8}{25}①^{-2} \\ \frac{1}{3} \end{bmatrix}. \quad (14)$$

Observe how its finite part has changed from $x^k_\delta = [\frac{1}{3}, \frac{1}{3}, \frac{1}{3}, 0]^T$ of the previous equilibrium to $x^k_\delta = [\frac{2}{5}, \frac{1}{5}, 0, \frac{2}{5}]^T$ of the current one.

Table 4 GM-S iterations for Game 3 of Eq. (13): row-player

It.	Base	$\tilde{x}$	$c^T\tilde{x}$
1	{5, 6, 7}	$\left[0, 0, 0, 0, \frac{1}{3}, \frac{1}{3}, \frac{1}{3}, 0, 0, 0\right]$	$-\textcircled{1}$
2	{5, 4, 7}	$\left[0, 0, 0, \frac{1}{4} - \frac{1}{4}\textcircled{1}^{-1} + \frac{1}{4}\textcircled{1}^{-2}, \frac{3}{8} - \frac{1}{8}\textcircled{1}^{-1} + \frac{1}{8}\textcircled{1}^{-2}, 0, \frac{3}{8} + \frac{3}{8}\textcircled{1}^{-1} - \frac{3}{8}\textcircled{1}^{-2}, 0, 0, 0\right]$	$-\frac{3}{4}\textcircled{1} - \frac{1}{2} + \frac{1}{2}\textcircled{1}^{-1} - \frac{1}{4}\textcircled{1}^{-2}$
3	{5, 4, 1}	$\left[\frac{3}{10} + \frac{3}{50}\textcircled{1}^{-1} + \frac{3}{250}\textcircled{1}^{-2}, 0, 0, \frac{2}{5} + \frac{2}{25}\textcircled{1}^{-1} + \frac{2}{125}\textcircled{1}^{-2}, \frac{3}{10} - \frac{7}{50}\textcircled{1}^{-1} - \frac{7}{250}\textcircled{1}^{-2}, 0, 0, 0, 0, 0\right]$	$-\frac{3}{10}\textcircled{1} - \frac{14}{25} - \frac{14}{125}\textcircled{1}^{-1} - \frac{3}{250}\textcircled{1}^{-2}$
4	{2, 4, 1}	$\left[\frac{2}{5} + \frac{4}{75}\textcircled{1}^{-1} - \frac{16}{25}\textcircled{1}^{-2}, \frac{1}{5} - \frac{28}{75}\textcircled{1}^{-1} + \frac{56}{125}\textcircled{1}^{-2}, 0, \frac{2}{5} + \frac{8}{25}\textcircled{1}^{-1} - \frac{48}{125}\textcircled{1}^{-2}, 0, 0, 0, 0, 0, 0\right]$	-1

Table 5 GM-S iterations for Game 3 of Eq. (13): column-player

It.	Base	$\tilde{y}$	$c^T\tilde{y}$
1	{4, 5, 6, 7}	$\left[0, 0, 0, \frac{1}{4}, \frac{1}{4}, \frac{1}{4}, \frac{1}{4}, 0, 0, 0, 0\right]$	$-\textcircled{1}$
2	{4, 3, 6, 7}	$\left[0, 0, \frac{2}{9} - \frac{17}{5}\textcircled{1}^{-1} + \frac{27}{50}\textcircled{1}^{-2}, \frac{4}{9} + \frac{1}{5}\textcircled{1}^{-1} - \frac{31}{100}\textcircled{1}^{-2}, 0, \frac{2}{9} + \frac{1}{10}\textcircled{1}^{-1} - \frac{3}{20}\textcircled{1}^{-2}, \frac{1}{9} + \frac{1}{25}\textcircled{1}^{-1} - \frac{2}{25}\textcircled{1}^{-2}, 0, 0, 0, 0\right]$	$-\frac{7}{9}\textcircled{1} + \frac{253}{450} + \frac{394}{100}\textcircled{1}^{-1} - \frac{27}{50}\textcircled{1}^{-2}$
3	{4, 3, 6, 1}	$\left[\frac{1}{10} + \frac{1}{25}\textcircled{1}^{-1} - \frac{8}{125}\textcircled{1}^{-2}, 0, \frac{3}{10} - \frac{7}{25}\textcircled{1}^{-1} + \frac{56}{125}\textcircled{1}^{-2}, \frac{1}{2} + \frac{2}{5}\textcircled{1}^{-1} - \frac{6}{25}\textcircled{1}^{-2}, 0, \frac{1}{10} - \frac{4}{25}\textcircled{1}^{-1} - \frac{18}{125}\textcircled{1}^{-2}, 0, 0, 0, 0, 0\right]$	$-\frac{3}{5}\textcircled{1} - \frac{16}{25} + \frac{78}{125}\textcircled{1}^{-1} - \frac{48}{125}\textcircled{1}^{-2}$
4	{4, 3, 2, 1}	$\left[\frac{1}{3} - \frac{4}{3}\textcircled{1}^{-1} + \frac{16}{3}\textcircled{1}^{-2}, \frac{1}{3} - \frac{8}{3}\textcircled{1}^{-1} + \frac{40}{3}\textcircled{1}^{-2}, \frac{1}{3} - \frac{8}{3}\textcircled{1}^{-2}, 4\textcircled{1}^{-1} - 16\textcircled{1}^{-2}, 0, 0, 0, 0, 0, 0, 0\right]$	$-5 + 20\textcircled{1}^{-1} - 16\textcircled{1}^{-2}$
5	{7, 3, 2, 1}	$\left[\frac{1}{3} - \frac{2}{9}\textcircled{1}^{-1} + \frac{4}{27}\textcircled{1}^{-2}, \frac{1}{3} - \frac{2}{9}\textcircled{1}^{-1} + \frac{4}{27}\textcircled{1}^{-2}, \frac{1}{3} - \frac{2}{9}\textcircled{1}^{-1} + \frac{4}{27}\textcircled{1}^{-2}, 0, 0, 0, \frac{2}{3}\textcircled{1}^{-1} - \frac{4}{9}\textcircled{1}^{-2}, 0, 0, 0, 0\right]$	$-\frac{5}{3} + \frac{10}{9}\textcircled{1}^{-1} - \frac{4}{9}\textcircled{1}^{-2}$
6	{10, 3, 2, 1}	$\left[\frac{1}{3} - \frac{8}{15}\textcircled{1}^{-2}, \frac{1}{3} - \frac{8}{15}\textcircled{1}^{-1} + \frac{8}{15}\textcircled{1}^{-2}, \frac{1}{3} - \frac{8}{39}\textcircled{1}^{-1}, 0, 0, 0, 0, 0, 0, \frac{4}{5}\textcircled{1}^{-1}, 0\right]$	$-1 + \frac{4}{5}\textcircled{1}^{-1}$

$$
\begin{aligned}
&(-3.58904 + 103.344\,g + 720.946\,g^{2} - 9651.52\,g^{3} - 338238.\,g^{4} - 3.18695\times10^{6}\,g^{5} + 1.24018\times10^{7}\,g^{6} + 2.8101\times10^{8}\,g^{7} + \\
&\quad 1.72011\times10^{9}\,g^{8} + 6.24568\times10^{9}\,g^{9} + 1.47068\times10^{10}\,g^{10} + 1.81561\times10^{10}\,g^{11} - 1.86775\times10^{10}\,g^{12} - 1.728\times10^{11}\,g^{13} - \\
&\quad 5.42735\times10^{11}\,g^{14} - 1.15499\times10^{12}\,g^{15} - 1.77031\times10^{12}\,g^{16} - 1.66508\times10^{12}\,g^{17} + 4.72222\times10^{11}\,g^{18} + 6.3175\times10^{12}\,g^{19} + \\
&\quad 1.73155\times10^{13}\,g^{20} + 3.38623\times10^{13}\,g^{21} + 5.46924\times10^{13}\,g^{22} + 7.68805\times10^{13}\,g^{23} + 9.65928\times10^{13}\,g^{24} + 1.10328\times10^{14}\,g^{25} + \\
&\quad 1.16008\times10^{14}\,g^{26} + 1.13484\times10^{14}\,g^{27} + 1.04198\times10^{14}\,g^{28} + 9.04283\times10^{13}\,g^{29} + 7.44483\times10^{13}\,g^{30} + 5.81056\times10^{13}\,g^{31} + \\
&\quad 4.27046\times10^{13}\,g^{32} + 2.91627\times10^{13}\,g^{33} + 1.80782\times10^{13}\,g^{34} + 9.75863\times10^{12}\,g^{35} + 4.15519\times10^{12}\,g^{36} + \\
&\quad 8.94097\times10^{11}\,g^{37} - 6.2536\times10^{11}\,g^{38} - 1.05118\times10^{12}\,g^{39} - 9.33968\times10^{11}\,g^{40} - 6.43649\times10^{11}\,g^{41} - 3.73419\times10^{11}\,g^{42} - \\
&\quad 1.87453\times10^{11}\,g^{43} - 8.23227\times10^{10}\,g^{44} - 3.16971\times10^{10}\,g^{45} - 1.07216\times10^{10}\,g^{46} - 3.19584\times10^{9}\,g^{47} - 8.59424\times10^{8}\,g^{48} - \\
&\quad 2.18065\times10^{8}\,g^{49} - 5.66965\times10^{7}\,g^{50} - 1.49953\times10^{7}\,g^{51} - 3.77275\times10^{6}\,g^{52} - 694785.\,g^{53} - 78327.8\,g^{54} + \\
&\quad 6045.51\,g^{55} - 923.935\,g^{56} - 1043.41\,g^{57} - 443.538\,g^{58} - 16.0318\,g^{59} + 30.3366\,g^{60} - 5.42663\,g^{61} + 0.175217\,g^{62}) / \\
&((0.993515 + 0.482686\,g + 1.\,g^{2})\ (3.27553 + 3.44271\,g + 4.46498\,g^{2} + 1.31183\,g^{3} + 1.\,g^{4}) \\
&\quad (-0.780488 + 16.1408\,g + 72.0581\,g^{2} + 80.2571\,g^{3} + 101.998\,g^{4} + 34.5331\,g^{5} + 18.3454\,g^{6} - 7.65728\,g^{7} + 1.\,g^{8}) \\
&\quad (-1.18922 - 13.4137\,g - 182.973\,g^{2} - 267.794\,g^{3} - 95.7229\,g^{4} + 702.883\,g^{5} + 1889.16\,g^{6} + 2617.11\,g^{7} + 2567.49\,g^{8} + \\
&\qquad 1513.39\,g^{9} + 483.063\,g^{10} - 235.831\,g^{11} - 342.028\,g^{12} - 225.883\,g^{13} - 84.4367\,g^{14} - 20.2516\,g^{15} + 1.\,g^{16}) \\
&\quad (-1.20679 + 22.8092\,g + 169.934\,g^{2} - 117.669\,g^{3} + 1673.5\,g^{4} + 20154.3\,g^{5} + 96848.8\,g^{6} + 359978.\,g^{7} + \\
&\qquad 970133.\,g^{8} + 2.21868\times10^{6}\,g^{9} + 4.12681\times10^{6}\,g^{10} + 6.67195\times10^{6}\,g^{11} + 9.10681\times10^{6}\,g^{12} + 1.08871\times10^{7}\,g^{13} + \\
&\qquad 1.1076\times10^{7}\,g^{14} + 9.80718\times10^{6}\,g^{15} + 7.23746\times10^{6}\,g^{16} + 4.47\times10^{6}\,g^{17} + 2.04326\times10^{6}\,g^{18} + \\
&\qquad 558717.\,g^{19} - 163762.\,g^{20} - 297648.\,g^{21} - 227335.\,g^{22} - 106240.\,g^{23} - 36877.5\,g^{24} - 5395.33\,g^{25} + \\
&\qquad 983.216\,g^{26} + 1276.05\,g^{27} + 343.541\,g^{28} + 53.5686\,g^{29} - 10.4184\,g^{30} - 2.82982\,g^{31} + 1.\,g^{32})),
\end{aligned}
$$

Fig. 1 First entry of x^* after symbolic computations in Mathematica (g here stands for ①). Please observe how the provided solution is very difficult to read

4.4 Experiment 4: High Dimensional Games

This experiment aims at challenging GM-S algorithm on high-dimensional games, comparing the efficacy with symbolic approaches one. The first test case is not a true high-dimensional game since it involves only 8 strategies per player. However, it is still a significant experiment since it already allows one to stress the limits of symbolic tools, such as Mathematica, which start to struggle even in this relatively low dimensional scenario. When game complexity grows, the gap between numerical and symbolic algorithms efficacy becomes more and more evident. The main reason is that, considering games in $\mathcal{NZG}$, symbolic tools search for Nash equilibria repeatedly applying Algorithm 2 with randomly chosen active strategies, as stated in [45], which is an unpractical and combinatorial NP-hard task. Moreover, when found, the Nash equilibrium readability would probably be quite low, as stressed by Fig. 1 which reports the first entry of the row player optimal strategy computed in Mathematica (the letter g stands for ①). On the other hand, the numerical GM-S algorithm is able to show the approximated equilibrium components in a very interpretable way, e.g., Eq. (15) where the entire solution of the 8 × 8 problem is shown. The payoff matrix and GM-S iterations can be found in the appendix of [10].

$$\tilde{x}^*_\delta = \begin{bmatrix} 0.18 - 0.36①^{-1} + 1.05①^{-2} \\ 0.15 + 0.01①^{-1} - 1.96①^{-2} \\ 0.26 - 0.66①^{-1} + 1.34①^{-2} \\ 0 \\ 0.22 + 0.3①^{-1} + 2.09①^{-2} \\ 0.08 + 0.36①^{-1} - 0.95①^{-2} \\ 0.11 + 0.34①^{-1} - 1.57①^{-2} \\ 0 \end{bmatrix}, \quad \tilde{y}^*_\delta = \begin{bmatrix} 0.01 + 0.54①^{-1} + 3.12①^{-2} \\ 0.14 - 0.15①^{-1} - 1.03①^{-2} \\ 0 \\ 0.21 + 0.37①^{-1} + 2.71①^{-2} \\ 0.27 + 0.34①^{-1} - 0.81①^{-2} \\ 0 \\ 0.1 - 0.81①^{-1} - 2.19①^{-2} \\ 0.27 - 0.29①^{-1} - 1.8①^{-2} \end{bmatrix} \quad (15)$$

Considering a game with a 10×10 payoff matrix instead, the symbolic tool struggles a lot in computing the game solution and its execution time grows drastically, passing from 1.28 s of the 8×8 game to 6.98. To stress even more this fact, a true high-dimensional non-Archimedean zero-sum game has been considered. This time, each player can choose among 60 different strategies, and the game has 5 levels of priority information, i.e., 60×60 payoff matrix is filled by random G-scalars having 4 infinitesimal components. While Mathematica does not output the result in a reasonable amount of time for practical purposes, GM-S took only 122.487 s to compute the Nash Equilibrium, a quite remarkable result. For the sake of brevity, the matrix and the associated Nash equilibrium for both the 10×10 and 60×60 games are omitted.

5 A Brief Overview of Applications

As in [10], to which we refer the reader interested in a more detailed illustration, also here we offer a compact perspective of the models and fields which plausibly feature the non-Archimedean zero-sum property. Whenever the latter is identified, the game under examination can be treated using GM and of course the GM-S algorithm.

The first example has a dual nature, depending on whether players are firms or political parties, and is traditionally known as the Hotelling/Downs game of spatial competition [20, 27].

In both cases, players are supposed to choose their respective locations along a finite linear segment along which consumers or voters are distributed. In absence of a price mechanism, this can describe the choice of electoral platforms, with parties trying to acquire political consensus. In such a case the game has a zero-sum nature because any additional vote gained by a party is lost by the other (leaving aside any degree of abstention), and the model naturally lends itself to a non-Archimedean interpretation.

If instead the game features a price stage appearing after the location choice and prices are strategically set by firms, the game is no longer a zero-sum one. Yet, it becomes so if prices are regulated, in which case firms' profits depend solely on locations and the resulting relative size of market shares. It is also worth stressing that in such a case the Hotelling version of the model is observationally equivalent to Downs's.

This example is particularly relevant as it concerns a game which simultaneously belongs to the backbone of large literatures in industrial economics and politics. But several other non-Archimedean zero-sum games can be figured out.

For instance, another setting which has a twofold interpretation is that in which an entrepreneur owning a firm in dire straits is forced to sell it but maintains a symbolic amount of share and the family name on the gate of the production plant. From a realistic point of view, this payoff may look almost immaterial, but this does not correspond to the original owner's perception. The same applies to a seemingly different scenario in which we may just replace property with sovereignty. This holds for very small States surrounded by larger ones and embedded in fully integrated regions like the EU (think of Montecarlo, Liechtenstein, Andorra and San Marino). Political independence, accompanied by economic integration, may still matter a lot, although these Countries are net importers and, say, their basketball or football clubs may happen to play in the leagues of larger European Countries.

One additional example relies on [34] and is also connected to Patrolling games. Here, the matter is about construction operation purposes. The non-Archimedean property arises if the quality criteria adopted to decide which project to pursue are defined in binary terms, in such a way that the presence of any relevant component (say, fire escapes) is valued 1 while its absence is valued 0.

6 Conclusions

The chapter introduced for the first time the non-Archimedean zero-sum games and a numerical algorithm able to compute approximations of their Nash equilibria, namely the GM-S algorithm. A lot of attention has been dedicated to characterized the type of strategy profile the algorithm is able to output, showing a strong correlation with the standard ε-Nash equilibrium. Both the algorithm implementation and the theoretical study leveraged Grossone Methodology to better reason about non-Archimedean quantities as well as execute numerical non-Archimedean operations. The chapter also illustrated several test cases, some of which high-dimensional, to verify the effectiveness of GM-S algorithm and to highlight the limits of symbolic tools. All the computations run over an Infinity Computer simulator implemented in software. Finally, Some possible real-world applications of this new modeling tool have been presented and discussed. It is right to say that such results would not be obtained without the previous achievements in non-Archimedean linear programming [6, 7, 9] and non-Archimedean Prisoner's Dilemmas [23–25].

References

1. Adler, I.: The equivalence of linear programs and zero-sum games. Int. J. Game Theory **42**, 165–177 (2013). Springer
2. Astorino, A., Fuduli, A.: Spherical separation with infinitely far center. Soft. Comput. **24**(23), 17751–17759 (2020)
3. Bazaraa, M. S., Jarvis, J. J., Sherali, H. D.: Linear Programming and Network Flows. Wiley (1990)
4. Bosch, S., Güntzer, U., Remmert, R.: Non-Archimedean Analysis: A Systematic Approach to Rigid Analytic Geometry. Springer, Berlin, Heidelberg (2012)
5. Cavoretto, R., De Rossi, A., Mukhametzhanov, M.S., Sergeyev, Y.D.: On the search of the shape parameter in radial basis functions using univariate global optimization methods. J. Global Optim. **79**(2), 305–327 (2021)
6. Cococcioni, M., Pappalardo, M., Sergeyev, Y.D.: Lexicographic multi-objective linear programming using grossone methodology: Theory and algorithm. Appl. Math. Comput. **318**, 298–311 (2018)
7. Cococcioni, M., Cudazzo, A., Pappalardo, M., Sergeyev, Y.D.: Grossone Methodology for Lexicographic Mixed-Integer Linear Programming Problems. In: Sergeyev, Y.D., Kvasov D.E. (eds) Numerical Computations: Theory and Algorithms. NUMTA 2019. Springer Lecture Notes in Computer Science, vol. 11974, pp. 337–345 (2020)
8. Cococcioni, M., Cudazzo, A., Pappalardo, M., Sergeyev, Y.D.: Solving the lexicographic multi-objective mixed-integer linear programming problem using branch-and-bound and grossone methodology. Commun. Nonlinear Sci. Numer. Simul. **84**, 105177 (2020)
9. Cococcioni, M., Fiaschi, L.: The Big-M method with the numerical infinite M. Optim. Lett. **15**, 2455–2468 (2021). https://doi.org/10.1007/s11590-020-01644-6
10. Cococcioni, M., Fiaschi, L., Lambertini, L.: Non-Archimedean zero-sum games. J. Comput. Appl. Math. **393**, 113483 (2021)
11. D'Alotto, L.: Infinite games on finite graphs using grossone. In: Sergeyev, Y.D., Kvasov D.E. (eds) Numerical Computations: Theory and Algorithms. NUMTA 2019. Springer Lecture Notes in Computer Science, vol. 11974, pp. 346–353 (2020)
12. D'Alotto, L.: Infinite games on finite graphs using grossone. Soft. Comput. **24**, 17509–17515 (2020)
13. Dantzig, G.B.: Programming in a Linear Structure. United Air Force, Washington, D.C. (1948)
14. Dantzig, G.B., Thapa, M.N.: Linear Programming 1: Introduction. Springer, New York (1997)

15. Dantzig, G.B., Thapa, M.N.: Linear Programming 2: Theory and Extensions. Springer, New York (2003)
16. De Cosmis, S., De Leone, R.: The use of grossone in mathematical programming and operations research. Appl. Math. Comput. **218**, 8029–8038 (2021)
17. De Leone, R., Egidi, N., Fatone, L.: The use of grossone in elastic net regularization and sparse support vector machines. Soft. Comput. **24**(23), 17669–917677 (2020)
18. De Leone, R., Fasano, G., Roma, M., Sergeyev, Y.D.: Iterative grossone-based computation of negative curvature directions in large-scale optimization. J. Optim. Theory Appl. **186**(2), 554–589 (2020)
19. De Leone, R., Fasano, G., Sergeyev, Y.D.: Planar methods and grossone for the conjugate gradient breakdown in nonlinear programming. Comput. Optim. Appl. **71**(1), 73–93 (2018)
20. Downs, A.: An Economic Theory of Democracy. Harper and Row (1957)
21. Falcone, A., Garro, A., Mukhametzhanov, M.S.S., Sergeyev, Y.D.: Representation of grossone-based arithmetic in simulink for scientific computing. Soft. Comput. **24**, 17525–17539 (2020)
22. Falcone, A., Garro, A., Mukhametzhanov, M.S., Sergeyev, Y.D.: A Simulink-based software solution using the Infinity Computer methodology for higher order differentiation. Appl. Math. Comput. **409**, 125606 (2021)
23. Fiaschi, L., Cococcioni, M.: Numerical asymptotic results in game theory using Sergeyev's arithmetic of infinity. Int. J. Unconvent. Comput. **14**, 1–25 (2018)
24. Fiaschi, L., Cococcioni, M.: Generalizing pure and impure iterated prisoner's dilemmas to the case of infinite and infinitesimal quantities. In: Sergeyev, Y.D., Kvasov D.E. (eds) Numerical Computations: Theory and Algorithms. NUMTA 2019. Springer Lecture Notes in Computer Science, vol. 11974, pp. 370–377 (2020)
25. Fiaschi, L., Cococcioni, M.: Non-Archimedean Game Theory: A Numerical Approach. Appl. Math. Comput. 125356 (2020)
26. Gaudioso, M., Giallombardo, G., Mukhametzhanov, M.: Numerical infinitesimals in a variable metric method for convex nonsmooth optimization. Appl. Math. Comput. **318**, 312–320 (2018)
27. Hotelling, H.: Stability in Competition. Econ. J. **39**, 41–57 (1929)
28. Kantorovič, L.V.: Mathematical methods of organizing and planning production. Publication House of the Leningrad State University (1939)
29. Levi-Civita, T.: Sugli infiniti ed infinitesimi attuali, quali elementi analitici. Tip. Ferrari (1893)
30. Levi-Civita, T.: Sui numeri transfiniti. Tipografia della R, Accademia dei Lincei (1898)
31. Myerson, R.B.: Game Theory. Harvard University Press (2013)
32. Nikaidô, H., Isoda, K.: Note on non-cooperative convex games. Pac. J. Math. **5**, 807–815 (1955)
33. Owen, G.: Game Theory. Academic (1995)
34. Peldschus, F.: Experience of the game theory application in construction management. Technol. Econ. Dev. Econ. **14**, 531–545 (2008)
35. Petrosyan, L.A., Zenkevich, N.A.: Game Theory. World Scientific Publishing Co. Pte. Ltd., 9824 (2016)
36. Radner, R.: Collusive behavior in noncooperative epsilon-equilibria of oligopolies with long but finite lives. J. Econ. Theory **22**, 136–154 (1980)
37. Rizza, D.: Numerical methods for infinite decision-making processes. Int. J. Unconv. Comput. **14**, 139–158 (2019)

38. Robinson, A.: Non-standard Analysis. Princeton University Press (2016)
39. Schneider, P.: Nonarchimedean Functional Analysis. Springer Science & Business Media (2013)
40. Sergeyev, Y.D.: Computer system for storing infinite, infinitesimal, and finite quantities and executing arithmetical operations with them. USA patent 7,860,914 (2010)
41. Sergeyev, Y.D.: The Olympic Medals Ranks, lexicographic ordering and numerical infinities. Math. Intell. **37**, 4–8 (2015)
42. Sergeyev, Y.D.: Numerical infinities and infinitesimals: methodology, applications, and repercussions on two Hilbert problems. EMS Surv. Math. Sci. **4**, 219–320 (2017)
43. Sergeyev, Y.D.: Independence of the grossone-based infinity methodology from non-standard analysis and comments upon logical fallacies in some texts asserting the opposite. Found. Sci. **24**, 153–170 (2019)
44. Sergeyev, Y.D., Nasso, M.C., Mukhametzhanov, M.S., Kvasov, D.: Novel local tuning techniques for speeding up one-dimensional algorithms in expensive global optimization using Lipschitz derivatives. J. Comput. Appl. Math. **383**, 113134 (2021)
45. Shapley, L.S., Snow, R.N.: Basic Solutions of Discrete Games. Ann. Math. Stud. (1950)
46. Soleimani-Damaneh, M.: Modified Big-M method to recognize the infeasibility of linear programming models. Knowl.-Based Syst. **21**, 377–382 (2008)
47. Vanderbei, R.J. et al.: Linear Programming. Springer (2015)
48. Washburn, A.R.: Two-Person Zero-Sum Games. Springer (2014)

Modeling Infinite Games on Finite Graphs Using Numerical Infinities

Louis D'Alotto

Abstract In his seminal work, Robert McNaughton (see [14] and [10]) developed a model of infinite games played on finite graphs. Here is presented a new model of infinite games played on finite graphs using the Grossone paradigm. The new Grossone model provides certain advantages such as allowing for draws, which are common in board games, and a more accurate and decisive method for determining the winner.

1 Introduction

The theory of grossone has been developed as an extension of the Cantor theory to better understand the infinite. Here grossone is applied to investigate games of infinite duration. The games investigated occur on finite graphs and are those with *perfect information*. That is, and typically, a perfect information game is played on a board where a player moves pieces subject to a given set of rules and each player knows everything important to the game that has previously occurred.

This chapter is dedicated in loving memory to my wife Zana, who has and will always be my motivation and my inspiration.

L. D'Alotto (✉)
York College, The City University of New York, Jamaica, Queens, NY 11451, USA
e-mail: ldalotto@york.cuny.edu

The Graduate Center, The City University of New York, 356 Fifth Avenue, New York, NY 10016, USA

Y. D. Sergeyev and R. De Leone (eds.), *Numerical Infinities and Infinitesimals in Optimization*, Emergence, Complexity and Computation 43,
https://doi.org/10.1007/978-3-030-93642-6_12

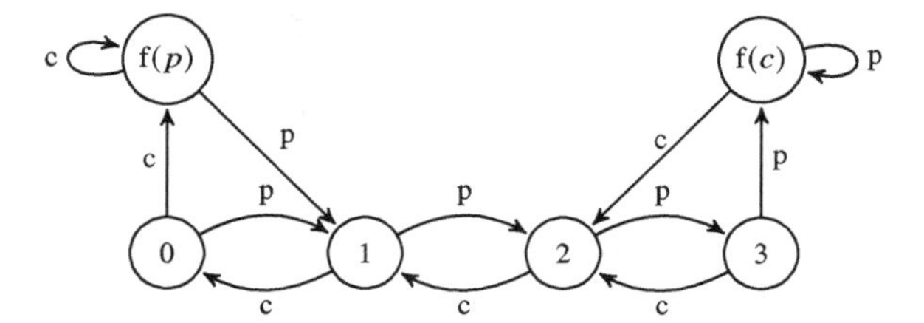

Fig. 1 The consumer-producer problem

Finite board games such as *tic-tac-toe*, *chess*, *checkers*, and *go* are common and most of us have a good understanding of them. These are games of strategy, once the specific positions are known. Of course we must exclude all games of chance and card games where players do not reveal their hands, since these are not games with perfect information. A board game will have a configuration (a state or a state of play) and it must be made precise to include all information about any situation in the game. The configuration describes the current state or standing of the game. Of significant importance, the configuration will dictate which player is to move next. Hence, in board games, the play moves go from one player to the other. A board game such as *tic-tac-toe* has only a very small number of configurations. Here we can easily compute (via computer search techniques) all the configurations and hence this game is not very interesting. However, and on the other hand, games of *checkers*, *chess* and *go* have an extremely large number of configurations and command a lot of attention from computer scientists and mathematicians.

Finite board games that are played to infinity may sound like science or mathematical fiction. Indeed, following the traditional Turing machine model, a computation is complete when it **halts** and produces some type of result. However when a game is played to infinity, it is implied that the game continues for an indefinite period and the play continues without bound. A typical application that can be considered an infinite game is the operating system of a computer (a multiprogramming machine). The operating system has to manage multiple processes (or users on a server) without termination. When one process (or user) is satisfied, there are others waiting for system resources to be processed. Hence process-oriented theory is an application of infinite games to computer science (see [14]). Running of a business may also be modeled as an infinite game. For example, the consumer-producer process (see Fig. 1) where the producer makes items, stores them in a warehouse, and the consumer purchases (consumes) them. In Fig. 1, for simplicity it is assumed the warehouse can store three items. Here the objective is to maintain the business and keep both the consumer and producer satisfied. If the warehouse is full and the producer still produces items, then the consumer fails the business game since they fail to satisfy the producer. If the warehouse is empty and the consumer still wants to buy, then the producer fails to satisfy the consumer. Here the objective is to keep items in the warehouse for the consumer to purchase but not overfill the warehouse. The game continues on as an infinite process. With respect to the theory of Büchi automata, the failing states, $f(c)$ and $f(p)$, can be entered finitely many times as long as the states 0 to 3 (call these the favorable states) are entered infinitely many times. However, with the current vague notion of infinity, suppose the favorable states are entered infinitely many times but the failing states are also entered infinitely many times. The consumer and the producer are satisfied infinitely many times but also are unsatisfied infinitely many times. Hence the game is failed by one (or both) of the players but the game also entered the favorable states infinitely many times. This dichotomy exists due to our vague notion of infinity.

Another important infinite game (or infinite process) is the famed *Dining Philosopher's Problem* and deals with synchronization. In the classic situation, there are five philosophers sitting at a round table. In front of each philosopher is a plate of spaghetti and a single fork on both the left and right sides of each philosopher. Each philosopher, not being very dexterous, requires two forks to eat their spaghetti. Hence only non-adjacent philosophers can eat and at most two at a time. Each of these philosophers can either eat or think. If a philosopher is not eating then they are thinking. If someone is thinking then that philosopher is not eating. So thinking and eating are mutually exclusive. If each philosopher grabs a single fork, then they are deadlocked and no one can eat. The game continues indefinitely, when a philosopher eats, the others (who are not eating) think. Of course, the objective is for the philosophers to avoid starvation.

2 The Infinite Unit Axiom and Grossone

Applying the following new paradigm facilitates us to better understand the notion of infinite games on graphs and update networks. The problem of better understanding the notion of computing with infinity was approached beginning in 2003 by Yaroslav D. Sergeyev (see [19–23]). In these works, a new unit of measure on the set of natural numbers $\mathbb{N}$ is defined. Thus, the following axiom evolves the idea of the infinite unit.

Axiom Infinite Unit axiom. The number of elements in the set $\mathbb{N}$ of natural numbers is equal to the infinite unit denoted as ① and called grossone. □

There are a few postulates that follow the *Infinite Unit axiom*:

1. Infinity: For any finite natural number n, $n < ①$.
2. Identity: The following relationships hold and are extended from the usual identity relationships of the natural numbers:

$$a)\ 0 \cdot ① = ① \cdot 0 = 0,\ b)\ ① - ① = 0,\quad c)\ \frac{①}{①} = 1,\quad d)\ ①^0 = 1,\quad e)\ 1^{①} = 1.$$

3. Divisibility: For any finite natural number n, the numbers

$$①, \frac{①}{2}, \frac{①}{3}, \frac{①}{4}, ..., \frac{①}{n}, ...$$

are defined as the number of elements in the nth part of $\mathbb{N}$.[1]

The divisibility property will be of significant importance in determining a winner of an infinite game. Indeed, determining a winner will result by determining the number of elements in a sequence. It is important to mention, with the introduction of the Infinite Unit axiom and grossone, ①, we list the natural numbers as

$$\mathbb{N} = \{1, 2, 3, 4, ..., ① - 2, ① - 1, ①\}.$$

As a consequence of this new paradigm, we have the following important theorem.

Theorem 1 *The number of elements of any infinite sequence is less or equal to ①.*

Proof See [20] or [25].

Recently there has been a large amount of research activity on the logical theory and applications of grossone. To name a few, see [2–8, 11, 13,

---

[1] In [19], Sergeyev formally presents the divisibility axiom as saying for any finite natural number n sets $\mathbb{N}_{k,n}$, $1 \leq k \leq n$, being the nth parts of the set $\mathbb{N}$, have the same number of elements indicated by the numeral $\frac{①}{n}$ where

$$\mathbb{N}_{k,n} = \{k, k+n, k+2n, k+3n, ...\},\ 1 \leq k \leq n,\ \bigcup_{k=1}^{n} \mathbb{N}_{k,n} = \mathbb{N}$$

and illustrates this with examples of the odd and even natural numbers.

15–18, 22, 26, 28]. As a result of the ① numeral system, we have the mathematical tools to express the number of elements in infinite sets. It must be stressed, however, that the grossone-based methodology is not related to non-standard analysis, see [27]. This next section will describe a new application of grossone to infinite games.

3 Infinite Games

Formally, an *infinite graph game* is defined on a finite bipartite directed graph whose set, Q, of vertices are partitioned into two sets: **R**, the set of vertices from which Red moves, and **B**, the set of vertices from which Blue moves. The game has a place marker which is moved from vertex to vertex along the directed edges. The place marker signifies the progress of the play. When the marker is on a vertex of **R**, it is Red's move to move to a vertex in **B**. When the marker is on a vertex of **B**, it is Blue's turn to move to a vertex of set **R**. The play continues in this fashion, Red moves to a Blue vertex and Blue moves to a Red vertex.

Definition 1 An infinite game, G, is a 6-tuple

$$G = (Q, B, R, E, W(B), W(R))$$

where,

1. Q is the finite set of positions (vertices).
2. B and R are subsets of Q, such that $B \cup R = Q$ and $B \cap R = \emptyset$.
3. E is a set of directed edges between B and R such that:
    a. for each $b \in B$ there exists $r \in R$ such that $(b, r) \in E$.
    b. for each $r \in R$ there exists $b \in B$ such that $(r, b) \in E$.
4. $W(B)$ is called the winning set for Blue.
5. $W(R)$ is called the winning set for Red.
6. $W(B) \cap W(R) = \emptyset$.

At this time it should be noted that the winning sets for each player are not limited to vertices of the player's color.

Definition 2 A play that begins from position q is a complete[2] infinite sequence $p = q_1 q_2 q_3 q_4 \ldots q_{①-1} q_{①}$ such that $q = q_1$ and $(q_i, q_{i+1}) \in E, \ \forall i \in$

[2] Here we use the notion of complete taken from [19], that is the sequence contains ① elements.

$\mathbb{N}$, E is the edge relation. Hence a play is a sequence of states of the game. That is,

$$p : \mathbb{N} \to Q.$$

To determine how a player can win, let p be a play and consider the set of all vertices that occur infinitely often. We now have the following definition.

Definition 3 $In(p)$ is the set of vertices, in play p, that occur infinitely often, called the **infinity set** of p.

We now have the following cases to determine a win:

1. $W(B) \subset In(p)$ and $W(R) \not\subset In(p)$, then Blue wins.
2. $W(B) \not\subset In(p)$ and $W(R) \subset In(p)$, then Red wins.
3. $W(B) \not\subset In(p)$ and $W(R) \not\subset In(p)$, then Draw.
4. $W(B) \subset In(p)$ and $W(R) \subset In(p)$, then the frequencies of occurrence of the elements in each set must be considered.

Cases 1 and 2 above are the result that whatever winning set a player chooses, all vertices must occur infinitely often for a player to have a chance of winning (this concept is consistent with the ideology presented in [24]). All vertices must occur infinitely often also prevents a player from choosing too many vertices for their winning set.[3] Next we look at a simple example to analyze the situation when a player chooses the empty set.

Example 1 Suppose Blue chooses $\emptyset$ as their winning set (this is consistent with the premise that no choice is also a choice). That is, $W(B) = \emptyset$. The reason for Blue's choice is clear. $\emptyset \subset In(p)$, hence Blue is hoping that $W(R) \not\subset In(p)$ and Blue wins the game (the same can be true for Red, if Red chooses the empty set). Of course the situation can arise if both players choose $\emptyset$. In that case, the game will result in a draw. However, to show this we first need to define more machinery.

It is necessary to define a frequency function to count the number of occurrences of a given vertex in a play sequence. This gives rise to the next two definitions.

Definition 4 Given $Q = \{q_1, q_2, ..., q_n\}$ is the finite set of states and let D be a subset of Q. Let p be an infinite sequence of states, from a play, define a new sequence by the function

$$\psi_{D,p} : \mathbb{N} \to \{0, 1\}$$

[3] It is noted here that, as is usual, the $\subset$ symbol can also imply equality.

where,

$$\psi_{D,p}(i) = \begin{cases} 1 \; if \;\; p(i) \in D \\ 0 \quad otherwise \end{cases}$$

$\forall i \in \mathbb{N}$.

Definition 5 Define the frequency function, $freq_p$, as

$$freq_p(D) = \sum_{i=1}^{①} \psi_{D,p}(i).$$

These definitions are valid in general, however here they are applied to the winning sets for Blue and Red, respectively $W(B)$ and $W(R)$.

With the previous definitions, if both winning sets are subsets of the infinity set (the elements of both player's winning sets occur infinitely often) a winner can be determined. If the frequency of the elements in $W(B)$ is greater than the frequency of the elements in $W(R)$, then Blue is the winner. If the frequency of the elements in $W(R)$ is greater than the frequency of the elements in $W(B)$, then Red is the winner. If the frequencies are equal, then a draw results. This is a key advancement as a result of the grossone theory. As an immediate consequence from the above definitions, the following propositions are true.

Proposition 1 *For any sequence p,* $freq_p(\emptyset) = 0$.

Proof $p(i) \notin \emptyset \; \forall i \in \mathbb{N}$. Hence $\psi_D(i) = 0 \; \forall i \in \mathbb{N}$. Therefore $freq_p(\emptyset) = 0$.

Proposition 2 *If both players choose the empty set as their winning set, then the game is a draw.*

Proof By Proposition 1, $freq_p(W(B)) = freq_p(W(R)) = freq_p(\emptyset) = 0$.

4 Strategies

As with all games, a strategy is important for winning the game or to prevent your opponent from winning (in the latter case, at least strive for a draw). Hence a strategy allows a player to give a definition of the winner (or draw) from a given position in the game. Generally, a strategy for a player is a rule that specifies the next move of the player, given the history of the past moves. To give a more formal notion of a strategy, let

$$p = q_1, q_2, q_3, ..., q_{①-1}, q_①$$

be a play in the game G. Then the histories of this play are

$$q_1, \quad q_1q_2, \quad q_1q_2q_3, \quad q_1q_2q_3q_4, \quad ....$$

It is immediately apparent that there are infinitely many histories in game G. Therefore we will look at two sets of histories, the set $H(B)$ that contains all histories whose last positions are positions where Blue makes a move, and the set $H(R)$ that contains all the histories whose last positions are positions where Red makes a move. Hence, a strategy for Blue is a function f where

$$f : H(B) \to G$$

such that $\forall\ v = q_1...q_n \in H(B)$, $(q_n, f(v)) \in E$.

Hence, if f is a strategy for a player and q is a state position in the game, then all plays that begin from q, in which the player follows the strategy f can be considered. These plays are called consistent with f.

Definition 6 Call the strategy f of a player a **winning strategy from a position** q if all plays consistent with f that begin from q are won by the player. In this case it can be said that the **player wins the game from** q.

5 Examples and Results

Example 2 Referring to the game in Fig. 2, assume that $W(B) = \{b1\}$ and $W(R) = \{r1\}$. Then Blue is always the winner, no matter where the game begins. If $W(B) = \{b1\}$ and if $W(R) = \{r1, r2\}$, then Blue's winning strategy would be to move to either r1 or r2 finitely many times and the other infinitely times. Therefore $W(R) \not\subset In(p)$.

For instance, if the following sequence is played

$$p = r1, b1, r2, b1, r1, b1, r2, b1, r1, b1, r1, b1, r1, b1, r1, ...$$

then

$$In(p) = \{b1, r1\}$$

and $W(R) \not\subset In(p)$, however $W(B) \subset In(p)$, which implies Blue wins the game. If $W(B) = \{r1\}$ and $W(R) = \{r2\}$ (as mentioned previously, a player does not have to choose their color as their winning set) then Blue wins the game. The winning strategy for Blue consists of moving to $r2$ finitely many times. Actually Blue can move to $r2$ infinitely many times, however it must

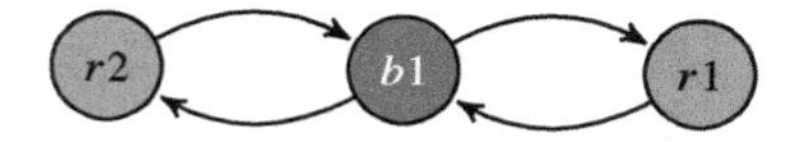

Fig. 2 A simple game

be less than ①/4 times. The following theorem and corollaries provide a better understanding of the frequency function.

Theorem 2 *For any set A and play p, the frequency function,* $freq_p(A) \leq$ ①.

Proof This follows directly from the properties of ① and Theorem 1.

Corollary 1 *For any game, the frequency of occurrence of any single vertex is* $\leq$ ①/2.

Proof Follows from Theorem 2 and the definition of a game, since there are two players.

Corollary 2 *For any game where Q is the set of vertices,* $freq_p(Q) =$ ①.

Proof Using the premise of a complete sequence, the corollary directly follows from Theorems 1 and 2.

This next example will illustrate this new application of the grossone paradigm to infinite games.

Example 3 Referring again to the game in Fig. 2, suppose the play goes as follows:

$$p = r2, b1, r1, b1, r2, b1, r1, b1, \overbrace{r1}^{r2\ skip}, b1, r1, b1, r2, b1, r1, b1, r2, b1, r1, b1, ...$$

Here the $In(p) = \{b1, r1, r2\}$. Hence, the frequency of occurrence for each vertex in the $In(p)$ is:

$$freq(\{b1\}) = ①/2,$$
$$freq(\{r1\}) = ①/4 + 1,$$
$$freq(\{r2\}) = ①/4 - 1.$$

Using the winning sets $W(B) = \{r1\}$ and $W(R) = \{r2\}$, Blue wins the game.

Example 4 In Fig. 3, if Blue chooses $b4$, that is $W(B) = \{b4\}$, a strategy for Red would be to choose $\emptyset$. Then from $r3$, Red can always move to $b3$ an infinite number of times or move to $b4$ a finite number of times. In this case, the best Blue can hope for is a draw.

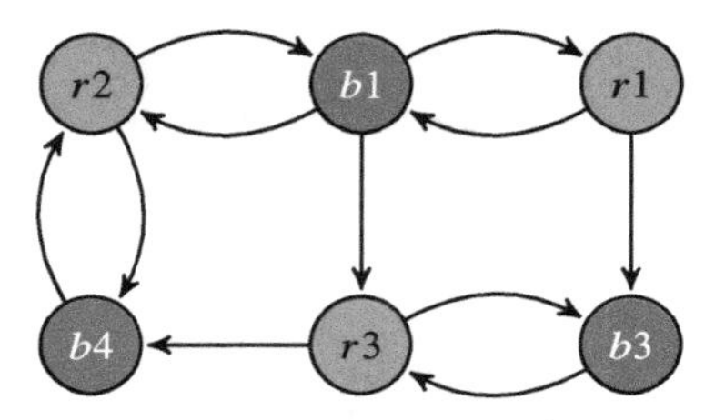

Fig. 3 A more complex game

Example 5 Referring again to Fig. 3, if each node is visited once in the 6 node outside cycle, that is via edges $(r1, b3)$, $(b3, r3)$, $(r3, b4)$, $(b4, r2)$, $(r2, b1)$, $(b1, r1)$, then the frequency of each vertex occurrence is ①/6. The sequence that will ensure this is:

$$p = r1, b3, r3, b4, r2, b1, r1, b3, r3, b4, r2, b1, r1, b3...$$

If player Blue chooses their winning sets $W(B) = \{b1, b3\}$, then Red can choose $W(R) = \{r2, r3\}$ and Red has a winning strategy. When Blue lands on vertex $b1$, Blue must move to $r1$ to get to $b3$ (part of Blue's winning set). The play continues and can follow the outside cycle. However, at some point, Red moves from $r2$ back to $b4$ a finite number of times. For instance, a play can follow:

$$p = r1, b3, r3, b4, r2, b4, r2, b4, r2, b4, r2, b1, r1, b3, r3, b4, r2, b1, r1, ...$$

hence

$$freq_p(\{b1, b3\}) = \frac{①}{3} - 2,$$

$$freq_p(\{r2, r3\}) = \frac{①}{3} + 1,$$

and Red wins the game.

So far the counting arguments presented seem quite straightforward. However, some counting arguments can be a bit more complicated but nevertheless still computable, see [25]. The following example in Fig. 4 shows that the number of moves does not straightforwardly and necessarily compute to a natural number. However applying the integer floor function, a natural number is obtained.

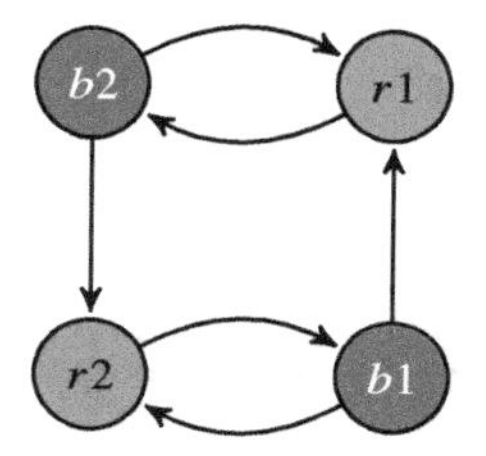

Fig. 4 A simple game where the number of occurrences is counted with a log function

Here it is supposed that a player chooses the set $\{b1\}$ as their winning set. Hence the number of occurrences of $b1$ will be counted. Consider the following sequence:

$$\underbrace{b1}_{1}, r2, \underbrace{b1}_{3}, r1, b2, r2, \underbrace{b1}_{7}, r1, b2, ..., r1, b2, r2, \underbrace{b1}_{15}, r1, b2, ..., r2, \underbrace{b1}_{31}, ...$$

The sequence repeats $r1$, $b2$ then $r2$ to $b1$. Continuing this pattern, the $b1$ vertex occurs at the next power of 2 away from the previous $b1$ vertex. This is the case where $k = -1$.

There are four vertices and hence three other cases:
$k = 0$

$$r2, \underbrace{b1}_{2}, r2, \underbrace{b1}_{4}, r1, b2, r2, \underbrace{b1}_{8}, r1, b2, ..., r1, b2, r2, \underbrace{b1}_{16}, r1, b2, ..., \underbrace{b1}_{32}, ...$$

$k = 1$

$$b2, r2, \underbrace{b1}_{3}, r2, \underbrace{b1}_{5}, r1, b2, r2, \underbrace{b1}_{9}, r1, b2, ..., r1, b2, r2, \underbrace{b1}_{17}, r1, ..., \underbrace{b1}_{33}, ...$$

$k = 2$

$$r1, b2, r2, \underbrace{b1}_{4}, r2, \underbrace{b1}_{6}, r1, b2, r2, \underbrace{b1}_{10}, r1, b2, ..., r1, b2, r2, \underbrace{b1}_{18}, ..., \underbrace{b1}_{34}, ...$$

The number of occurrences of $b1$ in each sequence is:

$$\lfloor \log_2(n - k) \rfloor \ \forall n \in \mathbb{N}.$$

Analogously, the set

$$\{k + 2^i \ : \ -1 \leq k \leq 2, \ i \in \mathbb{N}, \ k + 2^i \leq ①\}$$

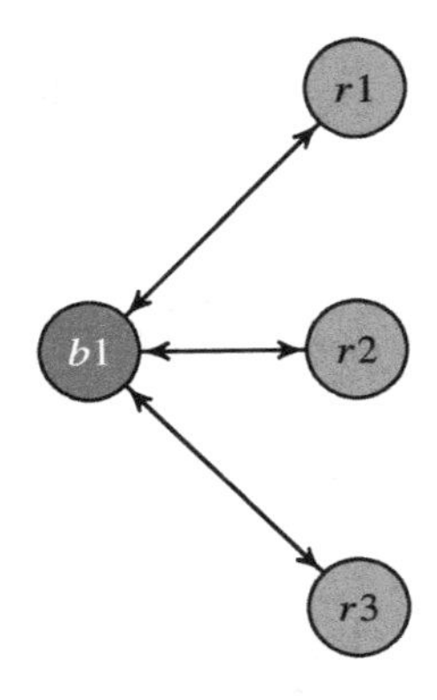

Fig. 5 An update game

for the maximum possible i, has the following number of elements

$$\lfloor \log_2(① - k) \rfloor.$$

6 An Application: Update Games and Networks

Communication network problems can be modeled using the concept of infinite games presented herein this paper. For example, in a distributed database system a problem of redundancy of data can be resolved by sharing information between all nodes of the system. This can be accomplished by a packet of current data continuously running through all nodes of the distributed system. Hence this is a solution (theoretically) without termination.

In an update game, one player is called the **control** player (color). The control player is the player who is responsible for updating all the vertices (nodes) in the game. The following definition can now be stated.

Definition 7 An **update game** is a game $G = (Q, B, R, E, W(B), W(R))$ for which Blue (or Red) is the control player, $Q = W(B)$ (or $Q = W(R)$) and $W(R) = \emptyset$ (respectively, $W(B) = \emptyset$).

Definition 8 An update game is an **update network** if the control player has a winning strategy to win the update game from any position of the game.

Figure 5, with Blue as the control player, is an update network. As is seen, Blue can update every vertex from $b1$ as follows:

$$p = b1, r1, b1, r2, b1, r3, b1, r1, b1, r2, b1, r3, b1, r1...$$

Here the set of vertices $Q = \{b1, r1, r2, r3\} = W(B) = In(p)$. It is obvious that $freq_p(\{b1\}) = ①/2$ and hence, by Corollary 2, $freq_p(\{r1, r2, r3\}) = ①/2$. From b1, Blue must move to one of the Red vertices. The Red vertices must be visited an infinite number of times in order for Blue to win and make this an update network, however their frequencies do not have to be equal. Hence note that Fig. 5 is **not** an update network with Red as the control player. Of course this update network does not have to be limited to three Red vertices. Hence the following proposition is obvious and stated without proof.

Proposition 3 *An update game with color players Blue (B) and Red (R), where the control color B has only one position, is an update network if* $\forall r \in R$, *there exists a bi-directional edge from the control color to each* r.

7 Conclusion

This paper has presented a new model of infinite games played on finite graphs by applying the theory of grossone and the Infinite Unit axiom. In his original work, McNaughton (see [14]) presented and developed a model of infinite games played on finite graphs using traditional methods of dealing with infinity. This paper has extended that work to count the number of times vertices in a board game are visited, although vertices can be visited an infinite number of times. Indeed, two players choose their winning sets and the player whose winning set is visited more frequently wins the game. With this new paradigm, as is common in the usual finite duration board games (chess, checkers, go), a draw can result. This was not the case in McNaughton's original work. Hence a more finer decision process is used in determining the winner or if the game results in a draw.

References

1. Cantor, G.: Contributions to the Founding of the Theory of Transfinite Numbers. Dover Publications, New York (1955)
2. Caldarola, F.: The exact measures of the Sierpiński d-dimensional tetrahedron in connection with a Diophantine nonlinear system. Commun. Nonlinear Sci. Numer. Simul. **63**, 228–238 (2018)
3. Cococcioni, M., Fiaschi, L., Lambertini, L. Non-Archimedean zero-sum games. J. Comput. Appl. Math. **393**, article 113483 (2021)
4. Cococcioni, M., Fiaschi, L. Non-Archimedean game theory: a numerical approach. Appl. Math. Comput. **409**, article 125356 (2021)

5. De Leone, R., Fasano, G., Sergeyev, Y.D.: Planar methods and grossone for the Conjugate Gradient breakdown in nonlinear programming. Comput. Optim. Appl. **71**(1), 73–93 (2018)
6. De Leone, R.: Nonlinear programming and grossone: quadratic programming and the role of constraint qualifications. Appl. Math. Comput. **318**, 290–297 (2018)
7. D'Alotto, L.: A classification of one-dimensional cellular automata using infinite computations. Appl. Math. Comput. **255**, 15–24 (2015)
8. Fiaschi, L., Cococcioni, M.: Numerical asymptotic results in game theory using Sergeyev's infinity computing. Int. J. Unconv. Comput. **14**(1), 1–25 (2018)
9. Gordon, P.: Numerical cognition without words: evidence from Amazonia. Science **306**, 496–499 (2004)
10. Khoussainov, B., Nerode, A., Automata Theory and its Applications. Birkhauser, (2001)
11. Iudin, D.I., Sergeyev, Y.D., Hayakawa, M.: Infinity computations in cellular automaton forest-fire model. Commun. Nonlinear Sci. Numer. Simul. **20**(3), 861–870 (2015)
12. Lolli, G.: Infinitesimals and infinites in the history of mathematics: a brief survey. Appl. Math. Comput. **218**(16), 7979–7988 (2012)
13. Lolli, G.: Metamathematical investigations on the theory of Grossone. Appl. Math. Comput. **255**, 3–14 (2015)
14. McNaughton, R.: Infinite games played on finite graphs. Ann. Pure Appl. Logic **65**, 149–184 (1993)
15. Margenstern, M.: Using Grossone to count the number of elements of infinite sets and the connection with bijections, p-Adic Numbers. Ultrametr. Anal. Appl. **3**(3), 196–204 (2011)
16. Montagna, F., Simi, G., Sorbi, A.: Taking the Pirahã seriously. Commun. Nonlinear Sci. Numer. Simul. **21**(1–3), 52–69 (2015)
17. Rizza, D.: Numerical methods for infinite decision-making processes. Int. J. Unconv. Comput. **14**(2), 139–158 (2019)
18. Rizza, D.: How to make an infinite decision. Bull. Symbol. Logic **24**(2), 227 (2018)
19. Sergeyev, Y.D.: Arithmetic of Infinity. Edizioni Orizzonti Meridionali, Italy (2003)
20. Sergeyev, Y.D.: A new applied approach for executing computations with infinite and infinitesimal quantities. Informatica **19**(4), 567–596 (2008)
21. Sergeyev, Y.D., Counting systems and the first Hilbert problem. Nonlinear Anal. Ser. A: Theory, Methods Appl. **72**(3–4), 1701–1708 (2010)
22. Sergeyev, Y.D.: Numerical computations and mathematical modeling with infinite and infinitesimal numbers. J. Appl. Math. Comput. **29**, 177–195 (2009)
23. Sergeyev, Y.D. Computations with grossone-based infinities. In: Calude, C.S., Dinneen, M.J. (eds.), Proceedings of the 14th International Conference "Unconventional Computation and Natural Computation", Lecture Notes in Computer Science, vol. 9252, pp. 89–106. Springer (2015)
24. Sergeyev, Y.D.: The Olympic medals ranks, lexicographic ordering and numerical infinities. Math. Intell. **37**(2), 4–8 (2015)
25. Sergeyev, Y.D.: Numerical infinities and infinitesimals: methodology, applications, and repercussions on two Hilbert problems. EMS Surv. Math. Sci. **4**(2), 219–320 (2017)
26. Sergeyev, Y.D., Garro, A.: Observability of turing machines: a refinement of the theory of computation. Informatica **21**(3), 425–454 (2010)
27. Sergeyev, Y.D.: Independence of the grossone-based infinity methodology from non-standard analysis and comments upon logical fallacies in some texts asserting the opposite. Found. Sci. **24**, 153–170 (2019)

28. Tohmé, F., Caterina, G., Gangle, R.: Computing truth values in the topos of infinite Peirce's α-existential graphs. Appl. Math. Comput. **385**, article 125343 (2020)
29. Zhigljavsky, A.: Computing sums of conditionally convergent and divergent series using the concept of grossone. Appl. Math. Comput. **218**, 8064–8076 (2012)

Adopting the Infinity Computing in Simulink for Scientific Computing

Alberto Falcone, Alfredo Garro, Marat S. Mukhametzhanov, and Yaroslav D. Sergeyev

Abstract Numerical computing represents a critical aspect of conventional computer architecture. Traditional computers adopt the IEEE 754-1985 binary floating-point standard to represent and work with real numbers. Due to the architectural limitations of traditional computers, it is impossible to handle infinite and infinitesimal quantities numerically. This chapter is devoted to the Infinity Computer, a supercomputer that permits to execute numerical computation with finite, infinite, and infinitesimal numbers. The accessible software simulator of the Infinity Computer is adopted in many industrial and research domains for addressing important real-world issues, where precision plays a crucial aspect. However, the Infinity Computer simulator is not suitable for handling problems in control theory and dynamics, where visual programming environments like Simulink are commonly used. In this context, the chapter presents the *Simulink-based Solution for the Infin-*

A. Falcone · A. Garro (✉) · M. S. Mukhametzhanov · Y. D. Sergeyev
Department of Informatics, Modeling, Electronics and Systems Engineering (DIMES), University of Calabria, via P. Bucci 41/C, 87036 Rende, Italy
e-mail: alfredo.garro@unical.it

A. Falcone
e-mail: alberto.falcone@dimes.unical.it

M. S. Mukhametzhanov
e-mail: m.mukhametzhanov@dimes.unical.it

Y. D. Sergeyev
e-mail: yaro@dimes.unical.it

Y. D. Sergeyev
Institute of Information Technology, Mathematics and Mechanics (IITMM), Lobachevsky
State University of Nizhny Novgorod, 603950 Nizhny Novgorod, Russia

Y. D. Sergeyev and R. De Leone (eds.), *Numerical Infinities and Infinitesimals in Optimization*, Emergence, Complexity and Computation 43,
https://doi.org/10.1007/978-3-030-93642-6_13

ity Computer, a novel solution that allows one to exploit the Infinity Computer arithmetic within the Simulink environment.

1 Introduction

To represent and deal with numbers, traditional computers adopt the IEEE 754-1985 binary floating-point standard (see [1]). Despite computers can deal with finite numbers, numerical computations involving infinite and infinitesimal quantities are impossible due to the presence of indeterminate forms and the inability of putting an infinite representation of a number into the finite computer memory (see [50]).

The Infinity Computer is a new type of supercomputer that can operate with finite, infinite, and infinitesimal numbers numerically. The already available software simulator of the Infinity Computer was implemented in the C++ language and is used to tackle complex real-world problems in a variety of industrial and research domains, including mathematics and physics (see [50] and references given therein). However, due to implementation issues related to extending and integrating the C++ source code of arithmetic and elementary operations in well-established environments like Simulink, the software simulator is not sophisticated enough to handle problems in control theory and dynamical systems.

To address these challenges, the chapter presents an innovative solution that integrates the Infinity Computer arithmetic within the Simulink environment, a well-known graphical programming environment for exploring and analyzing dynamic systems produced by MathWorks (see [21]). Simulink adopts a graphical block diagramming notation that is tightly integrated with Matlab. Simulink is widely used in the modeling and simulation domain, including distributed simulation, Co-Simulation of Cyber-Physical Systems (CPS) and Model-Based design (see [5, 14, 29, 38, 39]).

The chapter is organized as follows. Section 2 briefly presents the Infinity Computer and Matlab Simulink environment. The Simulink-based solution for operating with the Infinity Computing concepts within the Simulink environment is presented in Sect. 3. A set of numerical experiments is described in Sect. 4 to show the practicability and validity of the proposed solution. Finally, conclusions and future works are outlined in Sect. 5.

2 Background

The chapter employs notions and concepts from the Infinity Computing and its representation of numbers, as well as related algebraic operations, and the MATLAB Simulink environment, as described in the following subsections.

2.1 The Infinity Computing and Representation of Numbers

In the Infinity Computing (see [50]), numbers are represented using the positional numeral system with the infinite radix ① introduced as the number of elements of the set of natural numbers:

$$C = d_0①^{p_0} + d_1①^{p_1} + \cdots + d_n①^{p_n}, \tag{1}$$

where quantities d_i, $i = 0, ..., n$, are finite (positive or negative) floating-point numbers called *grossdigits*, and p_i, $i = 0, ..., n$, are called *grosspowers* and can be finite, infinite and infinitesimal (positive or negative), n is the number of grosspowers used in computations (can be fixed or variable for all computations).[1] Due to Simulink constraints (e.g., difficulties in working with variable-sized matrices in algebraic loops) and for simplicity, this chapter only considers finite floating-point *grosspowers*. It should be highlighted that this methodology is not related to non-standard analysis (see [51] for details).

In the Infinity Computer, a finite floating-point number A can be easily expressed using just one *grosspower* $p_0 = 0 : A = A①^0$. Furthermore, different infinite and infinitesimal numbers can be expressed: e.g., the numbers ①, $①^2$, $-1.5①^{2.5}$, $-1.2①^{3.2} - 1.2①^0 + 2.3①^{-1.2}$ are infinite, because they contain at least one finite positive *grosspower*; whereas, the numbers $①^{-1} = \frac{1}{①}$, $2.5①^{-1.5}$, $1.3①^{-2.2} - 1.7①^{-3.1}$ are infinitesimal, because they contain only finite negative *grosspowers*. Let us refer to the numbers given in the form (1) as *grossnumbers* hereinafter.

The Infinity Computer has previously been utilized effectively to solve real-world engineering issues and practical mathematics, such as handling ill-conditioning (see [34, 53]), optimization (see [10, 11, 15–17, 34, 53, 58]), Turing machines and infinite series (see [52, 57]), game theory and

---

[1] It should be noted that in the literature dedicated to the Infinity Computer, a different notation with $p_0 = 0$ is generally used, but here we adopted this matrix notation, since it better represents the details of our implementation.

probability (see [9, 12, 30, 41, 43]), fractals and cellular automata (see [7, 8, 13, 47, 49]), numerical differentiation and ordinary differential equations (see [2, 26, 35, 46, 48, 55]), binary spherical classifiers (see [3]), Spencer Brown's Calculus of Indications (see [31]), etc.

2.2 The MATLAB/Simulink Environment

Simulink is a simulation environment for MATLAB developed by MathWorks (see [37]). It enables engineers to model, simulate and analyze dynamic systems using the block diagram language before deploying on hardware. Furthermore, Simulink provides graphical tools for displaying the progress of a simulation, which considerably improves understanding of the system's behavior.

Simulink has the potential to significantly increase productivity (see [21, 37]). In the past, the traditional approach for developing a system was to begin with its components and describe their logic using blocks. The so-obtained blocks were then translated into the corresponding source code using a target programming language (for example, C/C++). Because the system had to be described twice, once in block notation and then in a programming language, this method required duplication of effort.

This approach exposes the translation process, from blocks to source code, to accuracy risks making the debugging stage difficult due to errors that could happen in the design (block diagram level), in the programming (programming level), and/or in the translation process. This approach is no longer required with Simulink because blocks are the "program".

Simulink is widely used in research and industry to investigate and assess design alternatives for complex systems so as to determine the optimal configuration that meets the requirements. The multi-domain nature of Simulink may be used by research teams to collaboratively simulate the behavior of the system's components, each of which developed by a team, also to evaluate how components influence the behaviour of the entire system [18, 22–24].

Simulink allows one to: (i) minimize the cost of prototypes by testing the system under risky and/or time-consuming conditions; (ii) verify and validate the system design with hardware-in-the-loop testing and rapid prototyping; and, (iii) maintain requirement traceability from design down to source code.

3 The Simulink-Based Solution for the Infinity Computer (SSIC)

This section presents the *Simulink-based Solution for the Infinity Computer (SSIC)*, a Simulink-based solution for operating with the Infinity Computing [25, 26]. It allows one to rapidly develop simulation models by exploiting the Infinity Computing arithmetic within the Simulink environment [37]. Furthermore, as demonstrated in several publications (see [45, 50] and references given therein), the Infinity Computer executes computations numerically using the IEEE 754 floating-point numbers, rather than symbolically.

In particular, it has been demonstrated in the papers [36, 40] that the Infinity Computer is significantly faster than symbolic computations, allowing one to achieve exact[2] solutions for various real-life problems as symbolic computations do. It should be emphasized that this methodology has nothing to do with non-standard analysis (see [51]).

The following subsections are devoted to describe in detail the proposed solution. Specifically, Sect. 3.1 describes how *grossnumbers* are represented in SSIC; whereas, Sect. 3.2 presents the architecture of SSIC along with the provided four functional modules, i.e., *Arithmetic Blocks Module (ABM)*, *Elementary Blocks Module (EBM)*, *Utility Blocks Module (UBM)*, and *Differentiation Blocks Module (DBM)*. For each module, the offered blocks are described together with examples showing their applicability.

3.1 Representation of Grossnumbers in SSIC

A *grossnumber* x is represented in SSIC through a standard Simulink *Constant block* as a variable sized vector (1-D array) or matrix (2-D array) depending on the dimensionality of the "Constant value" parameter (see [37]). The output has the same dimensions and elements as the "Constant value" parameter. If "Constant value" is a vector and the "Interpret vector parameters as 1-D" setting is enabled, Simulink considers the output as a 1-D array; otherwise, the output is handled as a 2-D array. Regardless of the output size, the first column indicates the *grossdigits*, while the second one specifies the *grosspowers* of a number expressed in the form (1): the number (1) is represented by the following 2-D array:

---

[2] Since the computations on the Infinity Computer are numerical and all the numbers and operations belong to the IEEE 754 floating-point standard [1], then the word "exact" means "up to machine precision".

$$C = \begin{bmatrix} d_0 & p_0 \\ d_1 & p_1 \\ \cdots & \\ d_n & p_n \end{bmatrix}. \quad (2)$$

For instance, the number 6 is represented in SSIC through the 1-D array $\begin{bmatrix} 6 & 0 \end{bmatrix}$, whereas the number $6①^0 + 1①^{-1}$ is represented by the 2-D array $\begin{bmatrix} 6 & 0 \\ 1 & -1 \end{bmatrix}$.

3.2 Architecture

SSIC brings the power of the Infinity Computer into the Simulink Graphical Programming Environment (GPE). The solution has been defined to facilitate the modeling and simulation of dynamic systems by allowing engineers to focus on the specific aspects of their system's components rather than the low-level functionalities provided by the Infinity Computer Arithmetic C++ library (*ICA-lib*).

The proposed solution is general-purpose and domain-independent and was created according to the *Modular Programming Paradigm* (see [4]), which emphasizes separating the functionalities of a software into independent modules, each of which has a group of closely related blocks and resources to perform a given function (e.g., *sqrt*, *cos*, and *sin*) (see [5]). The modular architecture of SSIC enables one to exploit the large set of Infinity Computer blocks in conjunction with the Simulink standard ones. Because modules have independent concerns, the proposed solution is easy to maintain and upgrade; this means that, it is possible to obtain a module and make the necessary changes without causing issues to the other modules.

Specialists can easily model, simulate, and evaluate hybrid systems using SSIC to face important real-world problems, where precision and accuracy in the calculations are critical aspects (e.g., Modern Complex Engineered Systems (CES), Automation, and Aerospace (see [6, 20, 28, 32, 33])).

SSIC has been developed by using the standard Simulink Blocks and S-Functions, which allows engineers to jointly exploit the benefits of the Infinity Computer with the already available Simulink functionalities. Figure 1 presents an overview of the proposed solution and its integration with the Matlab/Simulink environments. SSIC is placed between the *Simulink Graphical Programming Environment* and *Matlab Environment* layers.

The *Simulink Graphical Programming Environment* represents the Simulink environment used for modeling, simulating, and analyzing dynamic systems

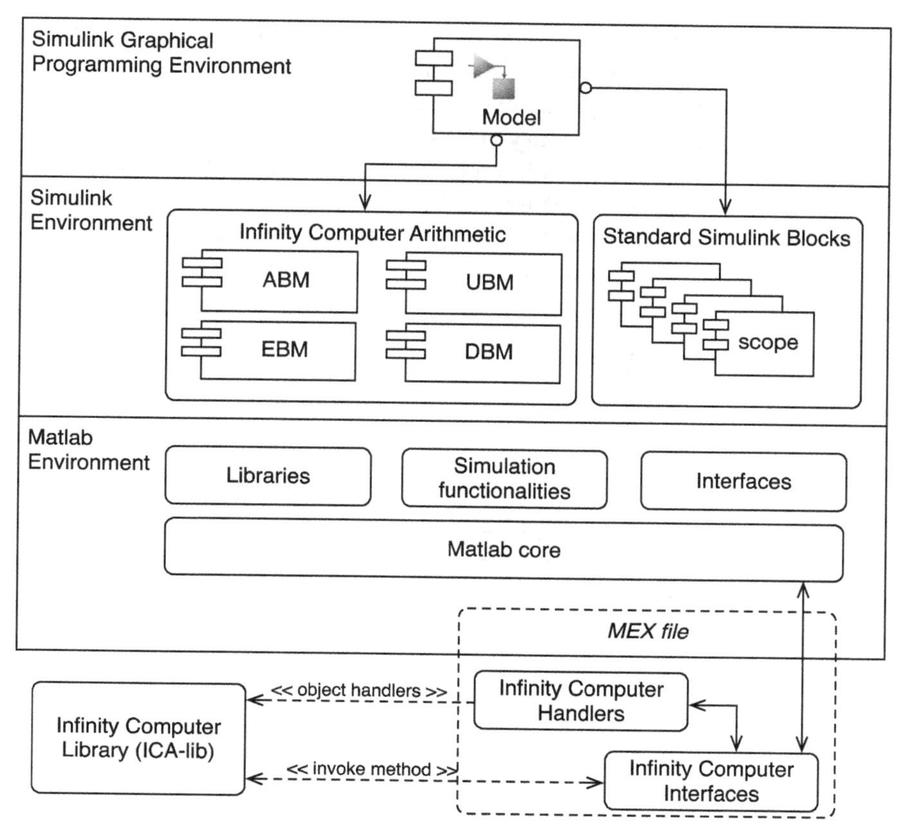

Fig. 1 The Simulink-based Solution for the Infinity Computer (SSIC) [27]

through the graphical block diagramming tool according to the Model-Based Design (MBD) paradigm (see [19]). MBD is an effective method for dealing with issues concerning the design and execution of complex systems, signal processing equipment, and communication components. MBD offers a common framework that can be used to create models with advanced functionalities by combining continuous-time and discrete-time blocks. The obtained models can be simulated in Simulink under different operational conditions leading to rapid prototyping, testing, and verification of the system's requirements and performances.

The *Simulink Environment* layer includes all the standard Simulink blocks as well as those offered by SSIC. Specifically, SSIC provides a collection of functional blocks to manage computations on infinite, finite, and infinitesimal quantities represented in the form (2). Each functional block accepts as input infinite, finite, and infinitesimal numbers that are forwarded to the appropriate

S-Function to perform the computation by interacting with *ICA-lib*. The SSIC functional block modules are depicted in Fig. 2.

The *Matlab Environment* represents the infrastructure where the Infinity Computer arithmetic C++ library (ICA-lib) has been integrated to perform infinite, finite, and infinitesimal computations. The ICA-lib was integrated into Simulink through a Matlab executable file (MEX) that serves as an interface between the involved parts. When the MEX file is compiled, Simulink dynamically loads it and allows to invoke the Infinity Computer arithmetic functions as if they were natively built-in.

Finally, the *ICA-lib* provides a collection of services, each of which offers C++ classes and interfaces that implement specific functionalities to handle infinite, finite, and infinitesimal quantities, as well as associated computations.

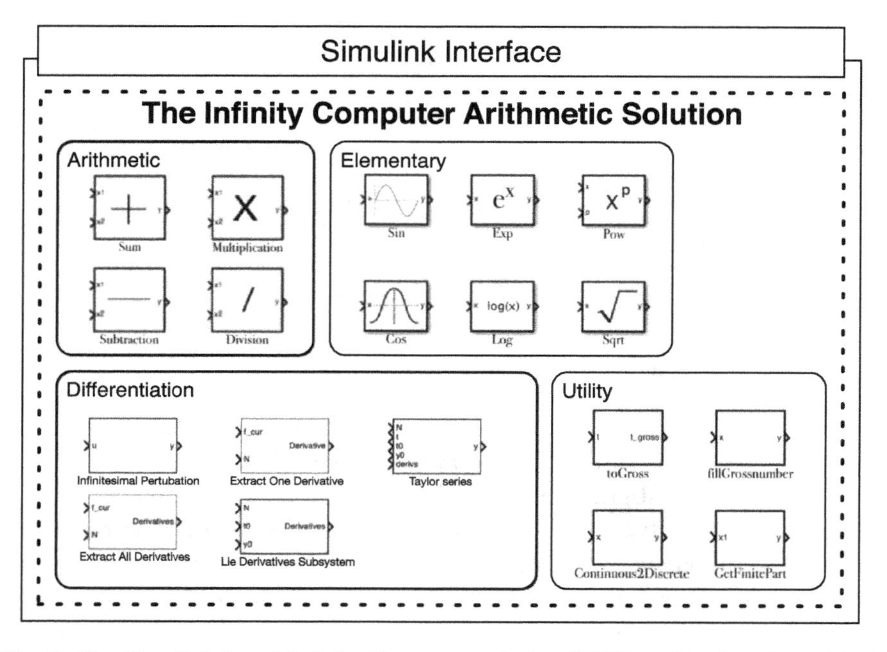

Fig. 2 The Simulink-based Infinity Computer solution (SSIC) and its functional block modules

3.3 Arithmetic Blocks Module

The *Arithmetic Blocks Module (ABM)* is described in this section. It includes different blocks specifically defined to perform arithmetic operations on infinite, finite, and infinitesimal numbers, such as *Sum*, *Subtraction*, *Multiplication* and *Division*. All the ABM blocks accept as input two arguments x, y, which are specified as follow:

$$x = \begin{bmatrix} x_0 & p_0 \\ x_1 & p_1 \\ \dots & \\ x_N & p_N \end{bmatrix}, \; y = \begin{bmatrix} y_0 & q_0 \\ y_1 & q_1 \\ \dots & \\ y_N & q_N \end{bmatrix}, \tag{3}$$

where N is the maximum precision used to perform arithmetic operations. This number determines the number of rows in the matrix representation (2) of each grossnumber (1). It is a configuration parameter defined within each block, and its value is set to 20 by default.

The precision of the Infinity Computer is specified through the parameter n, $n \leq N$, which can be configured in the "n_configuration.m" file.

The x, y arguments represent the *grossnumbers*

$$\begin{aligned} x &= x_0\text{①}^{p_0} + x_1\text{①}^{p_1} + \dots + x_N\text{①}^{p_N}, \\ y &= y_0\text{①}^{q_0} + y_1\text{①}^{q_1} + \dots + y_N\text{①}^{q_N}. \end{aligned} \tag{4}$$

The result of the applied operation is a matrix $z \in \mathbb{R}^{N \times 2}$:

$$z = z_0\text{①}^{\gamma_0} + z_1\text{①}^{\gamma_1} + \dots + z_N\text{①}^{\gamma_N}, \tag{5}$$

where z has dimension and number of elements according to the Infinity Computer algebra (see [45]), where the first n rows are significant, whereas the remaining $N - n$ ones are null.

For each ABM block, the specific function and an example showing its application are presented, hereinafter.

Sum. This block adds the values of its inputs. The Simulink model depicted in Fig. 3a adds the grossnumbers $A = 4\text{①}^0 + 2\text{①}^{-1}$ and $B = 1\text{①}^1 + 3\text{①}^0 + 4\text{①}^{-1}$. After running the simulation, the result $C = 1\text{①}^1 + 7\text{①}^0 + 6\text{①}^{-1}$ is displayed using the standard *Display block*.

The grossnumbers A, B, and C are defined as:

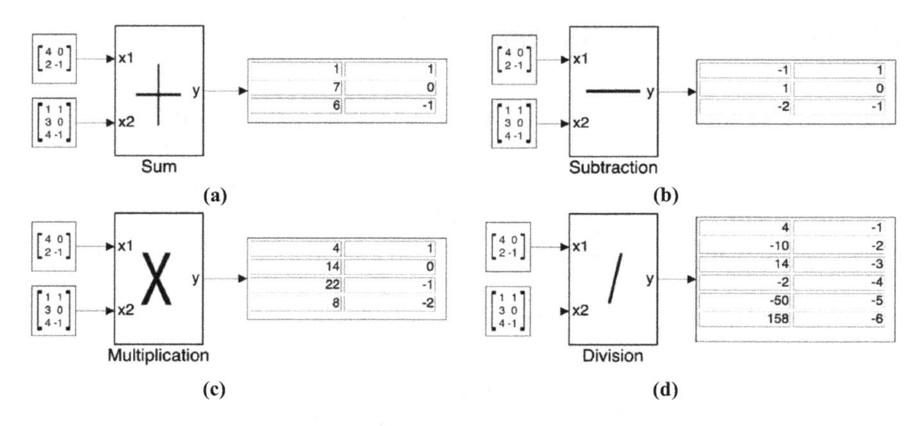

Fig. 3 Simulink models that perform addition (**a**), subtraction (**b**), multiplication (**c**), and division (**d**) of the *grossnumbers* $A = 4①^0 + 2①^{-1}$ and $B = 1①^1 + 3①^0 + 4①^{-1}$

$$A = \sum_{i=1}^{K} a_{k_i} ①^{k_i},\ B = \sum_{j=1}^{M} b_{m_j} ①^{m_j},\ C = \sum_{i=1}^{L} c_{l_i} ①^{l_i}. \tag{6}$$

According to the Infinity Computer arithmetic (see [50]), the result C is determined by including both the items of A, $a_{k_i}①^{k_i} : k_i \neq m_j$ with $1 \leq j \leq M$ and the ones of B, $b_{m_j}①^{m_j} : m_j \neq k_i$ with $1 \leq i \leq K$, and the terms with the same grosspower $(a_{l_i} + b_{l_i})①^{l_i}$.

Subtraction. This block subtracts the *grossnumbers* provided in input. This operation is a direct result of the *Sum* block explained above. Figure 3b depicts a Simulink model that executes the subtraction of $A = 4①^0 + 2①^{-1}$ and $B = 1①^1 + 3①^0 + 4①^{-1}$. By using the *Display block*, the result $C = -1①^1 + 1①^0 - 2①^{-1}$ is displayed.

Multiplication. This block multiplies the values of its inputs. Figure 3c depicts a Simulink model that performs the multiplication operation between the *grossnumbers* $A = 4①^0 + 2①^{-1}$ and $B = 1①^1 + 3①^0 + 4①^{-1}$. The *Display block* displays the outcome of the multiplication $C = 4①^1 + 14①^0 + 22①^{-1} + 8①^{-2}$, which is defined as follow:

$$C = \sum_{j=1}^{M} C_j, \qquad 1 \leq j \leq M, \tag{7}$$

where $C_j = b_{m_j}①^{m_j} \cdot A = \sum_{i=1}^{K} a_{k_i} b_{m_j} ①^{k_i + m_j}$.

Division. This block divides the values of its inputs. The Simulink model depicted in Fig. 3d carry out the division operation between the *grossnumbers* $A = 4①^0 + 2①^{-1}$ and $B = 1①^1 + 3①^0 + 4①^{-1}$. The *grossnumber*

$C = 4①^{-1} - 10①^{-2} + 14①^{-3} - 2①^{-4}$... that is the result of the operation is shown by a *Display block*.

The division operation $C = A/B$ yields to a result C and a reminer R, with the first *grossdigits* $c_{k_K} = a_{l_L}/b_{m_M}$ and the maximum exponent $k_K = l_L - m_M$.

The first partial reminder R^* is calculated as: $R^* = A - c_{k_K}①^{k_K} \cdot B$. The calculation ends when either $R^* = 0$ or the default accuracy is reached; otherwise, the number A is replaced with R^* and the computation process starts again (see [50]).

3.4 Elementary Blocks Module

The *Elementary Blocks Module (EBM)* provides common elementary functions, such as *sine*, *cosine*, *exponential*, and *logarithm*.

The truncated Taylor series has been used to implement each elementary function $f(x)$:

$$f(x) = f(x_0) + \sum_{i=1}^{N} \frac{d^i f(x_0)}{dx^i} \frac{(x - x_0)^i}{i!}, \tag{8}$$

where x_0 denotes a finite floating-point number. For each elementary function $f(x)$, where $f(x) \in \{\sin(x), \cos(x), \exp(x), \log(x), x^p$ (p is finite), $\sqrt{x}\}$, its Taylor expansion (8) is known as well as the analytical formulae for the respective derivatives $\frac{d^i f(x_0)}{dx^i}$ that is simple to implement. For each $f(x)$, except x^p and $\log(x)$, the value x_0 is selected as the finite part of the input x even if this finite part is equal to 0. Since the functions x^p and $\log(x)$ are not differentiable at the point $x0 = 0$, the value $x0$ was chosen as the finite part of x, if it is different from 0, and 0.1 otherwise (the number 0.1 was chosen to be neither too small nor too large and to keep the computations also in the case, when $x0 = 0$). Since the number x and the numbers $(x - x_0)$ are grossnumbers, then the computations in this Taylor expansion are performed using the arithmetic operations implemented in the Infinity Computer library. The value N is the same as in (1), since the resulting value $f(x)$ at a grossnumber x is also a grossnumber of the form (1).

All the EBM blocks accept one argument x as input (except the block Pow, which implements the function x^p and takes the second input p specified as a standard floating-point number), which is defined as follows:

$$x = \begin{bmatrix} x_0 & p_0 \\ x_1 & p_1 \\ \dots & \\ x_N & p_N \end{bmatrix}, \tag{9}$$

where N is the configuration parameter that has the same meaning as previously. The real precision of the Infinity Computer is also defined by the parameter n (i.e., the rows in (9) starting from the $(n+1)$-th contain only zeros).

In this solution, the expansions (8) are only relevant if the input x is not infinite; otherwise, the Taylor series diverges. If these elementary functions should be evaluated also in the infinite points x, then other implementations should be exploited (e.g., Newton's method).

Sin. The trigonometric sine of an argument x given as a *grossnumber* is computed by this block. The standard $C++$ library *math.h* is used to calculate the values $\pm\sin(x_0)$ and $\pm\cos(x_0)$, which are the corresponding derivatives used in the Taylor formula (8). Figure 4a depicts a Simulink model that calculates $sin(4①^0 + 2①^{-1})$. The result $-0.7568①^0 - 1.307①^{-1} + 1.514①^{-2}$... is shown through a *Display block*.

Cos. The trigonometric cosine of an argument x given as a *grossnumber* is computed by this block. The values $\pm\sin(x_0)$ and $\pm\cos(x_0)$ that represent the derivatives used in the Taylor formula (8) are calculated using the library *math.h*. The computation $cos(4①^0 + 2①^{-1})$ is performed by the Simulink model depicted in Fig. 4b, where the result of the computation $-0.6536①^0 + 1.514①^{-1} + 1.307①^{-2}$... is shown by a *Display block*.

Exp. This block calculates the base-e exponential function of a *grossnumber* x, which is e raised to the power $x : e^x$. The value $\exp(x_0)$, which represents the respective derivative used in the Taylor formula (8), is calculated using the library *math.h*. A Simulink model that performs the computation $e^{(4①^0+2①^{-1})}$ is shown in Fig. 4c, where the result $54.6①^0 + 109.2①^{-1} + 109.2①^{-2}$... is shown by a *Display block*.

Log. The natural logarithm of a *grossnumber* x is calculated by this block. The values $\log(x_0)$ and x_0^p, where p is finite, used for the computation of the respective derivatives in the Taylor formula (8) are calculated using the library *math.h*. Figure 4d shows a Simulink model that executes the computation $log(4①^0 + 2①^{-1})$. A *Display block* displays the result $1.386①^0 + 0.5①^{-1} - 0.125①^{-2} + 0.0416①^{-3}$... of the computation.

Pow. This block uses the function x^p that returns the base x raised to the power p. The values x_0^q, where q are finite, are computed using the library *math.h* for the computation of the respective derivatives in the Taylor formula (8). The value p is specified as a standard floating-point number. Figure 4e

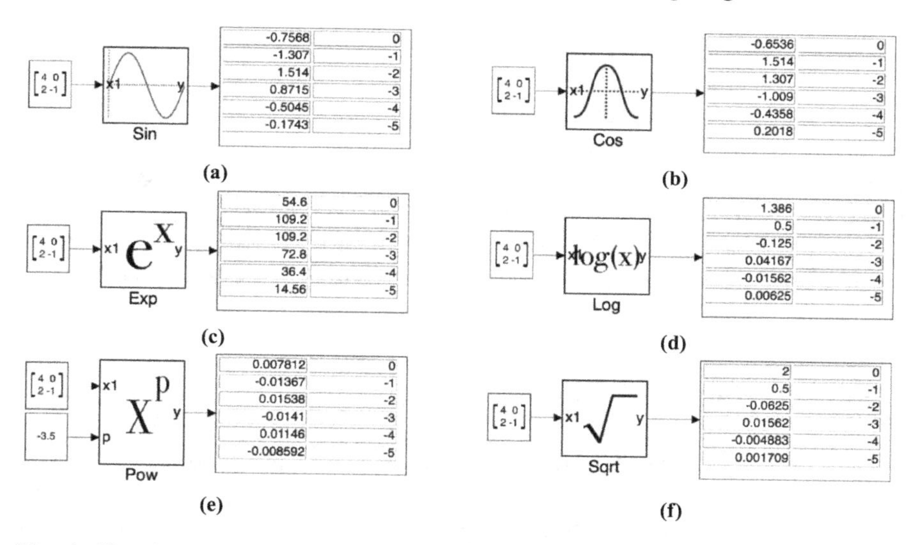

Fig. 4 Simulink models that perform the trigonometric function $\sin x$ (**a**), the trigonometric function $\cos x$ (**b**), the base-e exponential function e^x (**c**), the natural logarithm $\log\ x$ (**d**), the base x to the exponent power p, x^p, where $p = -3.5$ (**e**), and the square root $\sqrt{x}$ (**f**) of the *grossnumber* $x = 4①^0 + 2①^{-1}$

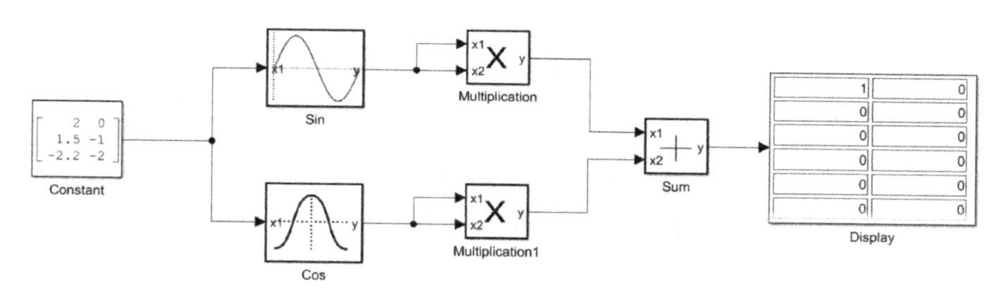

Fig. 5 Implementation of the Pythagorean trigonometric identity $\sin^2 x + \cos^2 x = 1$ using the presented blocks. The identity is satisfied even on matrix-type input of the form (2)

depicts a Simulink model that performs the computation $(4①^0 + 2①^{-1})^{-3.5}$. A *Display block* shows the result $0.0078①^0 - 0.0136①^{-1} + 0.0153①^{-2} - 0.0141①^{-3}$ of the computation.

Sqrt. This block has been added for convenience in calculations. It exploits the block *Pow* to return the square root of a *grossnumber* x : $\sqrt{x} = x^{1/2}$. Figure 4f shows a Simulink model that performs the computation $\sqrt{(4①^0 + 2①^{-1})}$. A *Display block* shows the result $2①^0 + 0.5①^{-1} - 0.0625①^{-2}$... of the computation.

An example of using of the above mentioned blocks is presented in Fig. 5. Here, an illustrative example with the Pythagorean trigonometric identity $\sin^2 x + \cos^2 x = 1$ is provided. One can see that the implementation of the sine and cosine functions is correct and maintains the main identity property even on matrix-type input of the form (2). The result of the operations $\sin^2 x + \cos^2 x$ is also given in the form (2) and consists of the finite part only: [1 0] (additional zero rows are added just to maintain the fixed size of the inputs/outputs and can be removed, if the output is variable-size).

3.5 Utility Blocks Module

The *Utility Blocks Module (UBM)* offers common utility function blocks necessary for enabling the realization of models according to the Infinity Computer solution and making them compatible with the Simulink environment. The provided blocks are: *Continuous2Discrete*, *fillGrossnumber*, *toGross*, and *getFinitePart*.

Continuous2Discrete. This block allows one to set the *Sample Time* to *Discrete* for all variable size blocks and Signals. It enables the sampling of time as a discrete numerical value. The generated value is used to update the blocks' internal states throughout the simulation execution.

fillGrossnumber. This block adds zero rows to the matrix representation of its input in order to fix the size of all variables (by default, the size of all variables expressed as *grossnumbers* is 20 by 2, i.e. the number N from (2) is set to 20). More in detail, given a matrix $M \in \mathbb{R}^{n\times m}$ with $m = 2$ and $i < n$ significant rows, the function adds an k-by-m rows of zeros, where $k = n - i$ to fill M.

This block is required, for example, to handle Simulink models containing algebraic loops where variable size variables are not allowed.

toGross. This block converts a finite floating-point number x into its matrix representation $\left[x\ 0\right]$ from (2). The resulting matrix is compatible with SSIC and therefore can be used as input for the other provided blocks.

getFinitePart. This block returns the finite part of a *grossnumber* x as a standard floating-point number. For instance, given $x = \begin{bmatrix} 3 & 1 \\ 6 & 0 \\ 4.1 & -5 \end{bmatrix}$, this block returns the value 6, while for $y = \begin{bmatrix} 3 & 2 \\ 3.1 & -1 \end{bmatrix}$ the result is 0.

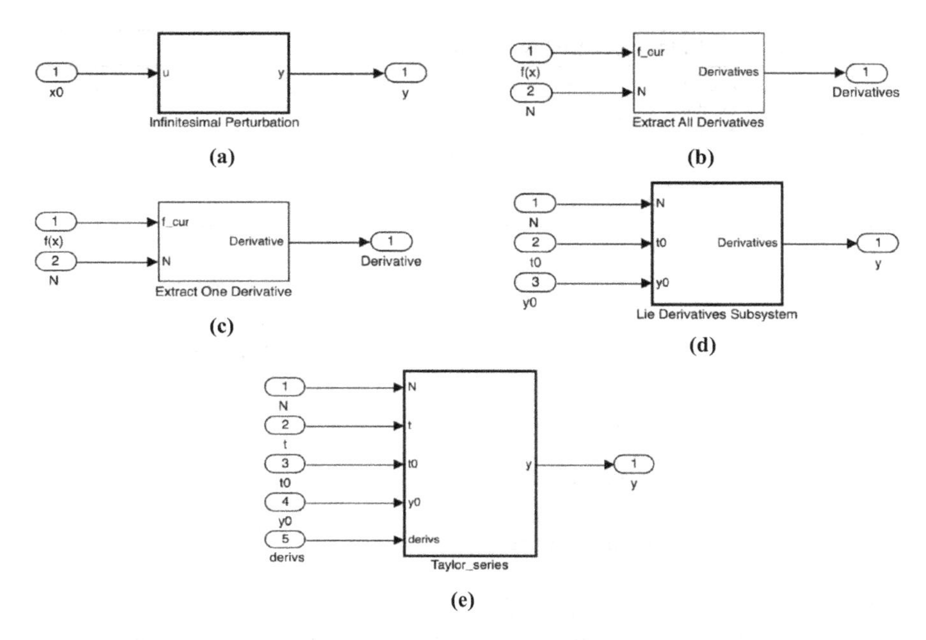

Fig. 6 Simulink models that perform the infinitesimal pertubation of x (**a**), the extraction of the derivatives from the value of the function $f(x_0 + ①^{-1})$ (**b**), the extraction of a specific $N-$th derivative from the value $f(x_0 + ①^{-1})$ (**c**), the computation of the Lie derivatives for ODE (**d**), and the Taylor series at the point t around the point $t0$ using the value $y0 = f(t0)$ and the first N derivatives specified in the vector *derivs* (**e**) [27]

3.6 Differentiation Blocks Module

The *Differentiation Blocks Module (DBM)* offers a collection of blocks and subsystems for handling higher order differentiation of a function implemented through Simulink standard blocks. Specifically, DBM provides five blocks that are: (i) *Infinitesimal Pertubation*, (ii) *Extract All Derivatives*, (iii) *Extract One Derivative*, (iv) *Lie Derivatives Subsystem*, and (v) *Taylor Series*.

Infinitesimal Pertubation. This block adds the infinitesimal pertubation $①^{-1}$ to its input $x : y = x + ①^{-1}$ whether x is specified; otherwise the block produces as output $y = ①^{-1}$. The structure of this block is depicted in Fig. 6a, where the input and output are both supplied as grossnumbers in matrix form. The input $x0$ is not mandatory; this means that, if it is not provided the output returns the grossnumber $①^{-1}$ in matrix form.

Algorithm 13.1: Extract the first N derivatives of $f(x)$

1: $F \leftarrow f(x_0 + ①^{-1})$;
2: **for** $i \leftarrow 1$ to N **do**
3: $F \leftarrow i \cdot ① \cdot F$;
4: $f^{(i)}(x_0) \leftarrow GetFinitePart(F)$;
5: **end for**
6: **return** $f'(x_0), f''(x_0), ..., f^{(N)}(x_0)$

Extract All Derivatives. This block extracts derivatives from the value of the function $f(x_0 + ①^{-1})$ (see, Fig. 6b). It takes as input two arguments: f_cur and N. The first one is a *grossnumber* defined according to Eq. 2; whereas N represents the number of derivatives to calculate. The output is the vector of the first N derivatives

$$[f(x_0),\ f'(x_0),\ f''(x_0),\ f^{(3)}(x_0), ...,\ f^N(x_0)],$$

where each element of the vector is an IEEE 754 binary64 floating-point number (see [1]).

The computation of the first N derivatives of $f(x)$, at the point x_0, is performed by this block as follows. First, the value of the function $f(x)$ at the point $x = x_0 + ①^{-1}$ is calculated:

$$f(x_0 + ①^{-1}) = f(x_0) + f'(x_0) \cdot ①^{-1} + \cdots + \frac{f^{(N)}(x_0)}{N!} \cdot ①^{-N} + O(①^{-N-1}), \tag{10}$$

where the coefficient of $①^{-i}$ provides the exact i−th derivative of $f(x)$ divided by $i!$ ($O(\sim)$ represents the standard Big O notation).

$$f^{(i)}(x_0) = GetFinitePart(f(x_0 + ①^{-1}) \cdot ①^i \cdot i!), \tag{11}$$

All the operations are performed using the *ABM* module and the utility block $GetFinitePart$. Since all derivatives of order i, $i = 1, 2, ..., N$, are extracted, the computational procedure has been designed in order to avoid the computation of the factorials $i!$ at each stage as presented in Algorithm 13.1 (see [27, 46] for details).

Figure 7 shows a Simulink model that computes the first N derivatives of a univariate function $f(x)$ at a finite point $x0$ using the blocks *Infinitesimal Perturbation* and *Extract All Derivatives* presented above. The model receives the input x_0 as an IEEE 754 binary64 floating-point number, which is converted by the $toGross$ block to the corresponding *grossnumber*. After that, the infinitesimal quantity $①^{-1}$ is added to the value of x_0 using the block "Infinitesimal Perturbation", yielding the value $y = x_0 + ①^{-1}$. The result y of this addition operation is given as input to the objective function $f(x)$, which is designed using the SSIC blocks. Then, the result $f(y)$ with

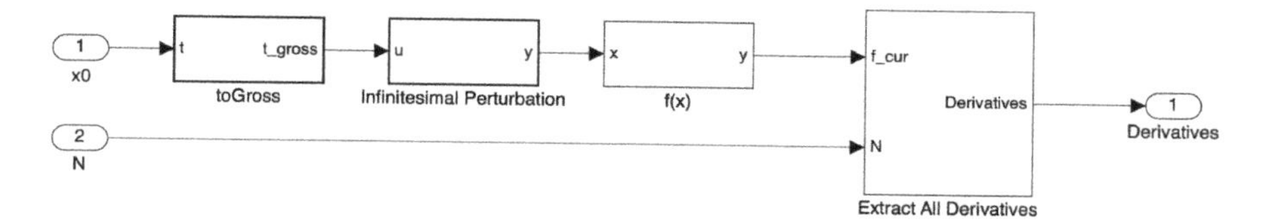

Fig. 7 The Simulink model for computation of the first N derivatives of a univariate function $f(x)$ at the finite point $x0$. The input $x0$ is given as a finite IEEE 754 binary64 floating-point number, the output of the $objective\_function$ block is represented in the matrix form as in [25, 26], the output $Derivatives$ is given as a finite floating-point vector $[f(x_0),\ f'(x_0),\ f''(x_0),\ f^{(3)}(x_0), ...,\ f^N(x_0)]$ [27]

the number of the desired derivatives N is moved to the block "Extract All Derivatives", where the values of the N derivatives are calculated according to (11).

Extract One Derivative. This block is similar to the *Extract All Derivatives* one, except that the *Extract One Derivative* block extracts just one derivative of order N, which is specified through the input argument N, from the result of the function $f(x_0 + ①^{-1})$. The structure of this block is shown in Fig. 6c. The block's inputs are the same as for the *Extract All Derivatives*, while the output is a finite IEEE 754 binary64 floating-point number.

Lie Derivatives Subsystem. This subsystem calculates Lie derivatives in the field of numerical solution to ODEs, where the first derivative of the unknown function $y(t)$ is given by the function $f(t, y(t))$. More in detail, this block allows one to calculate the first N total derivatives of the function $f(t, y(t))$, which are the first $N + 1$ derivatives of $y(t)$ at the point t_0. The structure of the block is depicted in Fig. 6d along with its inputs, which are supplied as standard finite IEEE 754 binary64 floating-point numbers. The output of this block is a vector containing the first N Lie derivatives of $f(t, y)$ at (t_0, y_0)

$$[D_0 f(t_0, y_0),\ D_1 f(t_0, y_0), ...,\ D_N f(t_0, y_0)],$$

where each element of the vector is an IEEE 754 binary64 floating-point number [1].

Because the computing method for the Lie derivatives is more complex to implement w.r.t. the other differentiation techniques, this subsystem was built differently w.r.t. the previous blocks. Specifically, the user who wishes to apply this subsystem should specify not only the required inputs, but also the function $f(t, y)$ in the appropriate computational block. The detailed description of this block is given in Fig. 8, where the Simulink models for

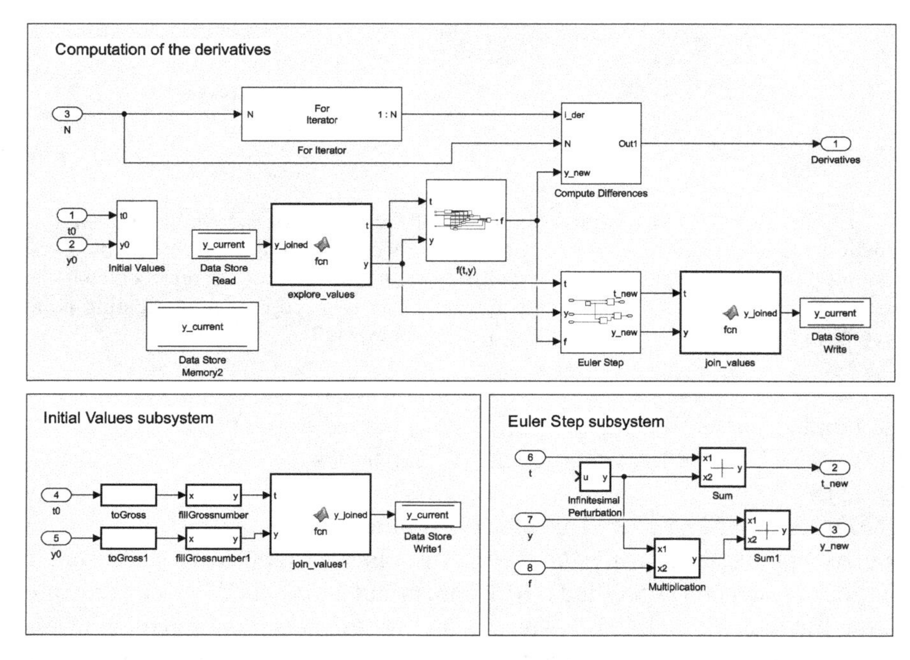

Fig. 8 The Simulink models based on SSIC for computation of the first N Lie derivatives of a function $f(t_0, y_0)$. Before the main computational system (shown at the top of the figure), the initial values should be generated and written to the global variable $y\_current$ (see the subsystem on the left and bottom). The subsystem $Euler\ Step$ performs the Euler steps (12) and is described at the bottom and right of the figure [27]

computation of the first N Lie derivatives of the function $f(t, y)$ (given by the block $f(t, y)$)[3] at the initial value (t_0, y_0) are presented.

First, the initial conditions $t0$ and $y0$ being the first two inputs of this computational system are transformed to the *grossnumbers* and stored in the Data Store Memory "*y_current*" (see the bottom and left of Fig. 8). The value $f(t_0 + k \cdot ①^{-1}, y_k)$ is then computed in the block $f(t, y)$ within the loop $for\ k = 1 : N$ block. Following that, the values y_k are computed at the subsystem $Euler\ Step$ (described at the bottom and right of the figure):

$$y_k = y_{k-1} + ①^{-1} f(y_{k-1}),\ \ k = 1, ..., N. \tag{12}$$

At the same time, $f(t_0 + k \cdot ①^{-1}, y_k)$ are transferred to the *Compute Differences* block, where these values are stored and the associated forward differences and Lie derivatives $D_k f(y_0)$ are calculated (see [27, 35, 36]):

[3] The function f is considered as non-autonomous to avoid unnecessary computations of the higher order derivatives w.r.t. t.

$$D_k f(y_0) = y^{(k+1)}(t_0) = ①^k \cdot F^k_{①^{-1}}(f(t_0, y_0), ..., f(t_k, y_k)) + O(①^{-1}),\ k = 1, ..., N, \tag{13}$$

$$F^k_{①^{-1}}(f(t_0, y_0), ..., f(t_k, y_k)) = \sum_{i=0}^{k} (-1)^i \binom{k}{i} f(t_{k-i}, y_{k-i}). \tag{14}$$

When the for loop finishes its work, the *Compute Differences* block returns the values of the Lie derivatives. Matlab functions $join\_values$ ($explore\_values$) are needed only to join separate variables t and y to one vector $[t, y]$ (and to explore the vector $[t, y]$ to two separate values t and y, respectively). The submodule *Compute Differences* contains Matlab functions, which calculate the respective backward difference and return the values of the derivatives. When the for loop ends, the *Compute Differences* block returns the values of the Lie derivatives. The Matlab functions $join\_values(explore\_values)$ are required to join the single variables t and y into a single vector $[t, y]$ (and to explore the vector $[t, y]$ to two independent values t and y, respectively). The submodule *Compute Differences* provides Matlab functions to compute the respective backward difference and return the values of the derivatives.

Taylor Series. This block computes the Taylor series of a function $f(t)$ at the point t around the point $t0$ using the function's value $y0 = f(t0)$ and the first N derivatives of the function $f(t)$ (calculated, e.g., using the previously defined differentiation blocks). The structure of this block is depicted in Fig. 6e. This block has five inputs and one output that are:

- N. Number of the used derivatives;
- t. The point at which the Taylor series should be calculated;
- $t0$. The initial point around which the Taylor series is calculated;
- $y0$. The value of the function to be approximated at the point $t0$;
- $derivs$. The vector of the first N derivatives;
- y. The output of the block as a finite IEEE 754 binary64 floating-point number.

4 Assessment and Evaluation

Three series of numerical experiments have been carried out to study and evaluate the proposed solution. All the experiments have been performed on a computer with the Windows 10 operating system, an i7-8550U processor, 8 GB of RAM, and the Matlab version 2016b. All the arithmetical operations have been calculated according to the IEEE 754-1985 binary floating-point standard using double precision.

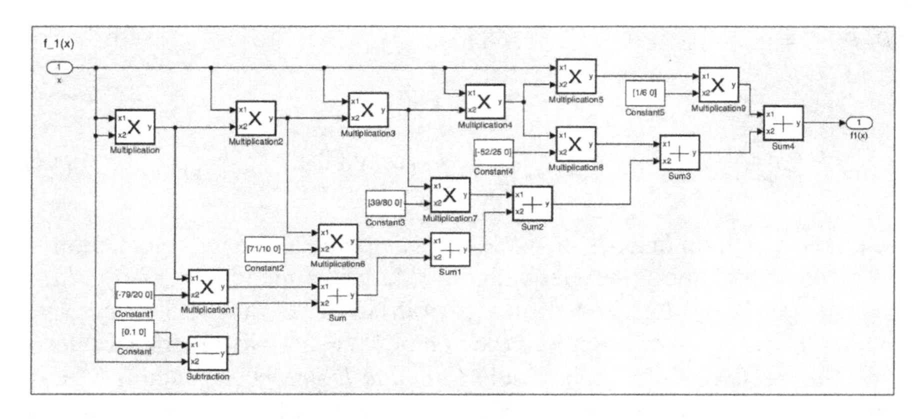

Fig. 9 The Simulink submodules of the functions $f_1(x)$ from (16) defined by using SSIC [27]

4.1 Differentiation of a Univariate Function

In the first series of the experiments introduced in [27], the first derivative has been calculated through SSIC using (11) and numerically using the central difference with finite steps h implemented manually (let us recall that in Simulink, only the backward difference w.r.t. the time variable t is implemented internally):

$$f'(x) \approx F_c^h(x) = \frac{f(x+h) - f(x-h)}{2h}, \tag{15}$$

The first three test functions $f_i(x),\ i = 1, 2, 3$ delineated in [54] and the function $f_4(x) = \frac{f_3(x)}{1+f_2(x)}$ have been considered:

$$\begin{aligned} f_1(x) &= \tfrac{1}{6}x^6 - \tfrac{52}{25}x^5 + \tfrac{39}{80}x^4 + \tfrac{71}{10}x^3 - \tfrac{79}{20}x^2 - x + 0.1, \\ f_2(x) &= \sin(x) + \sin(\tfrac{10x}{3}), \\ f_3(x) &= -\textstyle\sum_{i=1}^{5} i \sin[(i+1)x + i], \\ f_4(x) &= \tfrac{f_3(x)}{1+f_2(x)}. \end{aligned} \tag{16}$$

The Simulink implementations of these functions implemented by using SSIC are shown in Figs. 9 and 10 (for the functions $f_1(x)$ and $f_2(x)$, respectively) and Fig. 11 (for the function $f_3(x)$). Concerning the function $f_4(x)$, its Simulink implementation is omitted, since it follows from the implementations of the functions $f_2(x)$ and $f_3(x)$ and includes only two basic operations between them.

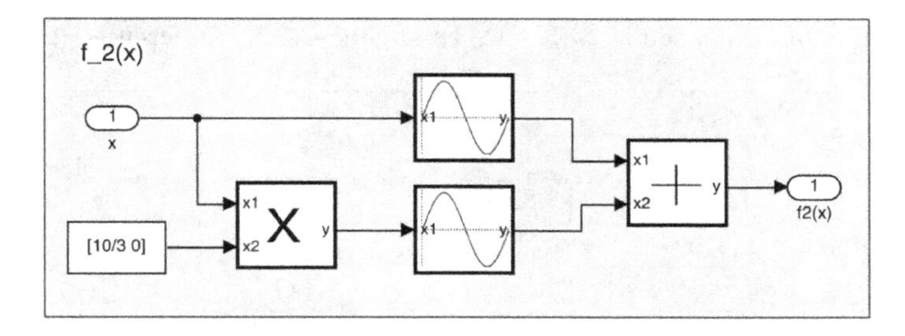

Fig. 10 The Simulink submodules of the function $f_2(x)$ from (16) defined by using SSIC [27]

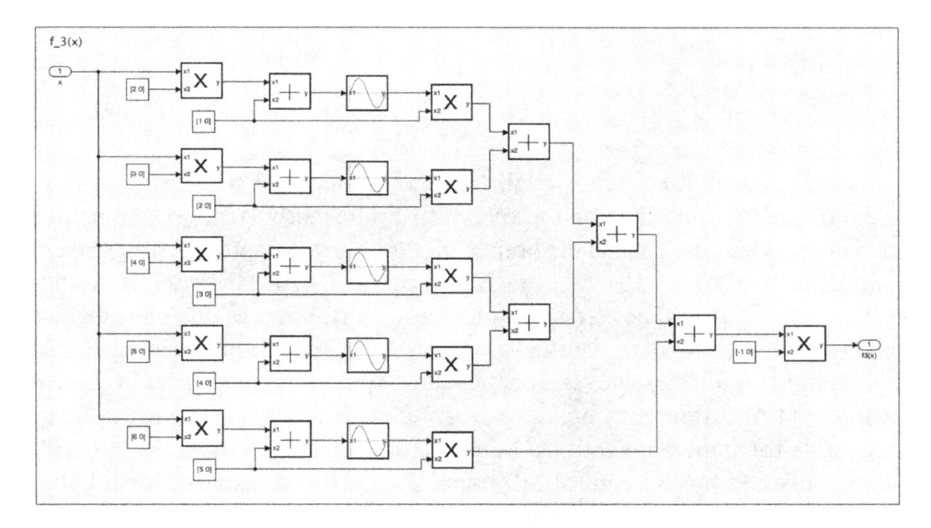

Fig. 11 The Simulink subsystem of the function $f_3(x)$ from (16) defined through SSIC [27]

For each test function, the following values of the steps h have been used: 10^{-1}, 10^{-2},...,10^{-15}. Then, the first derivative of each test function has been computed by both the Infinity Computer and numerically using (15) at each point $x_i = i \cdot 10^{-3}$, $i = 0, \ldots, 10^3$. Following that, the Normalized Root Mean Square Errors (NRMSE) have been determined for each test problem as reported below:

$$NRMSE = \frac{\sqrt{\frac{1}{N_{trials}} \sum_{i=1}^{N_{trials}} (y_i^{approx} - y_i^{exact})^2}}{\max\limits_{i=1,\ldots,N_{trials}} y_i^{exact} - \min\limits_{i=1,\ldots,N_{trials}} y_i^{exact}}, \tag{17}$$

where y_i^{approx} is the approximation of the first derivative at the point x_i obtained by SSIC (let us call them $y'_{IC}(x_i)$, hereinafter) and by the central

Table 1 NRMSE's obtained by SSIC (11) and by the Central Difference (15) for each test function $f_i(x),\ i = 1, 2, 3, 4$

Differentiation using SSIC				
	$f_1'(x)$	$f_2'(x)$	$f_3'(x)$	$f_4'(x)$
	2.1e-16	5.8e-17	1.6e-16	1.6e-16
Differentiation using (15) with the stepsize h				
h	$f_1'(x)$	$f_2'(x)$	$f_3'(x)$	$f_4'(x)$
10^{-5}	7.7e-11	6.4e-11	2.4e-10	3.4e-10
10^{-6}	2.5e-11	1.3e-11	3.6e-11	3.1e-11
10^{-7}	3.1e-10	1.3e-10	3.5e-10	3.1e-10

differences from (15) (let us call them $y'_{NUM}(x_i)$, hereinafter). The obtained NRMSE values[4] are reported in Table 1.

As reported in Table 1, SSIC allows one to calculate the first derivative of the presented functions on the interval [0, 1] exactly (i.e., up to machine precision), while the central difference did not allow to obtain errors smaller than $2.5 \cdot 10^{-11}$, $1.3 \cdot 10^{-11}$, $3.6 \cdot 10^{-11}$, and $3.1 \cdot 10^{-11}$ for the functions $f_1(x)$, $f_2(x)$, $f_3(x)$, and $f_4(x)$, respectively. Furthermore, one can see also that errors started to grow for the values $h < 10^{-6}$ due to numerical (mainly cancellation) issues. However, for these test functions, the errors produced by the central differences are always smaller than 10^{-1}, which might be acceptable for some applications. Figure 12 shows the graphs of the NRMSE values obtained by the central differences w.r.t. the stepsize h for a better visualization of the obtained results.

4.2 *Higher Order Differentiation of a Univariate Function*

In the second series of experiments introduced in [27], the higher order derivatives calculated with SSIC (11) are compared with those obtained numerically using the consecutive application of the internal Simulink block *Derivative*. The first five Pintér's functions $f_i^{pinter}(x)$ from [42] commonly used to evaluate global optimization algorithms have been used as test functions. Figure 13 shows the Simulink subsystem to evaluate the Pintér's functions.

[4] Only the results with the smallest errors obtained with $h = 10 - 5, 10 - 6, 10 - 7$ for the central difference are shown in Table 1, while the remaining values are displayed in Fig. 12.

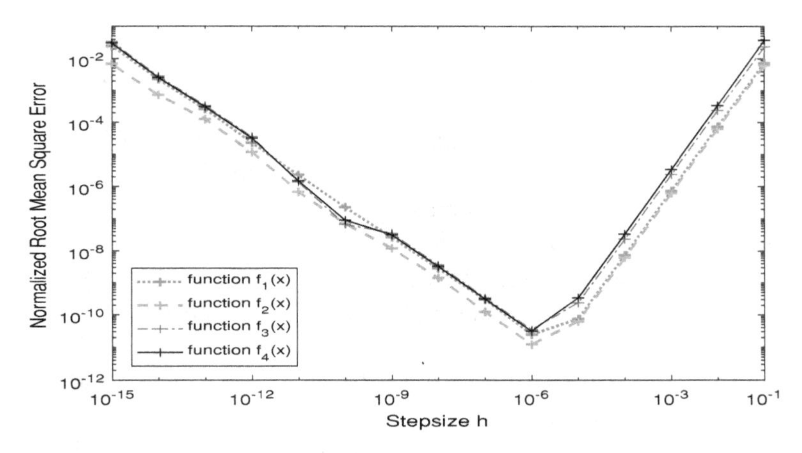

Fig. 12 Graphs of the NRMSE's obtained by the Central Difference (15) for each test function $f_i(x)$, $i = 1, 2, 3, 4$, w.r.t. the stepsize h [27]

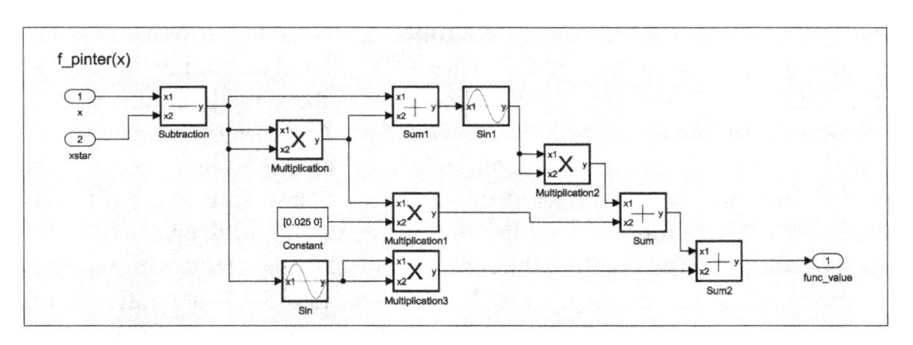

Fig. 13 The Simulink subsystem based on SSIC of the Pintér's functions $f_i^{pinter}(x)$ from [27, 42]

Table 2 Average NRMSE's over the first 5 Pintér's test functions obtained by computation of the first 5 derivatives on SSIC and by the consecutive application of the Simulink blocks *Derivative*

	$\Delta t = 10^{-1}$		$\Delta t = 10^{-2}$		$\Delta t = 10^{-3}$		$\Delta t = 10^{-4}$		$\Delta t = 10^{-5}$	
	SSIC	Num	SSIC	Num	SSIC	Num	SSIC	Num	SSIC	Num
f'	4.0e-17	2.5e-01	5.0e-17	2.6e-02	5.0e-17	2.6e-03	4.9e-17	2.6e-04	4.9e-17	2.6e-05
f''	6.6e-17	4.1e-01	6.5e-17	4.9e-02	6.3e-17	4.9e-03	6.3e-17	4.9e-04	6.4e-17	4.9e-05
f'''	4.9e-17	4.7e-01	5.1e-17	7.3e-02	4.9e-17	7.2e-03	4.9e-17	7.2e-04	4.9e-17	1.9e-03
$f^{(4)}$	7.5e-17	4.9e-01	6.8e-17	8.8e-02	7.2e-17	8.9e-03	7.1e-17	4.3e-03	7.1e-17	2.8e+01
$f^{(5)}$	6.8e-17	6.3e-01	6.8e-17	1.1e-01	6.3e-17	1.1e-02	6.3e-17	1.2e+01	6.3e-17	7.4e+05

For each test function, the first five derivatives have been calculated by using SSIC and the consecutive application of the Simulink *Derivative* blocks at the points $x_i = t_0 + i \cdot \Delta t,\ \ i = 0, 1, ..., N_{trials}$, where $t_0 = 0$, $N_{trials} = \frac{1}{\Delta t}$, and Δt is the Simulink's sample time value (fixed sample times have been used). The sample time Δt have been set to the following values: 10^{-1}, 10^{-2},...,10^{-5}, whereas the values of x_i have been calculated through the Simulink internal block *clock*.

For each $s-$th test function and for each $j-$th derivative $f_s^{(j)}(x_i)$, i=1, ..., N_{trials}, s, $j = 1, ..., 5$, the values $NRMSE_{s,j}$ have been calculated according to (17). Then, for each $j-$th derivative, the average NRMSE's have been calculated over the five test functions: $NRMSE_{avg,j} = \frac{1}{5}\sum_{s=1}^{5} NRMSE_{s,j}$. Table 2 and Fig. 14 show the obtained results.

As reported in Table 2, SSIC allows one to compute the higher order derivatives of the Pintér's functions up to machine precision, and the result is independent of the sample time (for each derivative, the results obtained with SSIC are different for various sample times Δt, since the grids x_i on which the derivatives' values are calculated are different and depend on the sample time Δt: $x_i = t_0 + i \times \Delta t$, but the error order is the same in all the cases).

In their turn, the errors produced by the Simulink *Derivative* block are sensitive to the choice of the sample time Δt. Since the *Derivative* block employs the first order backward difference, it is known that the differentiation error is of order 1, implying that the error should decrease approximately with the same rate that the sample time Δt does. However, the error decreases at the same rate for the 3−rd, 4−th, and 5−th derivatives until the value $\Delta t = 10^{-4}$ for the 3−rd derivative and $\Delta t = 10^{-3}$ for the 4−th and 5−th derivatives. If Δt decreases after these values, the error does not decrease with the same rate or even starts to increase due to numerical (mostly cancellation) issues.

For example, with Δt less than or equal to 10^{-4}, the computation of the fifth derivative becomes not significant due to large errors, while the computation of the fourth derivative becomes not significant with Δt less than or equal to 10^{-5}. Furthermore, for the first five derivatives, numerical differentiation using the block *Derivative* did not allow to obtain errors less than 2.6×10^{-5}, 4.9×10^{-5}, 7.2×10^{-4}, 4.3×10^{-3}, and 1.1×10^{-2}, respectively.

Another crucial issue is the loss of information caused by the *Derivative* blocks at the first k points when computing the $k-$th derivative. This is explained by the backward difference formula, which cannot be used at the first points, while SSIC is able to calculate derivatives at any finite point without loss of information. The graphs of these average NRMSE's are depicted in Fig. 14.

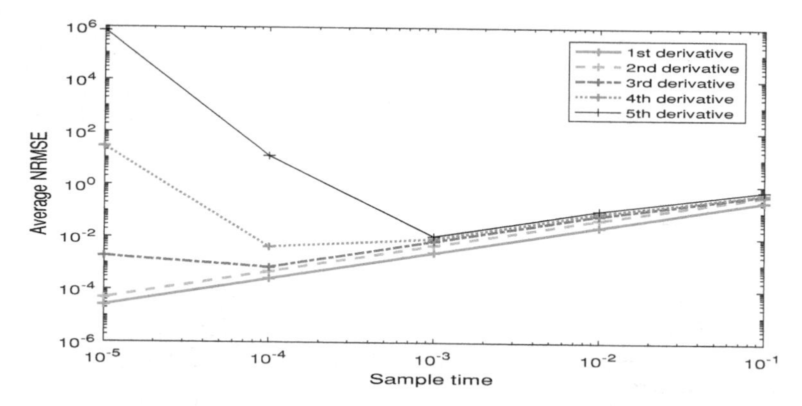

Fig. 14 Graphs of the average NRMSE's over the first 5 Pintér's test functions obtained by computation of the first 5 derivatives by the consecutive application of the Simulink blocks *Derivative* w.r.t. the sample time Δt [27]

It is also evident that the error coming from the *Derivative* blocks can be reduced using different Simulink tools as, e.g., *variable-step* methods instead of *fixed sample times*. However, these tools do not overcome the numerical issues discussed above; rather, they allow improving the accuracy but do not ensure to obtain results up to machine precision.

4.3 Computation of the Lie Derivatives for ODEs

In the third series of the experiments introduced in [27], the first five test problems from [55] have been considered. Each test problem is made up of an ODE of the form:

$$\begin{cases} y'(t) = f(y(t)), \\ y(t_0) = y_0, \end{cases} \tag{18}$$

where $y \in \mathbb{R}^n$, $f : \mathbb{R}^n \to \mathbb{R}^n$, $t \in \mathbb{R}$.

The solution $y(t)$ of the corresponding ODE is a univariate function continuously differentiable infinitely many times at the finite points t. Problems 3 and 5 are autonomous, while the remaining ones are not (in this cases, let us refer to the function $f(y)$ from the right side of the Eq. (18) as the function $f(t, y)$, since $y(t)$ is always univariate in these problems). The Simulink models are presented in Fig. 15.

Since the initial conditions delineated in [55] for each test problem are too simple (specifically, all the conditions are $y(0) = 1$), then in this series

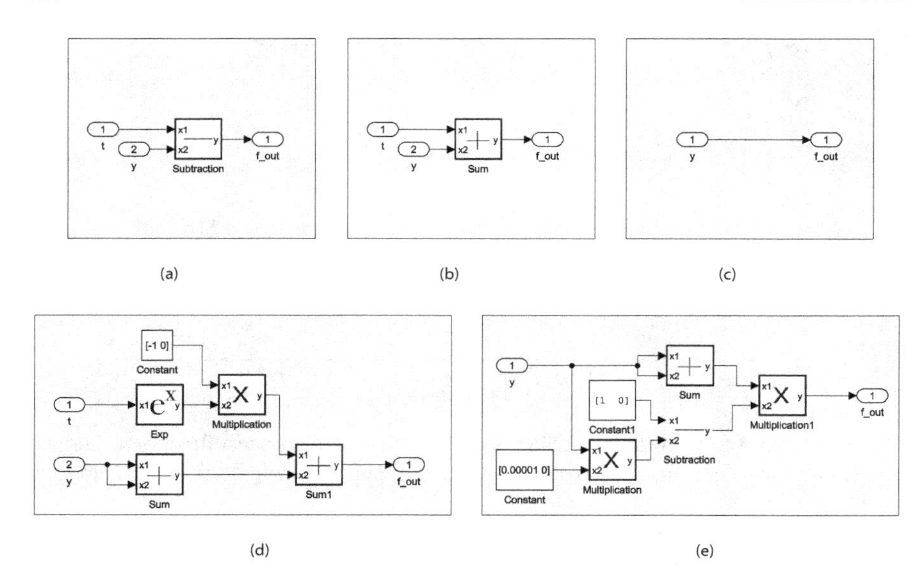

Fig. 15 The Simulink submodules of the ODE's right-side functions $f_i(t, y)$, $i = 1, ..., 5$, from [55] defined through SSIC. Test functions $f_3(t, y)$ and $f_5(t, y)$ (shown in **c** and **e**, respectively) are autonomous (i.e.,they do not depend on t), while the functions $f_1(t, y)$, $f_2(t, y)$, and $f_4(t, y)$ (shown in **a**, **b**, and **d**, respectively) are not [27]

of the experiments, the initial conditions were generated at the point $t_0 = 1$, maintaining the same solution of each ODE as in [55].

Traditionally, there are numerous numerical methods accessible in Matlab for computing higher order Lie derivatives of the right-side functions $f(y)$ from (18) (e.g., using Matlab's Symbolic computations or other available automated differentiation algorithms). For example, in the recent work [36], Matlab's symbolic toolbox and ADiGator (see [56]) have been investigated. It has been shown that these methods are inefficient, since they require the use of higher order tensors and, as a consequence, a large amount of processing resources, particularly, in the case of high dimensions and complex functions. Thus, because there is no method in Simulink for computing exact Lie derivatives without using complicated formulas involving higher order tensors (see [36] for details), then with the aim of showing the potential and the advantages of SSIC in differentiation of the solution to the initial value problems, we do not compare the presented computational method with another techniques. Instead, let us compare the *Taylor for the Infinity Computer* (TIC) method proposed in [55], which is the simplest method that uses the exact values of the derivatives, with the standard numerical methods available in Simulink that use fixed stepsizes: Runge-Kutta of order 4 (ODE4) and

Table 3 Relative errors obtained by the internal Simulink numerical methods Runge-Kutta (ODE4) and Dormand-Prince (ODE5 and ODE8) and by the TIC using 4, 5, 8, and 15 derivatives, respectively, at the points $t_i = 1.05,\ 1.1,\ 1.15$, and 1.2 for the first 5 test problems from [55]

N	t	E_{ODE4}	E_{TIC4}	E_{ODE5}	E_{TIC5}	E_{ODE8}	E_{TIC8}	E_{TIC15}
1	1.05	4.4e-16	2.5e-09	3.0e-16	2.1e-11	3.0e-16	1.5e-16	1.5e-16
	1.10	5.8e-16	7.9e-08	5.8e-16	1.3e-09	5.8e-16	2.9e-15	2.9e-16
	1.15	7.1e-16	5.8e-07	8.5e-16	1.5e-08	8.5e-16	9.8e-14	2.8e-16
	1.20	1.2e-15	2.4e-06	5.5e-16	7.9e-08	5.5e-16	1.3e-12	1.4e-16
2	1.05	7.3e-16	3.9e-09	2.4e-16	3.2e-11	2.4e-16	1.2e-16	1.2e-16
	1.10	2.0e-15	1.2e-07	4.5e-16	2.0e-09	4.5e-16	3.5e-15	3.4e-16
	1.15	2.1e-15	8.5e-07	2.1e-16	2.1e-08	2.1e-16	1.4e-13	0.0e+00
	1.20	3.0e-15	3.4e-06	4.0e-16	1.1e-07	4.0e-16	1.8e-12	4.0e-16
3	1.05	1.6e-16	2.5e-09	1.6e-16	2.1e-11	1.6e-16	1.6e-16	1.6e-16
	1.10	1.0e-15	7.7e-08	5.9e-16	1.3e-09	5.9e-16	2.5e-15	0.0e+00
	1.15	8.4e-16	5.6e-07	0.0e+00	1.4e-08	0.0e+00	9.3e-14	0.0e+00
	1.20	1.6e-15	2.3e-06	4.0e-16	7.5e-08	4.0e-16	1.2e-12	2.7e-16
4	1.05	3.7e-15	2.5e-09	1.6e-16	2.1e-11	1.6e-16	1.6e-16	1.6e-16
	1.10	7.5e-15	7.7e-08	5.9e-16	1.3e-09	5.9e-16	2.5e-15	0.0e+00
	1.15	1.2e-14	5.6e-07	0.0e+00	1.4e-08	0.0e+00	9.3e-14	0.0e+00
	1.20	1.6e-14	2.3e-06	4.0e-16	7.5e-08	4.0e-16	1.2e-12	2.7e-16
5	1.05	1.3e-14	7.6e-08	0.0e+00	1.3e-09	0.0e+00	2.2e-15	2.2e-16
	1.10	2.6e-14	2.3e-06	3.9e-16	7.5e-08	3.9e-16	1.1e-12	3.9e-16
	1.15	3.9e-14	1.6e-05	1.8e-16	7.8e-07	1.8e-16	4.0e-11	1.8e-16
	1.20	5.2e-14	6.1e-05	1.6e-16	4.0e-06	1.6e-16	4.8e-10	3.2e-16

Dormand-Prince of orders 5 and 8 (ODE5 and ODE8, respectively). All these methods have been selected only for illustrative reasons: see, for example, [2, 35, 55] for more complex algorithms employing exact higher order Lie derivatives and a further detailed comparison between them.

First, the derivatives $y^{(i)}(t_0),\ i = 1, \ldots, N$, have been computed for each test problem using (13). After that, the Taylor expansion has been reconstructed:

$$y_{TIC,N}(t) = y_0 + \sum_{i=1}^{N} \frac{y^{(i)}(t_0)}{i!}(t - t_0)^i. \tag{19}$$

Then, the sequence of points $t_i = t_0 + h \cdot i,\ i = 0, \ldots, N_{trials}$, has been generated, where $N_{trials} = (t_f - t_0)/h$, $t_f = 1.2$, and h is the finite stepsize that was set equal to 10^{-3}. At each point t_i, the values $y_{TIC,N}(t_i)$ have been calculated using $N = 4, 5, 8, 15$ (let us call the respective methods as TIC4, TIC5, TIC8, and TIC15, hereinafter). On the grid t_i, the methods ODE4, ODE5, and ODE8 have been also applied with the same initial condition (t_0, y_0). The values produced by these methods at each point t_i are denoted as $y_{ODE,N}(t_i)$, where $N = 4, 5, 8$ for the methods ODE4, ODE5, and ODE8, respectively. Finally, relative errors $E_{method}(t_i) = \frac{|y_{method}(t_i) - y_{exact}(t_i)|}{|y_{exact}(t_i)}|$ have been calculated at the points $t_i = 1.05, 1.1, 1.15, 1.2$ for each $method \in \{TIC4, TIC5, TIC8, TIC15, ODE4, ODE5, ODE8\}$. The results are presented in Table 3, where it is possible to observe that the error produced by the TIC4, TIC5, and TIC8 methods increases as the values of t_i. increase. However, for small time intervals, i.e., at the point $t = 1.05$, the error produced by TIC8 is the best among all methods, except for the last problem, where it was higher than the errors of ODE8 and TIC15. In general, the numerical methods ODE4, ODE5, and ODE8 yield less errors than the TIC methods of the same order (the errors of ODE4 are better than the errors of TIC4, the errors of ODE5 are better than the ones of TIC5, etc.). It is evident that more sophisticated than TIC methods should be exploited, when the search interval is large and/or just a few number of derivatives can be calculated. In this case, TIC methods are inappropriate, because other numerical methods of the same order are more "stable" (see Table 3).

It should be emphasized that the application of TIC methods is appropriate when higher order derivatives of the solution $y(t)$ can be calculated efficiently at the starting point t_0. In this case, one can use a high-order TIC method (e.g., order 15), for which the complexity is always constant: only a fixed number of derivatives should be calculated at the initial point. The so-obtained derivatives can be used to compute the Taylor series at any desired point t without re-evaluating them. Traditionally, it is difficult to built a numerical method of a high order (e.g., of order 15); therefore lower order methods are typically used with a smaller stepsize, requiring more computational resources for their implementation. As reported in Table 3, the method TIC15 gives the best error for all test problems, with the exception of the last one at $t = 1.2$, where the error is slightly greater than the machine precision. In this case, the TIC method generates a polynomial approximation that is continuously differentiable up to 15-th order. It may also be computed at any desired time t without having to re-evaluate the derivatives or system functions.

5 Conclusion

This chapter proposed the Simulink-based software solution to the Infinity Computer (SSIC), which allows one to use the Infinity Computer arithmetic within the Simulink environment for studying and analyzing dynamic systems. SSIC exploits the Infinity Computer software simulator, which was developed in $C++$ according to the patents (see [44]) and integrated in the Simulink environment through MEX-files for low-level functionalities. The proposed solution is user-friendly, general purpose, and domain independent, i.e., it can be used in any domain where high precision computations are required.

SSIC provides four functional modules, i.e., *Arithmetic Blocks Module (ABM)*, *Elementary Blocks Module (EBM)*, *Utility Blocks Module (UBM)*, and *Differentiation Blocks Module (DBM)*. The blocks provided by each module can be used in the same way as their internal Simulink equivalents; thus, no additional tools or sophisticated techniques are required.

The solution has been evaluated through three series of numerical experiments involving high order differentiation. It has been shown that the complexity of implementing functions using the proposed solution and traditional Simulink blocks is the same, whereas SSIC allows one to use the potentiality of the Infinity Computer in Simulink without having to refer to the Infinity Computer's low-level procedures.

Future research efforts will be directed to: (i) improve and extend the proposed solution to support a wider set of concepts and operations offered by the Infinity Computer; (ii) conduct experiments with the solution in different research domains.

References

1. IEEE Standard for Binary Floating-Point Arithmetic: ANSI/IEEE Std. **754–1985**, 1–20 (1985). https://doi.org/10.1109/IEEESTD.1985.82928
2. Amodio, P., Iavernaro, F., Mazzia, F., Mukhametzhanov, M.S., Sergeyev, Y.D.: A generalized Taylor method of order three for the solution of initial value problems in standard and infinity floating-point arithmetic. Math. Comput. Simul. **141**, 24–39 (2017)
3. Astorino, A., Fuduli, A.: Spherical separation with infinitely far center. Soft Comput. **24**(23), 17751–17759 (2020)
4. Baldwin, C.Y., Clark, K.B.: Modularity in the Design of Complex Engineering Systems, pp. 175–205. Springer, Berlin, Heidelberg (2006). https://doi.org/10.1007/3-540-32834-3_9
5. Bocciarelli, P., D'Ambrogio, A., Falcone, A., Garro, A., Giglio, A.: A model-driven approach to enable the simulation of complex systems on distributed architectures.

SIMULATION: Trans. Soc. Model. Simul. Int. **95**(12) (2018). https://doi.org/10.1177/0037549719829828
6. Bouskela, D., Falcone, A., Garro, A., Jardin, A., Otter, M., Thuy, N., Tundis, A.: Formal requirements modeling for cyber-physical systems engineering: an integrated solution based on form-l and modelica. Requirements Engineering (2021). https://doi.org/10.1007/s00766-021-00359-z
7. Caldarola, F.: The exact measures of the Sierpinski d-dimensional tetrahedron in connection with a diophantine nonlinear system. Commun. Nonlinear Sci. Numer. Simul. **63**, 228–238 (2018)
8. Caldarola, F., Maiolo, M.: On the topological convergence of multi-rule sequences of sets and fractal patterns. Soft Comput. **24**(23), 17737–17749 (2020)
9. Calude, C.S., Dumitrescu, M.: Infinitesimal probabilities based on grossone. SN Comput. Sci. (2020). https://doi.org/10.1007/s42979-019-0042-8
10. Cococcioni, M., Cudazzo, A., Pappalardo, M., Sergeyev, Y.D.: Solving the lexicographic multi-objective mixed-integer linear programming problem using branch-and-bound and grossone methodology. Commun. Nonlinear Sci. Numer. Simul. **84**, 105177 (2020). https://doi.org/10.1016/j.cnsns.2020.105177
11. Cococcioni, M., Pappalardo, M., Sergeyev, Y.D.: Lexicographic multi-objective linear programming using grossone methodology: theory and algorithm. Appl. Math. Comput. **318**, 298–311 (2018). https://doi.org/10.1016/j.amc.2017.05.058
12. D'Alotto, L.: Cellular automata using infinite computations. Appl. Math. Comput. **218**(16), 8077–8082 (2012)
13. D'Alotto, L.: Infinite games on finite graphs using grossone. Soft Comput. **24**(23), 17509–17515 (2020)
14. D'Ambrogio, A., Falcone, A., Garro, A., Giglio, A.: Enabling reactive streams in HLA-based simulations through a model-driven solution. In: 23rd IEEE/ACM International Symposium on Distributed Simulation and Real Time Applications, DS-RT 2019, Cosenza, Italy, October 7-9, 2019, pp. 1–8. Institute of Electrical and Electronics Engineers Inc. (2019). https://doi.org/10.1109/DS-RT47707.2019.8958697
15. De Leone, R.: Nonlinear programming and grossone: quadratic programming and the role of constraint qualifications. Appl. Math. Comput. **318**, 290–297 (2018)
16. De Leone, R., Egidi, N., Fatone, L.: The use of grossone in elastic net regularization and sparse support vector machines. Soft Comput. **24**(23), 17669–17677 (2020)
17. De Leone, R., Fasano, G., Sergeyev, Y.D.: Planar methods and grossone for the conjugate gradient breakdown in nonlinear programming. Comput. Optim. Appl. **71**(1), 73–93 (2018)
18. Falcone, A., Garro, A.: Using the HLA standard in the context of an international simulation project: The experience of the "smashteam". In: 15th International Conference on Modeling and Applied Simulation, MAS 2016, Held at the International Multidisciplinary Modeling and Simulation Multiconference, I3M 2016, Larnaca, Cyprus, September 26-28, 2016, pp. 121–129. Dime University of Genoa (2016)
19. Falcone, A., Garro, A.: A java library for easing the distributed simulation of space systems. In: 16th International Conference on Modeling and Applied Simulation, MAS 2017, Held at the International Multidisciplinary Modeling and Simulation Multiconference, I3M 2017, Barcelona, Spain, September 18-20, 2017, pp. 6–13. CAL-TEK S.r.l. (2017)
20. Falcone, A., Garro, A.: Reactive HLA-based distributed simulation systems with RxHLA. In: 22nd IEEE/ACM International Symposium on Distributed Simulation and Real Time Applications, DS-RT 2018, Madrid, Spain, October 15-17, 2018, pp.

1–8. Institute of Electrical and Electronics Engineers Inc. (2018). https://doi.org/10.1109/DISTRA.2018.8600936
21. Falcone, A., Garro, A.: Distributed co-simulation of complex engineered systems by combining the high level architecture and functional mock-up interface. Simul. Model. Pract. Theory **97**(August), 101967 (2019). https://doi.org/10.1016/j.simpat.2019.101967
22. Falcone, A., Garro, A., Anagnostou, A., Taylor, S.J.E.: An introduction to developing federations with the high level architecture (HLA). In: 2017 Winter Simulation Conference, WSC 2017, Las Vegas, NV, USA, December 3-6, 2017, pp. 617–631. Institute of Electrical and Electronics Engineers Inc. (2017). https://doi.org/10.1109/WSC.2017.8247820
23. Falcone, A., Garro, A., D'Ambrogio, A., Giglio, A.: Engineering systems by combining BPMN and HLA-based distributed simulation. In: 2017 IEEE International Conference on Systems Engineering Symposium, ISSE 2017, Vienna, Austria, October 11-13, 2017, pp. 1–6. Institute of Electrical and Electronics Engineers Inc. (2017). https://doi.org/10.1109/SysEng.2017.8088302
24. Falcone, A., Garro, A., D'Ambrogio, A., Giglio, A.: Using BPMN and HLA for engineering sos: lessons learned and future directions. In: 2018 IEEE International Conference on Systems Engineering Symposium, ISSE 2018, Rome, Italy, October 1-3, 2018, pp. 1–8. Institute of Electrical and Electronics Engineers Inc. (2018). https://doi.org/10.1109/SysEng.2018.8544399
25. Falcone, A., Garro, A., Mukhametzhanov, M.S., Sergeyev, Y.D.: Representation of grossone-based arithmetic in simulink for scientific computing. Soft Comput. **24**(23), 17525–17539 (2020). https://doi.org/10.1007/s00500-020-05221-y
26. Falcone, A., Garro, A., Mukhametzhanov, M.S., Sergeyev, Y.D.: A simulink-based infinity computer simulator and some applications. In: 3rd International Conference and Summer School 'Numerical Computations: Theory and Algorithms', NUMTA 2019, Le Castella, Crotone, Italy, June 15-21, 2019, pp. 362–369. Springer Nature Switzerland AG (2020). https://doi.org/10.1007/978-3-030-40616-5_31
27. Falcone, A., Garro, A., Mukhametzhanov, M.S., Sergeyev, Y.D.: A simulink-based software solution using the infinity computer methodology for higher order differentiation. Appl. Math. Comput. 125606 (2021). https://doi.org/10.1016/j.amc.2020.125606
28. Falcone, A., Garro, A., Taylor, S.J.E., Anagnostou, A.: Simplifying the development of HLA-based distributed simulations with the HLA development kit software framework (DKF). In: 21st IEEE/ACM International Symposium on Distributed Simulation and Real Time Applications, DS-RT 2017, Rome, Italy, October 18-20, 2017, pp. 216–217 (2017). https://doi.org/10.1109/DISTRA.2017.8167691
29. Falcone, A., Garro, A., Tundis, A.: Modeling and simulation for the performance evaluation of the on-board communication system of a metro train. In: 13th International Conference on Modeling and Applied Simulation, MAS 2014, Held at the International Multidisciplinary Modeling and Simulation Multiconference, I3M 2014, Bordeaux, France, September 10-12, 2014, pp. 20–29. Dime University of Genoa (2014)
30. Fiaschi, L., Cococcioni, M.: Numerical asymptotic results in game theory using Sergeyev's Infinity Computing. Int. J. Unconvent. Comput. **14**(1), 1–25 (2018)
31. Gangle, R., Caterina, G., Tohmé, F.: A constructive sequence algebra for the calculus of indications. Soft Comput. **24**(23), 17621–17629 (2020)

32. Garro, A., Falcone, A., Chaudhry, N.R., Salah, O., Anagnostou, A., Taylor, S.J.E.: A prototype HLA development kit: Results from the 2015 simulation exploration experience. In: 3rd ACM Conference on SIGSIM-Principles of Advanced Discrete Simulation, ACM SIGSIM PADS 2015, London, United Kingdom, June 10-12, 2015, pp. 45–46. Association for Computing Machinery Inc. (2015). https://doi.org/10.1145/2769458.2769489
33. Garro, A., Falcone, A., D'Ambrogio, A., Giglio, A.: A model-driven method to enable the distributed simulation of BPMN models. In: 27th IEEE International Conference on Enabling Technologies: Infrastructure for Collaborative Enterprises, WETICE 2018, Paris, France, June 27-29, 2018, pp. 121–126. Institute of Electrical and Electronics Engineers Inc. (2018). https://doi.org/10.1109/WETICE.2018.00030
34. Gaudioso, M., Giallombardo, G., Mukhametzhanov, M.: Numerical infinitesimals in a variable metric method for convex nonsmooth optimization. Appl. Math. Comput. **318**, 312–320 (2018)
35. Iavernaro, F., Mazzia, F., Mukhametzhanov, M., Sergeyev, Y.D.: Conjugate-symplecticity properties of Euler-Maclaurin methods and their implementation on the infinity computer. Appl. Numer. Math. **155**, 58–72 (2020). https://doi.org/10.1016/j.apnum.2019.06.011
36. Iavernaro, F., Mazzia, F., Mukhametzhanov, M.S., Sergeyev, Y.D.: Computation of higher order lie derivatives on the infinity computer. J. Comput. Appl. Math. **383**, 113135 (2021)
37. MathWorks: Simulink home page (2019). https://www.mathworks.com/products/simulink.html. Accessed 03 Dec 2019
38. Möller, B., Garro, A., Falcone, A., Crues, E.Z., Dexter, D.E.: Promoting a-priori interoperability of HLA-based simulations in the space domain: The SISO space reference FOM initiative. In: 20th IEEE/ACM International Symposium on Distributed Simulation and Real Time Applications, DS-RT 2016, London, UK, September 21-23, 2016, pp. 100–107. Institute of Electrical and Electronics Engineers Inc. (2016). https://doi.org/10.1109/DS-RT.2016.15
39. Möller, B., Garro, A., Falcone, A., Crues, E.Z., Dexter, D.E.: On the execution control of HLA federations using the SISO space reference FOM. In: 21st IEEE/ACM International Symposium on Distributed Simulation and Real Time Applications, DS-RT 2017, Rome, Italy, October 18-20, 2017, pp. 75–82. Institute of Electrical and Electronics Engineers Inc. (2017). https://doi.org/10.1109/DISTRA.2017.8167669
40. Mukhametzhanov, M.S., Sergeyev, Y.D.: The infinity computer vs. symbolic computations: First steps in comparison. In: AIP Conference Proceedings, vol. 2293, p. 420045. AIP Publishing LLC (2020)
41. Pepelyshev, A., Zhigljavsky, A.: Discrete uniform and binomial distributions with infinite support. Soft Comput. **24**(23), 17517–17524 (2020)
42. Pintér, J.D.: Global optimization: software, test problems, and applications. In: Pardalos, P.M., Romeijn, H.E. (eds.) Handbook of Global Optimization, vol. 2, pp. 515–569. Kluwer Academic Publishers, Dordrecht (2002)
43. Rizza, D.: Numerical methods for infinite decision-making processes. Int. J. Unconvent. Comput. **14**(2), 139–158 (2019)
44. Sergeyev, Y.D.: Computer system for storing infinite, infinitesimal, and finite quantities and executing arithmetical operations with them. USA patent 7,860,914 (2010), EU patent 1728149 (2009), RF patent 2395111 (2010)
45. Sergeyev, Y.D.: Arithmetic of Infinity. Edizioni Orizzonti Meridionali, CS (2003, 2nd ed. 2013)

46. Sergeyev, Y.D.: Higher order numerical differentiation on the infinity computer. Optim. Lett. **5**(4), 575–585 (2011)
47. Sergeyev, Y.D.: Using blinking fractals for mathematical modelling of processes of growth in biological systems. Informatica **22**(4), 559–576 (2011)
48. Sergeyev, Y.D.: Solving ordinary differential equations by working with infinitesimals numerically on the infinity computer. Appl. Math. Comput. **219**(22), 10668–10681 (2013)
49. Sergeyev, Y.D.: The exact (up to infinitesimals) infinite perimeter of the Koch snowflake and its finite area. Commun. Nonlinear Sci. Numer. Simul. **31**(1–3), 21–29 (2016)
50. Sergeyev, Y.D.: Numerical infinities and infinitesimals: methodology, applications, and repercussions on two Hilbert problems. EMS Surv. Math. Sci. **4**, 219–320 (2017)
51. Sergeyev, Y.D.: Independence of the grossone-based infinity methodology from non-standard analysis and comments upon logical fallacies in some texts asserting the opposite. Found. Sci. **24**(1), 153–170 (2019)
52. Sergeyev, Y.D., Garro, A.: Single-tape and multi-tape Turing machines through the lens of the Grossone methodology. J. Supercomput. **65**(2), 645–663 (2013)
53. Sergeyev, Y.D., Kvasov, D.E., Mukhametzhanov, M.S.: On strong homogeneity of a class of global optimization algorithms working with infinite and infinitesimal scales. Commun. Nonlinear Sci. Numer. Simul. **59**, 319–330 (2018)
54. Sergeyev, Y.D., Mukhametzhanov, M.S., Kvasov, D.E., Lera, D.: Derivative-free local tuning and local improvement techniques embedded in the univariate global optimization. J. Optim. Theory Appl. **171**(1), 186–208 (2016)
55. Sergeyev, Y.D., Mukhametzhanov, M.S., Mazzia, F., Iavernaro, F., Amodio, P.: Numerical methods for solving initial value problems on the Infinity Computer. Int. J. Unconvent. Comput. **12**(1), 3–23 (2016)
56. Weinstein, M., Rao, A.: Algorithm 984: Adigator, a toolbox for the algorithmic differentiation of mathematical functions in matlab using source transformation via operator overloading. ACM Trans. Math. Softw. **44**(2) (2017)
57. Zhigljavsky, A.: Computing sums of conditionally convergent and divergent series using the concept of grossone. Appl. Math. Comput. **218**(16), 8064–8076 (2012)
58. Žilinskas, A.: On strong homogeneity of two global optimization algorithms based on statistical models of multimodal objective functions. Appl. Math. Comput. **218**(16), 8131–8136 (2012)

Addressing Ill-Conditioning in Global Optimization Using a Software Implementation of the Infinity Computer

Marat S. Mukhametzhanov and Dmitri E. Kvasov

Abstract The present chapter studies the impact of scaling on global optimization algorithms. In particular, the notion of *strong homogeneity* is under study. A method is strongly homogeneous if it produces the same sequences of evaluation points independently both of multiplication of the objective function by a scaling constant and of adding a shifting constant. It is shown that even if a method possesses this property theoretically, numerically very small and large scaling constants can lead to ill-conditioning of the scaled problem. A new class of global optimization problems where the objective function can have not only finite but also infinite or infinitesimal Lipschitz constants is described. The strong homogeneity of several Lipschitz global optimization algorithms is then addressed within the Infinity Computing framework. It is finally shown that the usage of numerical infinities and infinitesimals can in certain cases avoid ill-conditioning produced by scaling.

1 Introduction

Global optimization problems are considered in this chapter. One of the desirable properties of numerical global optimization methods (see [5, 19, 62, 67, 74]) is their *strong homogeneity*: this means that a method produces the same sequences of trial points (i.e., points where the objective function $f(x)$ is evaluated) independently of both shifting $f(x)$ vertically and its multiplication by a scaling constant. Therefore, the optimization of a scaled function

$$g(x) = g(x; \alpha, \beta) = \alpha f(x) + \beta, \qquad \alpha > 0, \tag{1}$$

M. S. Mukhametzhanov · D. E. Kvasov (✉)
University of Calabria, Rende (CS), Italy
e-mail: m.mukhametzhanov@dimes.unical.it

D. E. Kvasov
Lobachevsky University of Nizhny Novgorod, Nizhny Novgorod, Russia
e-mail: kvadim@dimes.unical.it

Y. D. Sergeyev and R. De Leone (eds.), *Numerical Infinities and Infinitesimals in Optimization*, Emergence, Complexity and Computation 43,
https://doi.org/10.1007/978-3-030-93642-6_14

instead of the original objective function $f(x)$ can be useful. The concept of strong homogeneity was introduced in [74] where it was shown that both the P-algorithm (see [73]) and the one-step Bayesian algorithm (see [42]) are *strongly homogeneous*. The case $\alpha = 1, \beta \neq 0$ was considered in [19, 67] where a number of *homogeneous* methods enjoying this property were studied. Therefore, there exist global optimization methods that are homogeneous or strongly homogeneous and algorithms (see, for example, the DIRECT algorithm from [35] and a huge number of its modifications as surveyed in [34, 35]) that do not possess this useful property.

The above-mentioned methods were developed for solving practically important Lipschitz global optimization problems (see, e.g., [3, 27, 39, 43, 45, 55, 59, 67–69, 71, 72]) and belong to the class of "Divide-the-Best" algorithms as from [48] (see also [31, 59]). The so-called geometric and information algorithms from this class are considered in this chapter (see [59, 66–68]). The first type of algorithms is based on a geometrical interpretation of the Lipschitz condition and originates from [46] where a piecewise linear minorant for the objective function was constructed using the Lipschitz condition. The second approach is based on the information-statistical algorithm proposed in [67] and uses a stochastic model to calculate probabilities of locating global minimizers within subregions of the search domain (for other rich ideas in stochastic global optimization see [71, 72, 74]). In both the cases, different strategies to estimate global and local Lipschitz constants can be used (see, e.g., [46, 59, 63, 65–68]).

In this chapter, following the paper [62] (see also [37]) it will be shown that several univariate methods using the local tuning on the objective function behaviour (as surveyed, e.g., in [59, 63, 65]) enjoy the strong homogeneity property. Moreover, it will be shown that this property is valid for the considered methods not only for finite values of the constants α and β from (1) but for infinite and infinitesimal ones, as well. A class of global optimization problems with the objective function having infinite or infinitesimal Lipschitz constants is also described as in [62]. Numerical computations with functions that can assume infinite and infinitesimal values are executed using the Infinity Computing paradigm: its comprehensive description can be found in [49, 53] (it should be stressed that it is not-related to non-standard analysis, see [54]). This computational methodology has already been successfully applied in optimization and numerical differentiation [4, 9–11, 14–18, 20–22, 25, 32, 38, 51, 65] and in a number of other theoretical and applied research areas such as, e.g., cellular automata [12, 13], hyperbolic geometry [41], percolation [33], fractals [2, 6, 7, 52], infinite series [50, 70], probability theory [8, 44], logic [24, 40, 47], Turing machines [56, 57], numerical solution of ordinary differential equations [1, 64], etc. In particular, in the paper [25], grossone-based numerical infinities and infinitesimals were successfully used to handle ill-conditioning in a multidimensional optimization problem.

The importance to have the possibility to work with infinite and infinitesimal scaling/shifting constants α and β has an additional value due to the following fact (see [62]). It can happen that when a (theoretically) strongly homogeneous method is applied to the function $g(x)$ from (1), numerically very small and/or large finite constants α and β can lead to the ill-conditioning of the corresponding global optimization problem due to overflow and/or underflow during the construction of $g(x)$ from $f(x)$. Thus, global minimizers can change their locations and the values of global minima can change, as well. As a result, applying even strong homogeneous methods to solve these problems will lead to finding the changed values of minima related to $g(x)$ and not the desired global solution to the original function $f(x)$ we are interested in. As shown in [62], numerical infinities and infinitesimals and the Infinity Computing framework can help in this situation.

The rest of this chapter is structured as follows. Section 2 states the problem formally and discusses ill-conditioning induced by scaling. It is stressed that the introduction of numerical infinities and infinitesimals allows us to consider a new class of functions having infinite or infinitesimal Lipschitz constants. Section 3 presents geometric and information Lipschitz global optimization algorithms studied in this chapter and shows how an adaptive estimation of global and local Lipschitz constants can be performed. So far, the fact whether these methods are strongly homogeneous or not was an open problem even for finite constants α and β until [62]. Section 4 shows that these methods enjoy the strong homogeneity property for finite, infinite, and infinitesimal scaling and shifting constants. Section 5 describes how in certain cases the usage of numerical infinities and infinitesimals can avoid ill-conditioning produced by scaling and illustrates these results numerically. Finally, a brief conclusion is given in Sect. 6.

2 Problem Statement and Ill-Conditioning Induced by Scaling

2.1 Lipschitz Global Optimization and Ill-Conditioning Produced by Scaling

Let us consider a univariate global optimization problem where it is required to find the global minimum f^* and global minimizers x^* such that

$$f^* = f(x^*) = \min f(x), \qquad x \in D = [a, b] \subset \mathbb{R}. \tag{2}$$

It is assumed that the objective function $f(x)$ is multiextremal, non-differentiable, and Lipschitz continuous over the interval D, i.e.,

$$|f(x_1) - f(x_2)| \leq L|x_1 - x_2|, \qquad x_1, x_2 \in D, \tag{3}$$

where L is the Lipschitz constant, $0 < L < \infty$ (see, e.g., [23, 28, 30, 36, 43, 58, 59, 61, 66, 68, 72]). In many situations, it can be required to optimize a scaled function $g(x)$ from (1) instead of the original objective function $f(x)$ (see, e.g., [19, 67, 74]), thus giving rise to the study of strong homogeneity for global optimization algorithms (as introduced in Sect. 1).

As an illustration, let us consider the Pintér's class of univariate test problems from [45]. Each function $f_s(x)$, $1 \leq s \leq 100$, of this class is defined over the interval $[-5, 5]$ and has the following form:

$$f_s(x) = 0.025(x - x_s^*)^2 + \sin^2[(x - x_s^*) + (x - x_s^*)^2] + \sin^2(x - x_s^*), \quad (4)$$

where the global minimum f^* is equal to 0 for all the functions and the global minimizer x_s^*, $1 \leq s \leq 100$, is chosen randomly (and differently for each test function of the class) from the interval $[-5, 5]$ (we used the random number generator implemented in the GKLS-generator of multidimensional test classes from [26, 60]).

For example, let us consider the first test function $f_1(x)$ of the class (4) with $x_1^* = -3.9711481954$ and take $\alpha = 10^{-8}$ and $\beta = 10^8$, thus obtaining the scaled function

$$\hat{g}_1(x) = 10^{-8} f_1(x) + 10^8. \quad (5)$$

As can be seen in Fig. 1a and b, the functions $f_1(x)$ and $\hat{g}_1(x)$ have completely different minimizers due to the used scaling constants in $\hat{g}_1$. As a result, if we wish to invert the function $\hat{g}_1(x)$ trying to establish the original function $f_1(x)$, i.e., to compute the function $\hat{f}_1(x) = (\hat{g}_1(x) - 10^8)/10^{-8}$, then it will not coincide with $f_1(x)$. Figure 1c shows $\hat{f}_1(x)$ constructed from $\hat{g}_1(x)$ using piecewise linear approximations with step $h = 0.001$. Due to underflows taking place in commonly used numeral systems, the function $\hat{g}_1(x)$ degenerates in constant functions over several sub-intervals and many local minimizers disappear (see, e.g., local minimizers from the interval $[-3, -2]$ in Fig. 1a, b). On the other hand, due to overflows, several local minimizers become global ones for the scaled function $\hat{g}_1(x)$. In particular, one can find from Fig. 1b, c the following global solutions to the problems with the objective functions $\hat{g}_1(x)$ and $\hat{f}_1(x)$: $(x^*, \hat{g}_1^*) = (-5, 10^8)$ and $(x^*, \hat{f}_1^*) = (-5, 0)$, respectively, while it can be seen from Fig. 1a that the point $x = -5$ is not even a local minimizer.

This simple example shows that it does not make sense to talk about the strong homogeneity property in case of very huge or very small finite values of constants α and β in (1), since it is not possible to construct correctly the corresponding scaled functions on the traditional computers. Within the Infinity Computing paradigm not only finite, but also numerical infinite and infinitesimal values of α and β can be adopted and ill-conditioning of global optimization problems can be avoided in certain cases.

2.2 Functions with Infinite and Infinitesimal Lipschitz Constants

The introduction of the Infinity Computer paradigm allows us to consider univariate global optimization problems with the objective function $g(x)$ from (1) that can assume not only finite values, but also infinite and infinitesimal ones. Let us suppose

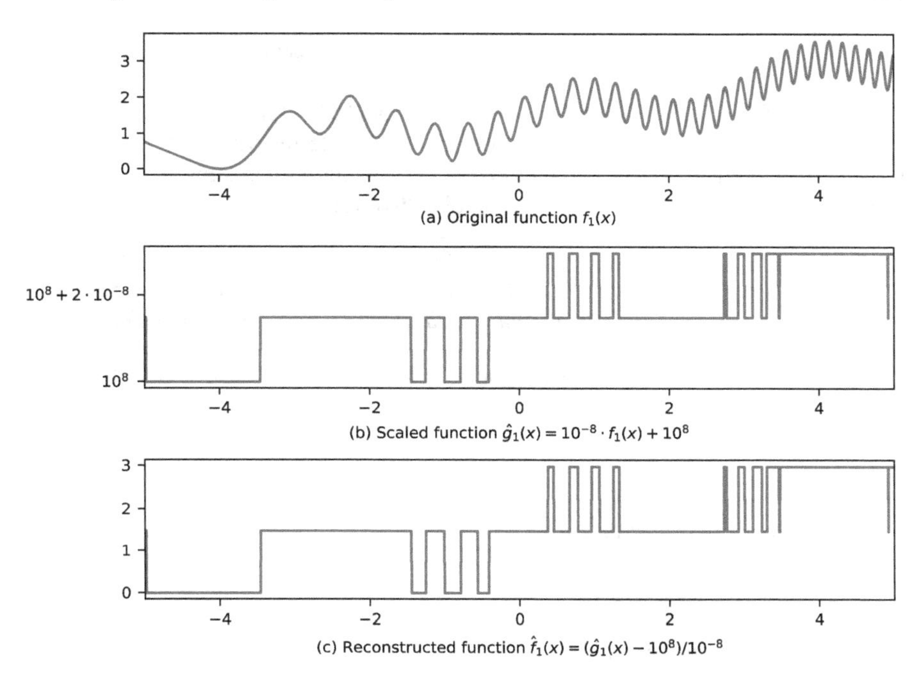

Fig. 1 Graphs of **a** the test function (4), **b** the scaled function $\hat{g}_1(x)$ from (5), **c** the inverted scaled function $\hat{f}_1(x) = (\hat{g}_1(x) - 10^8)/10^{-8}$. The form of the functions $\hat{g}_1(x)$ and $\hat{f}_1(x)$ is qualitatively different with respect to that of the original function $f_1(x)$ due to overflows and underflows

that the original function $f(x)$ assumes finite values only and satisfies condition (3) with a finite constant L. Since in (1) the scaling/shifting parameters α and β can be not only finite but also infinite and infinitesimal, the Infinity Computing framework is required to work with $g(x)$. Thus, the following optimization problem is considered

$$\min g(x) = \min \ (\alpha f(x) + \beta), \quad x \in D = [a, b] \subset \mathbb{R}, \quad \alpha > 0, \tag{6}$$

where the function $f(x)$ can be multiextremal, non-differentiable, and Lipschitz continuous with a finite value of the Lipschitz constant L from (3). In their turn, the values α and β can be finite, infinite, and infinitesimal numbers representable in the grossone-based numeral system.

The finiteness of the original Lipschitz constant L from (3) is the essence of the Lipschitz condition allowing people to construct optimization methods for traditional computers. The scaled objective function $g(x)$ from (6) can assume not only finite, but also infinite and infinitesimal values and, therefore, in these cases it is not Lipschitzian in the traditional sense. However, the Infinity Computer paradigm extends the space of functions that can be treated theoretically and numerically to functions assuming infinite and infinitesimal values. This fact allows us to extend the concept of Lipschitz functions to the cases where the Lipschitz constant can assume infinite/infinitesimal values.

The rest of the chapter will follow the results from [62]. Hereafter, we will indicate by "$\widehat{\ }$" all the values related to the function $g(x)$ while the terms without "$\widehat{\ }$" will indicate the values related to the function $f(x)$. The following lemma shows an important property of the Lipschitz constant for the objective function $g(x)$.

Lemma 1 *The Lipschitz constant $\widehat{L}$ of the function $g(x) = \alpha f(x) + \beta$, where $f(x)$ assumes only finite values and has the finite Lipschitz constant L over the interval $[a, b]$ and α, $\alpha > 0$, and β can be finite, infinite, and infinitesimal, is equal to αL.*

Proof The following relation can be obtained from the definition of $g(x)$ and the fact that $\alpha > 0$

$$|g(x_1) - g(x_2)| = \alpha|f(x_1) - f(x_2)|, \qquad x_1,\ x_2 \in [a, b].$$

Since L is the Lipschitz constant for $f(x)$, then

$$\alpha|f(x_1) - f(x_2)| \le \alpha L|x_1 - x_2| = \widehat{L}|x_1 - x_2|, \qquad x_1,\ x_2 \in [a, b],$$

and this inequality proves the lemma.

Thus, the new Lipschitz condition for the function $g(x)$ from (1), (6) can be written as

$$|g(x_1) - g(x_2)| \le \alpha L|x_1 - x_2| = \widehat{L}|x_1 - x_2|, \qquad x_1, x_2 \in D, \tag{7}$$

where the constant L from (3) is finite and the quantities α and $\widehat{L}$ can assume infinite and infinitesimal values. Notice that infinities and infinitesimals are expressed in ① numerals, and Lemma 1 describes the first property of this class.

Some geometric and information global optimization methods (see [45, 46, 59, 63, 66–68]) used for solving the traditional Lipschitz global optimization problem (2) are adopted hereinafter for solving the problem (6). A general scheme describing these methods is presented in the next section.

3 A General Scheme Describing Geometric and Information Algorithms

Methods studied in this chapter have a similar structure and belong to the class of "Divide-the-Best" global optimization algorithms from [48]. They can have the following differences in their computational schemes:

(i) Methods are either Geometric or Information (see [59, 67, 68] for detailed descriptions of these classes of methods);

(ii) Methods can use different approaches to estimate the Lipschitz constant: an a priori estimate, a global adaptive estimate, and local tuning techniques, for example, Maximum Local Tuning (MLT) and Maximum-Additive Local Tuning (MALT) (see [59, 63, 68] for detailed descriptions of these approaches).

The first difference, (i), consists of the choice of characteristics R_i [31] for subintervals $[x_{i-1}, x_i]$, $2 \le i \le k$, where the points x_i, $1 \le i \le k$, are called *trial points* and are points where the objective function $g(x)$ has been evaluated during previous iterations:

$$R_i = \begin{cases} \frac{z_i+z_{i-1}}{2} - l_i \frac{x_i - x_{i-1}}{2}, & \text{for geometric methods,} \\ 2(z_i + z_{i-1}) - l_i(x_i - x_{i-1}) - \frac{(z_i - z_{i-1})^2}{l_i(x_i - x_{i-1})}, & \text{for information methods,} \end{cases} \quad (8)$$

where $z_i = g(x_i)$ and l_i is an estimate of the Lipschitz constant for subinterval $[x_{i-1}, x_i]$, $2 \le i \le k$.

The second issue, (ii), is related to four different strategies used in this chapter to estimate the Lipschitz constant L. The first one is an a priori given estimate $\overline{L} > L$. The second way is to use an adaptive global estimate of the Lipschitz constant L during the search (the word *global* means that the same estimate is used for the whole region D). The global adaptive estimate $\overline{L}_k$ can be calculated as follows

$$\overline{L}_k = \begin{cases} r \cdot H^k, & \text{if } H^k > 0, \\ 1, & \text{otherwise,} \end{cases} \quad (9)$$

where $r > 0$ is a reliability parameter and

$$H^k = \max\{H_i : 2 \le i \le k\}, \quad (10)$$

$$H_i = \frac{|z_i - z_{i-1}|}{x_i - x_{i-1}}, \quad 2 \le i \le k. \quad (11)$$

Finally, the Maximum (MLT) and Maximum-Additive (MALT) local tuning techniques consist of estimating local Lipschitz constants l_i for each subinterval $[x_{i-1}, x_i]$, $2 \le i \le k$, as follows

$$l_i^{MLT} = \begin{cases} r \cdot \max\{\lambda_i, \gamma_i\}, & \text{if } H^k > 0, \\ 1, & \text{otherwise,} \end{cases} \quad (12)$$

$$l_i^{MALT} = \begin{cases} r \cdot \max\{H_i, \frac{\lambda_i+\gamma_i}{2}\}, & \text{if } H^k > 0, \\ 1, & \text{otherwise,} \end{cases} \quad (13)$$

where H_i is from (11), and λ_i and γ_i are calculated as follows

$$\lambda_i = \max\{H_{i-1}, H_i, H_{i+1}\}, \quad 2 \le i \le k, \quad (14)$$

$$\gamma_i = H^k \frac{(x_i - x_{i-1})}{X^{max}}, \tag{15}$$

with H^k from (10) and

$$X^{max} = \max\{x_i - x_{i-1} : 2 \le i \le k\}. \tag{16}$$

When $i = 2$ and $i = k$ only H_2, H_3, and H_{k-1}, H_k, should be considered, respectively, in (14).

After these preliminary descriptions we are ready to describe the General Scheme (GS) of algorithms studied in this chapter.

Step 0. *Initialization.* Execute first two trials at the points a and b, i.e., $x^1 := a$, $z^1 := g(a)$ and $x^2 := b$, $z^2 := g(b)$. Set the iteration counter $k := 2$. Suppose that $k \ge 2$ iterations of the algorithm have already been executed. The iteration $k + 1$ consists of the following steps.

Step 1. *Reordering.* Reorder the points $x^1, \ldots, x^k$ (and the corresponding function values $z^1, \ldots, z^k$) of previous trials by subscripts so that

$$a = x_1 < \ldots < x_k = b, \qquad z_i = g(x_i),\ 1 \le i \le k.$$

Step 2. *Estimates of the Lipschitz constant.* Calculate the current estimates l_i of the Lipschitz constant for each subinterval $[x_{i-1}, x_i]$, $2 \le i \le k$, in one of the following ways.

Step 2.1. *A priori given estimate.* Take an a priori given estimate $\overline{L}$ of the Lipschitz constant for the whole interval $[a, b]$, i.e., set $l_i := \overline{L}$.
Step 2.2. *Global estimate.* Set $l_i := \overline{L}_k$, where $\overline{L}_k$ is from (9).
Step 2.3. *"Maximum" local tuning.* Set $l_i := l_i^{MLT}$, where l_i^{MLT} is from (12).
Step 2.4. *"Maximum-Additive" local tuning.* Set $l_i := l_i^{MALT}$, where l_i^{MALT} is from (13).

Step 3. *Calculation of characteristics.* Compute for each subinterval $[x_{i-1}, x_i]$, $2 \le i \le k$, its characteristic R_i by using one of the following rules.

Step 3.1. *Geometric methods.*

$$R_i = \frac{z_i + z_{i-1}}{2} - l_i \frac{x_i - x_{i-1}}{2}. \tag{17}$$

Step 3.2. *Information methods.*

$$R_i = 2(z_i + z_{i-1}) - l_i(x_i - x_{i-1}) - \frac{(z_i - z_{i-1})^2}{l_i(x_i - x_{i-1})}. \tag{18}$$

Step 4. *Subinterval selection.* Determine an interval $[x_{t-1}, x_t]$, $t = t(k)$, for performing the next trial as follows

$$t = \min \arg \min_{2 \le i \le k} R_i. \tag{19}$$

Step 5. *Stopping rule*. **If**

$$x_t - x_{t-1} \le \varepsilon, \tag{20}$$

where $\varepsilon > 0$ is a given accuracy of the global search, **then Stop** and take as an estimate of the global minimum g^* the value $g_k^* = \min_{1 \le i \le k}\{z_i\}$ obtained at a point $x_k^* = \arg\min_{1 \le i \le k}\{z_i\}$.

Otherwise, go to Step 6.

Step 6. *New trial*. Execute the next trial $z^{k+1} := g(x^{k+1})$ at the point

$$x^{k+1} = \frac{x_t + x_{t-1}}{2} - \frac{z_t - z_{t-1}}{2l_t}. \tag{21}$$

Increase the iteration counter $k := k + 1$, and go to Step 1.

4 Strong Homogeneity of Algorithms Belonging to General Scheme with Finite, Infinite, and Infinitesimal Scaling and Shifting Constants

In this section, we address the strong homogeneity of the previously described algorithms both in the traditional and in the Infinity Computing frameworks. We show (as given in [62]) that methods belonging to the GS enjoy the strong homogeneity property for finite, infinite, and infinitesimal scaling and shifting constants. Recall that all the values related to the scaled function $g(x)$ are indicated by "$\widehat{\ }$" and the values related to the function $f(x)$ are written without "$\widehat{\ }$".

The following lemma establishes how the adaptive estimates of the Lipschitz constant $\widehat{\overline{L}}_k, \widehat{l}_i^{MLT}$, and $\widehat{l}_i^{MALT}$ that can assume finite, infinite, and infinitesimal values are related to the respective original estimates $\overline{L}_k$, l_i^{MLT}, and l_i^{MALT} that can be finite only.

Lemma 2 *Let us consider the function $g(x) = \alpha f(x) + \beta$, where $f(x)$ assumes only finite values and has a finite Lipschitz constant L over interval $[a, b]$ and α, $\alpha > 0$, and β can be finite, infinite and infinitesimal numbers. Then, the adaptive estimates $\widehat{\overline{L}}_k$, $\widehat{l}_i^{MLT}$ and $\widehat{l}_i^{MALT}$ from (9), (12) and (13) are equal to $\alpha\overline{L}_k$, αl_i^{MLT} and αl_i^{MALT}, respectively, if $H^k > 0$, and to 1, otherwise.*

Proof It follows from (11) that

$$\widehat{H}_i = \frac{|\widehat{z}_i - \widehat{z}_{i-1}|}{x_i - x_{i-1}} = \frac{\alpha|z_i - z_{i-1}|}{x_i - x_{i-1}} = \alpha H_i. \tag{22}$$

If $H^k \neq 0$, then $H^k = \max\limits_{2 \le i \le k} \frac{|z_i - z_{i-1}|}{x_i - x_{i-1}}$ and $H^k \ge H_i,\ 2 \le i \le k$. Thus, using (22) we obtain $\alpha H^k \ge \alpha H_i = \widehat{H}_i$, and, therefore, $\widehat{H}^k = \alpha H^k$ and from (9) it follows $\widehat{\overline{L}}_k = \alpha \overline{L}_k$. On the other hand, if $H^k = 0$, then both estimates for the functions $g(x)$ and $f(x)$ are equal to 1 (see (9)).

The same reasoning can be used to show the respective results for the local tuning techniques MLT and MALT (see (12) and (13))

$$\widehat{\lambda}_i = \max\{\widehat{H}_{i-1}, \widehat{H}_i, \widehat{H}_{i+1}\} = \alpha \max\{H_{i-1}, H_i, H_{i+1}\},$$

$$\widehat{\gamma}_i = \widehat{H}^k \frac{x_i - x_{i-1}}{X^{max}} = \alpha H^k \frac{x_i - x_{i-1}}{X^{max}} = \alpha \gamma_i,$$

$$\widehat{l}_i^{MLT} = \begin{cases} r \cdot \max\{\widehat{\lambda}_i, \widehat{\gamma}_i\}, \text{ if } \widehat{H}^k > 0, \\ 1, \qquad\qquad\qquad\quad \text{otherwise.} \end{cases}$$

$$\widehat{l}_i^{MALT} = \begin{cases} r \cdot \max\{\widehat{H}_i, \frac{\widehat{\lambda}_i + \widehat{\gamma}_i}{2}\}, \text{ if } \widehat{H}^k > 0, \\ 1, \qquad\qquad\qquad\qquad\quad \text{otherwise.} \end{cases}$$

Therefore, we can conclude that

$$\widehat{l}_i^{\{MLT,MALT\}} = \begin{cases} \alpha l_i^{\{MLT,MALT\}}, \text{ if } H^k > 0, \\ 1, \qquad\qquad\qquad\quad \text{otherwise.} \end{cases}$$

Lemma 3 *Suppose that characteristics $\widehat{R}_i$, $2 \le i \le k$, for the scaled objective function $g(x)$ are equal to an affine transformation of the characteristics R_i calculated for the original objective function $f(x)$*

$$\widehat{R}_i = \widehat{\alpha}_k R_i + \widehat{\beta}_k, \qquad 2 \le i \le k, \tag{23}$$

where the constants $\widehat{\alpha}_k$, $\widehat{\alpha}_k > 0$, and $\widehat{\beta}_k$ can be finite, infinite, or infinitesimal and possibly different for different iterations k. Then, the same interval $[x_{t-1}, x_t]$, $t = t(k)$, from (19) is selected at each iteration for the next subdivision during optimizing $f(x)$ and $g(x)$, i.e., $\widehat{t(k)} = t(k)$.

Proof Since, due to (19), $t = \arg\min_{2 \le i \le k} R_i$, then $R_t \le R_i$ and

$$\widehat{\alpha}_k R_t + \widehat{\beta}_k \le \widehat{\alpha}_k R_i + \widehat{\beta}_k, \qquad 2 \le i \le k.$$

Due to (23), this can be re-written as

$$\widehat{R}_t = \min_{2 \le i \le k} \widehat{R}_i = \widehat{\alpha}_k R_t + \widehat{\beta}_k.$$

Notice that if there are several values j such that $R_j = R_t$, then (see (19)) we have $t < j$, $j \neq t$, i.e., even in this situation it follows $\widehat{t(k)} = t(k)$. This observation concludes the proof.

The following Theorem shows that methods belonging to the GS enjoy the strong homogeneity property.

Theorem 1 *Algorithms belonging to the GS and applied for solving the problem (6) are strongly homogeneous for finite, infinite, and infinitesimal scales $\alpha > 0$ and β.*

Proof Two algorithms optimizing functions $f(x)$ and $g(x)$ will generate the same sequences of trials if the following conditions hold:

(i) The same interval $[x_{t-1}, x_t]$, $t = t(k)$, from (19) is selected at each iteration for the next subdivision during optimizing functions $f(x)$ and $g(x)$, i.e., it follows $\widehat{t(k)} = t(k)$.
(ii) The next trial at the selected interval $[x_{t-1}, x_t]$ is performed at the same point during optimizing functions $f(x)$ and $g(x)$, i.e., in (21) it follows $\widehat{x}^{k+1} = x^{k+1}$.

In order to prove assertions (i) and (ii), let us consider computational steps of the GS. For both functions, $f(x)$ and $g(x)$, Steps 0 and 1 of the GS work with the same interval $[a, b]$, do not depend on the objective function, and, as a result, do not influence (i) and (ii). Step 2 is a preparative one, it is responsible for estimating the Lipschitz constants for all the intervals $[x_{i-1}, x_i]$, $2 \leq i \leq k$ and was studied in Lemmas 1 and 2. Step 3 calculates characteristics of the intervals and, therefore, is directly related to the assertion (i). In order to prove it, we consider computations of characteristics $\widehat{R}_i$ for all possible cases of calculating estimates l_i during Step 2 and show that there always possible to indicate constants $\widehat{\alpha}_k$ and $\widehat{\beta}_k$ from Lemma 3.

Lemmas 1 and 2 show that for the a priori given finite Lipschitz constant L for the function $f(x)$ (see Step 2.1) it follows $\widehat{L} = \alpha L$. For the adaptive estimates of the Lipschitz constants for intervals $[x_{i-1}, x_i]$, $2 \leq i \leq k$, (see (9), (12), (13) and Steps 2.2–2.4 of the GS) we have $\widehat{l_i} = \alpha l_i$, if $H^k > 0$, and $\widehat{l_i} = l_i = 1$, otherwise (remind that the latter corresponds to the situation $z_i = z_1$, $1 \leq i \leq k$). Since Step 3 includes substeps defining information and geometric methods, then the following four combinations of methods with Lipschitz constant estimates computed at one of the substeps of Step 2 can take place:

(a) The value $\widehat{l_i} = \alpha l_i$ and the geometric method is used. From (17) we obtain

$$\widehat{R}_i = \frac{\widehat{z}_{i-1} + \widehat{z}_i}{2} - \widehat{l_i}\frac{x_i - x_{i-1}}{2} = \alpha(\frac{z_{i-1} + z_i}{2} - l_i\frac{x_i - x_{i-1}}{2}) + \beta = \alpha R_i + \beta.$$

Thus, in this case we have $\widehat{\alpha}_k = \alpha$ and $\widehat{\beta}_k = \beta$.
(b) The value $\widehat{l_i} = \alpha l_i$ and the information method is used. From (18) we get

$$\widehat{R}_i = 2(\widehat{z}_i + \widehat{z}_{i-1}) - \widehat{l}_i(x_i - x_{i-1}) - \frac{(\widehat{z}_i - \widehat{z}_{i-1})^2}{\widehat{l}_i(x_i - x_{i-1})} =$$

$$2\alpha(z_i + z_{i-1}) + 4\beta - \alpha l_i(x_i - x_{i-1}) - \frac{\alpha^2(z_i - z_{i-1})^2}{\alpha l_i(x_i - x_{i-1})} = \alpha R_i + 4\beta.$$

Therefore, in this case it follows $\widehat{\alpha}_k = \alpha$ and $\widehat{\beta}_k = 4\beta$.

(c) The value $\widehat{l}_i = l_i = 1$ and the geometric method is considered. Since in this case $z_i = z_1$, $1 \leq i \leq k$, then for the geometric method (see (17)) we have

$$\widehat{R}_i = \frac{\widehat{z}_{i-1} + \widehat{z}_i}{2} - \widehat{l}_i \frac{x_i - x_{i-1}}{2} = \widehat{z}_1 - \frac{x_i - x_{i-1}}{2} =$$

$$\alpha z_1 + \beta - \frac{x_i - x_{i-1}}{2} = R_i + \alpha z_1 - z_1 + \beta.$$

Thus, in this case we have $\widehat{\alpha}_k = 1$ and $\widehat{\beta}_k = z_1(\alpha - 1) + \beta$.

(d) The value $\widehat{l}_i = l_i = 1$ and the information method is used. Then, the characteristics (see (18)) are calculated as follows

$$\widehat{R}_i = 2(\widehat{z}_i + \widehat{z}_{i-1}) - \widehat{l}_i(x_i - x_{i-1}) - \frac{(\widehat{z}_i - \widehat{z}_{i-1})^2}{\widehat{l}_i(x_i - x_{i-1})} =$$

$$4\widehat{z}_1 - (x_i - x_{i-1}) = 4\alpha z_1 + 4\beta - (x_i - x_{i-1}) = R_i + 4\alpha z_1 - 4z_1 + 4\beta.$$

Therefore, in this case it follows $\widehat{\alpha}_k = 1$ and $\widehat{\beta}_k = 4(z_1(\alpha - 1) + \beta)$.

Let us show now that assertion (ii) also holds. Since for both the geometric and the information approaches the the same formula (21) for computing x^{k+1} is used, we should consider only two cases related to the estimates of the Lipschitz constant:

(a) If $\widehat{l}_t = \alpha l_t$, then it follows

$$\widehat{x}^{k+1} = \frac{x_t + x_{t-1}}{2} - \frac{\widehat{z}_t - \widehat{z}_{t-1}}{2\widehat{l}_t} = \frac{x_t + x_{t-1}}{2} - \frac{\alpha(z_t - z_{t-1})}{2\alpha l_t} = x^{k+1}.$$

(b) If $\widehat{l}_t = l_t = 1$, then $z_i = z_1$, $1 \leq i \leq k$, and we have

$$\widehat{x}^{k+1} = \frac{x_t + x_{t-1}}{2} - \frac{\widehat{z}_t - \widehat{z}_{t-1}}{2\widehat{l}_t} = \frac{x_t + x_{t-1}}{2} = x^{k+1}.$$

This result concludes the proof.

5 Numerical Illustrations

In order to illustrate the behavior of methods belonging to the GS in the Infinity Computer framework, the following four algorithms being examples of concrete implementations of the GS have been tested:

- **Geom-AL:** Geometric method with an a priori given overestimate of the Lipschitz constant. It is constructed by using Steps 2.1 and 3.1 in the GS.
- **Geom-GL:** Geometric method with the global estimate of the Lipschitz constant. It is formed by using Steps 2.2 and 3.1 in the GS.
- **Inf-GL:** Information method with the global estimate of the Lipschitz constant. It is formed by using Steps 2.2 and 3.2 in the GS.
- **Inf-LTMA:** Information method with the "Maximum-Additive" local tuning. It is built by applying Steps 2.4 and 3.2 in the GS.

The algorithm Geom-AL has one parameter, namely, an a priori given overestimate of the Lipschitz constant. In algorithms Inf-LTMA, Inf-GL, and Geom-GL, the Lipschitz constant is estimated during the search and the reliability parameter r is used. As in [62], the values of the Lipschitz constants of the functions $f(x)$ for the algorithm Geom-AL were taken from [36] (and multiplied by α for $g(x)$). The value of the parameter r for the Information algorithms Inf-LTMA and Inf-GL was set equal to 2.0, while for the geometric algorithm Geom-GL it was set equal to 1.1. The value $\epsilon = 10^{-4}(b-a)$ was used in the stopping criterion (20).

Recall that (see Sect. 2) huge or very small scaling/shifting constants can provoke ill-conditioning of the scaled function $g(x)$ in the traditional computational framework. In the Infinity Computing framework, the positional grossone-based numeral system allows us to avoid this ill-conditioning and to work safely with infinite and infinitesimal scaling/shifting constants if the respective grossdigits and grosspowers are not too large or too small. In order to illustrate this fact, two pairs of the values α and β were used in our experiments: $(\alpha_1, \beta_1) = (①^1, ①^{-1})$ and $(\alpha_2, \beta_2) = (①^{-2}, ①^5)$. The corresponding grossdigits and grosspowers involved in their grossone-based representation are, respectively: 1 and 1 for α_1; 1 and -1 for β_1; 1 and -2 for α_2; and 1 and 5 for β_2. These values do not provoke instability in numerical operations. Hereinafter, scaled functions constructed using constants (α_1, β_1) are indicated as $g(x)$ and functions using (α_2, β_2) are designated as $h(x)$.

The algorithms Geom-AL, Geom-GL, Inf-GL, and Geom-LTMA were tested on the Pintér's class of 100 univariate test problems and on the respective scaled functions $g(x)$ and $h(x)$ constructed from them. On all 100 test problems with infinite and infinitesimal constants (α_1, β_1) and (α_2, β_2), the results for the original functions $f(x)$ and for the scaled functions $g(x)$ and $h(x)$ coincided. To illustrate this fact, let us consider the first problem from the set of 100 tests (see Figs. 2a and 1a).

In Fig. 2b, the results for the scaled function

$$g_1(x) = ①^1 f_1(x) + ①^{-1},$$

are presented, and in Fig. 2c, the results for the scaled function

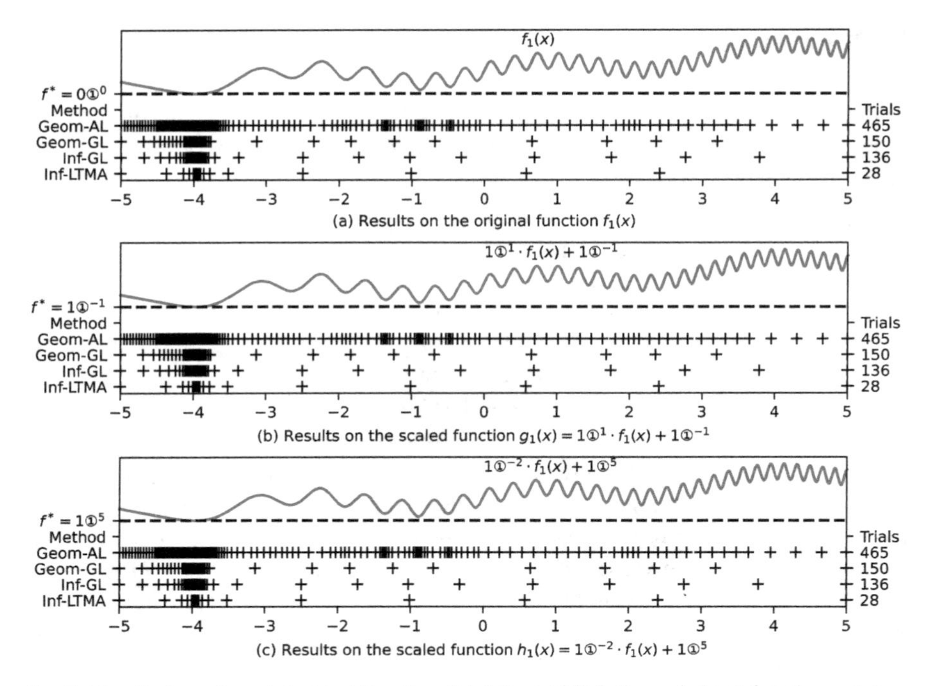

Fig. 2 Results for **a** the original test function $f_1(x)$ from [45], **b** the scaled test function $g_1(x) = ①^1 f_1(x) + ①^{-1}$, **c** the scaled test function $h_1(x) = ①^{-2} f_1(x) + ①^5$. Trial points are indicated by the signs "+" under the graphs of the functions and the number of trials for each method is indicated on the right. The results coincide for each method on all three test functions

$$h_1(x) = ①^{-2} f_1(x) + ①^5,$$

are shown. It can be seen that the results coincide for all four methods on all three test functions $f_1(x)$, $g_1(x)$, and $h_1(x)$.

Analogous results hold for the remaining test problems $f_i(x)$, $g_i(x)$, and $h_i(x)$, $i = 2, \ldots, 100$. In Fig. 3, the operational characteristics for each algorithm on each test class are presented. An operational characteristic, constructed on a class of 100 randomly generated test functions, is a graph showing the number of solved problems in dependence on the number of executed evaluations of the objective function (see [29]). Each test problem was considered to be solved if an algorithm generated a point in the ϵ-neighborhood of the global minimizer x^*. It can be seen from Fig. 3 that the operational characteristics of the algorithms perfectly coincide for the presented three test classes.

It can be seen from these experiments that even if the scaling constants α and β have a different order (e.g., when α is infinitesimal and β is infinite) the scaled problems continue to be well-conditioned (cf. discussion on ill-conditioning in the traditional framework with finite scaling/shifting constants, see Fig. 1). This fact suggests that even if finite constants of significantly different orders are required,

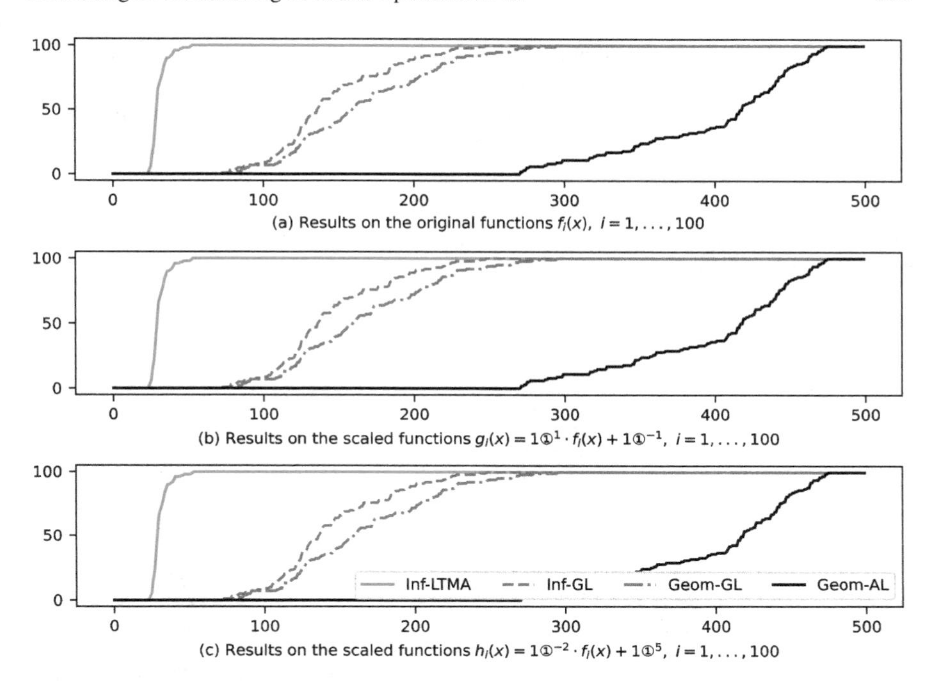

Fig. 3 Operational characteristics for **a** the original test class $f_i(x)$ from [45], **b** the scaled test class $g_i(x) = ①^1 f_i(x) + ①^{-1}$, **c** the scaled test class $h_i(x) = ①^{-2} f_i(x) + ①^5$. The results coincide for each method on all three test classes

① can also be used to avoid ill-conditioning by substituting very small constants by $①^{-1}$ and very huge constants by ①. If, for instance, α is too small (as, e.g., in (5), $\alpha = 10^{-8}$) and β is too large (as, e.g., in (5), $\beta = 10^8 \gg 10^{-8}$), the values $\alpha_1 = ①^{-1}$ and $\beta_1 = ①$ can be used in computations instead of $\alpha = 10^{-8}$ and $\beta = 10^8$ avoiding so the situations of underflow and overflow. When the optimization process is terminated, the global minimum f^* of the original function $f(x)$ can be easily extracted from the grossone-based solution $g^* = \alpha_1 f^* + \beta_1 = ①^{-1} f^* + ①$ of the scaled problem using $①^{-1}$ and ① while the original finite constants α and β can be then used to get the required value $g^* = \alpha f^* + \beta$ (in our case, $g^* = 10^{-8} f^* + 10^8$).

6 Concluding Remarks

The scaling issues in Lipschitz global optimization have been considered in this chapter by using the Infinity Computing framework. A class of global optimization problems has been particularly examined in which the objective function can have finite, infinite or infinitesimal Lipschitz constants. The strong homogeneity of geometric and information algorithms for solving univariate Lipschitz global optimization problems belonging to this class has been studied and illustrated numerically for finite, infinite, and infinitesimal scaling constants.

It has also been shown that when global optimization problems become ill-conditioned due to very huge and/or small scaling/shifting constants in the traditional computational frameworks working with finite numbers, the application of the Infinity Computing can help. It is useful in these situations to substitute finite constants provoking numerical problems by infinite and infinitesimal grossone-based numbers thus allowing one to avoid ill-conditioning of the scaled problems.

References

1. Amodio, P., Iavernaro, F., Mazzia, F., Mukhametzhanov, M.S., Sergeyev, Y.D.: A generalized Taylor method of order three for the solution of initial value problems in standard and infinity floating-point arithmetic. Math. Comput. Simul. **141**, 24–39 (2017)
2. Antoniotti, L., Caldarola, F., Maiolo, M.: Infinite numerical computing applied to Hilbert's, Peano's, and Moore's curves. Mediter. J. Math. **17**(3) (2020)
3. Archetti, F., Candelieri, A.: Bayesian Optimization and Data Science. Springer Briefs in Optimization. Springer, New York (2019)
4. Astorino, A., Fuduli, A.: Spherical separation with infinitely far center. Soft Comput. **24**, 17751–17759 (2020)
5. Audet, C., Caporossi, G., Jacquet, S.: Constraint scaling in the mesh adaptive direct search algorithm. Pacific J. Optim. **16**(4), 595–610 (2020)
6. Caldarola, F.: The Sierpinski curve viewed by numerical computations with infinities and infinitesimals. Appl. Math. Comput. **318**, 321–328 (2018)
7. Caldarola, F., Maiolo, M.: On the topological convergence of multi-rule sequences of sets and fractal patterns. Soft Comput. **24**, 17737–17749 (2020)
8. Calude, C.S., Dumitrescu, M.: Infinitesimal probabilities based on grossone. SN Comput. Sci. **1** (2020)
9. Cococcioni, M., Cudazzo, A., Pappalardo, M., Sergeyev, Y.D.: Solving the lexicographic multi-objective mixed-integer linear programming problem using branch-and-bound and grossone methodology. Commun. Nonlinear Sci. Numer. Simul. **84** (2020). Article 105177
10. Cococcioni, M., Fiaschi, L.: The Big-M method with the numerical infinite M. Optim. Lett. **15**(7), 2455–2468 (2021)
11. Cococcioni, M., Pappalardo, M., Sergeyev, Y.D.: Lexicographic multi-objective linear programming using grossone methodology: theory and algorithm. Appl. Math. Comput. **318**, 298–311 (2018)
12. D'Alotto, L.: Cellular automata using infinite computations. Appl. Math. Comput. **218**(16), 8077–8082 (2012)
13. D'Alotto, L.: A classification of one-dimensional cellular automata using infinite computations. Appl. Math. Comput. **255**, 15–24 (2015)
14. D'Alotto, L.: Infinite games on finite graphs using grossone. Soft Comput. **55**, 143–158 (2020)
15. De Cosmis, S., De Leone, R.: The use of grossone in mathematical programming and operations research. Appl. Math. Comput. **218**(16), 8029–8038 (2012)
16. De Leone, R.: Nonlinear programming and grossone: quadratic programming and the role of constraint qualifications. Appl. Math. Comput. **318**, 290–297 (2018)
17. De Leone, R., Fasano, G., Roma, M., Sergeyev, Y.D.: Iterative grossone-based computation of negative curvature directions in large-scale optimization. J. Optim. Theory Appl. **186**(2), 554–589 (2020)
18. De Leone, R., Fasano, G., Sergeyev, Y.D.: Planar methods and grossone for the conjugate gradient breakdown in nonlinear programming. Comput. Optim. Appl. **71**(1), 73–93 (2018)

19. Elsakov, S.M., Shiryaev, V.I.: Homogeneous algorithms for multiextremal optimization. Comput. Math. Math. Phys. **50**(10), 1642–1654 (2010)
20. Falcone, A., Garro, A., Mukhametzhanov, M.S., Sergeyev, Y.D.: A Simulink-based Infinity Computer simulator and some applications. In: LNCS, vol. 11974, pp. 362–369. Springer (2017)
21. Falcone, A., Garro, A., Mukhametzhanov, M.S., Sergeyev, Y.D.: Representation of Grossone-based arithmetic in Simulink and applications to scientific computing. Soft Comput. **24**, 17525–17539 (2020)
22. Falcone, A., Garro, A., Mukhametzhanov, M.S., Sergeyev, Y.D.: A Simulink-based software solution using the infinity computer methodology for higher order differentiation. Appl. Math. Comput. **409** (2021). Article 125606
23. Floudas, C.A., Pardalos, P.M. (eds.): Encyclopedia of Optimization (6 Volumes), 2nd edn. Springer (2009)
24. Gangle, R., Caterina, G., Tohmé, F.: A constructive sequence algebra for the calculus of indications. Soft Comput. **24**(23), 17621–17629 (2020)
25. Gaudioso, M., Giallombardo, G., Mukhametzhanov, M.S.: Numerical infinitesimals in a variable metric method for convex nonsmooth optimization. Appl. Math. Comput. **318**, 312–320 (2018)
26. Gaviano, M., Kvasov, D.E., Lera, D., Sergeyev, Y.D.: Algorithm 829: software for generation of classes of test functions with known local and global minima for global optimization. ACM Trans. Math. Softw. **29**(4), 469–480 (2003)
27. Gergel, V., Barkalov, K., Sysoev, A.: Globalizer: A novel supercomputer software system for solving time-consuming global optimization problems. Numer. Algebra Control Optim. **8**(1), 47–62 (2018)
28. Gergel, V.P., Grishagin, V.A., Gergel, A.V.: Adaptive nested optimization scheme for multidimensional global search. J. Glob. Optim. **66**, 35–51 (2016)
29. Grishagin, V.A.: Operating characteristics of some global search algorithms. Probl. Stoch. Search **7**, 198–206 (1978). In Russian
30. Grishagin, V.A., Israfilov, R.A., Sergeyev, Y.D.: Convergence conditions and numerical comparison of global optimization methods based on dimensionality reduction schemes. Appl. Math. Comput. **318**, 270–280 (2018)
31. Grishagin, V.A., Sergeyev, Y.D., Strongin, R.G.: Parallel characteristic algorithms for solving problems of global optimization. J. Global Optim. **10**(2), 185–206 (1997)
32. Iavernaro, F., Mazzia, F., Mukhametzhanov, M.S., Sergeyev, Y.D.: Computation of higher order Lie derivatives on the Infinity Computer. J. Comput. Appl. Math. **383** (2021)
33. Iudin, D.I., Sergeyev, Y.D., Hayakawa, M.: Infinity computations in cellular automaton forest-fire model. Commun. Nonlinear Sci. Numer. Simul. **20**(3), 861–870 (2015)
34. Jones, D.R., Martins, J.R.R.A.: The DIRECT algorithm: 25 years later. J. Glob. Optim. **79**, 521–566 (2021)
35. Jones, D.R., Perttunen, C.D., Stuckman, B.E.: Lipschitzian optimization without the Lipschitz constant. J. Optim. Theory Appl. **79**, 157–181 (1993)
36. Kvasov, D.E., Mukhametzhanov, M.S.: Metaheuristic vs. deterministic global optimization algorithms: the univariate case. Appl. Math. Comput. **318**, 245–259 (2018)
37. Kvasov, D.E., Mukhametzhanov, M.S., Sergeyev, Y.D.: Ill-conditioning provoked by scaling in univariate global optimization and its handling on the Infinity Computer. In: M.T.M. Emmerich et al. (ed.) Proceedings LEGO – 14th International Global Optimization Workshop, vol. 2070 (1). AIP Conference Proceedings (2019). Article 020011
38. Lai, L., Fiaschi, L., Cococcioni, M.: Solving mixed Pareto-Lexicographic manyobjective optimization problems: the case of priority chains. Swarm Evol. Comput. **55** (2020). Article 100687
39. Lera, D., Posypkin, M., Sergeyev, Y.D.: Space-filling curves for numerical approximation and visualization of solutions to systems of nonlinear inequalities with applications in robotics. Appl. Math. Comput. **390** (2021). Article 125660
40. Lolli, G.: Metamathematical investigations on the theory of grossone. Appl. Math. Comput. **255**, 3–14 (2015)

41. Margenstern, M.: Fibonacci words, hyperbolic tilings and grossone. Commun. Nonlinear Sci. Numer. Simul. **21**(1–3), 3–11 (2015)
42. Mockus, J.: Bayesian Approach to Global Optimization. Kluwer Academic Publishers, Dodrecht (1988)
43. Paulavičius, R., Žilinskas, J.: Simplicial Global Optimization. SpringerBriefs in Optimization. Springer, New York (2014)
44. Pepelyshev, A., Zhigljavsky, A.: Discrete uniform and binomial distributions with infinite support. Soft Comput. **24**, 17517–17524 (2020)
45. Pintér, J.D.: Global optimization: software, test problems, and applications. In: Pardalos, P.M., Romeijn, H.E. (eds.) Handbook of Global Optimization, vol. 2, pp. 515–569. Kluwer Academic Publishers, Dordrecht (2002)
46. Piyavskij, S.A.: An algorithm for finding the absolute extremum of a function. USSR Comput. Math. Math. Phys. **12**(4), 57–67 (1972)
47. Rizza, D.: Numerical methods for infinite decision-making processes. Int. J. Unconv. Comput. **14**(2), 139–158 (2019)
48. Sergeyev, Y.D.: On convergence of "divide the best" global optimization algorithms. Optimization **44**(3), 303–325 (1998)
49. Sergeyev, Y.D.: A new applied approach for executing computations with infinite and infinitesimal quantities. Informatica **19**(4), 567–596 (2008)
50. Sergeyev, Y.D.: Numerical point of view on Calculus for functions assuming finite, infinite, and infinitesimal values over finite, infinite, and infinitesimal domains. Nonlinear Anal. Ser. A: Theory Methods Appl. **71(12)**, e1688–e1707 (2009)
51. Sergeyev, Y.D.: Higher order numerical differentiation on the Infinity Computer. Optim. Lett. **5**(4), 575–585 (2011)
52. Sergeyev, Y.D.: Using blinking fractals for mathematical modelling of processes of growth in biological systems. Informatica **22**(4), 559–576 (2011)
53. Sergeyev, Y.D.: Numerical infinities and infinitesimals: methodology, applications, and repercussions on two Hilbert problems. EMS Surv. Math. Sci. **4**(2), 219–320 (2017)
54. Sergeyev, Y.D.: Independence of the grossone-based infinity methodology from non-standard analysis and comments upon logical fallacies in some texts asserting the opposite. Found. Sci. **24**(1), 153–170 (2019)
55. Sergeyev, Y.D., Candelieri, A., Kvasov, D.E., Perego, R.: Safe global optimization of expensive noisy black-box functions in the δ-Lipschitz framework. Soft Comput. **24**(23), 17715–17735 (2020)
56. Sergeyev, Y.D., Garro, A.: Observability of turing machines: a refinement of the theory of computation. Informatica **21**(3), 425–454 (2010)
57. Sergeyev, Y.D., Garro, A.: Single-tape and multi-tape Turing machines through the lens of the Grossone methodology. J. Supercomput. **65**(2), 645–663 (2013)
58. Sergeyev, Y.D., Grishagin, V.A.: A parallel method for finding the global minimum of univariate functions. J. Optim. Theory Appl. **80**(3), 513–536 (1994)
59. Sergeyev, Y.D., Kvasov, D.E.: Deterministic Global Optimization: An Introduction to the Diagonal Approach. Springer, New York (2017)
60. Sergeyev, Y.D., Kvasov, D.E., Mukhametzhanov, M.S.: Emmental-type GKLS-based multiextremal smooth test problems with non-linear constraints. In: LNCS, vol. 10556, pp. 383–388. Springer (2017)
61. Sergeyev, Y.D., Kvasov, D.E., Mukhametzhanov, M.S.: Operational zones for comparing metaheuristic and deterministic one-dimensional global optimization algorithms. Math. Comput. Simul. **141**, 96–109 (2017)
62. Sergeyev, Y.D., Kvasov, D.E., Mukhametzhanov, M.S.: On strong homogeneity of a class of global optimization algorithms working with infinite and infinitesimal scales. Commun. Nonlinear Sci. Numer. Simul. **59**, 319–330 (2018)
63. Sergeyev, Y.D., Mukhametzhanov, M.S., Kvasov, D.E., Lera, D.: Derivative-free local tuning and local improvement techniques embedded in the univariate global optimization. J. Optim. Theory Appl. **171**, 186–208 (2016)

64. Sergeyev, Y.D., Mukhametzhanov, M.S., Mazzia, F., Iavernaro, F., Amodio, P.: Numerical methods for solving initial value problems on the infinity computer. Int. J. Unconv. Comput. **12**(1), 3–23 (2016)
65. Sergeyev, Y.D., Nasso, M.C., Mukhametzhanov, M.S., Kvasov, D.E.: Novel local tuning techniques for speeding up one-dimensional algorithms in expensive global optimization using Lipschitz derivatives. J. Comput. Appl. Math. **383** (2021). Article 113134
66. Sergeyev, Y.D., Strongin, R.G., Lera, D.: Introduction to Global Optimization Exploiting Space-Filling Curves. Springer, New York (2013)
67. Strongin, R.G.: Numerical Methods in Multiextremal Problems: Information-Statistical Algorithms. Nauka, Moscow (1978). (In Russian)
68. Strongin, R.G., Sergeyev, Y.D.: Global Optimization with Non-Convex Constraints: Sequential and Parallel Algorithms. Kluwer Academic Publishers, Dordrecht (2000)
69. Žilinskas, A., Žilinskas, J.: A hybrid global optimization algorithm for non-linear least squares regression. Journal of Global Optimization **56**(2), 265–277 (2013)
70. Zhigljavsky, A.: Computing sums of conditionally convergent and divergent series using the concept of grossone. Appl. Math. Comput. **218**(16), 8064–8076 (2012)
71. Zhigljavsky, A., Žilinskas, A.: Bayesian and High-Dimensional Global Optimization. Springer Briefs in Optimization. Springer, New York (2021)
72. Zhigljavsky, A., Žilinskas, A.: Stochastic Global Optimization. Springer, New York (2008)
73. Žilinskas, A.: Axiomatic characterization of a global optimization algorithm and investigation of its search strategies. Oper. Res. Lett. **4**, 35–39 (1985)
74. Žilinskas, A.: On strong homogeneity of two global optimization algorithms based on statistical models of multimodal objective functions. Appl. Math. Comput. **218**(16), 8131–8136 (2012)

CPSIA information can be obtained
at www.ICGtesting.com
Printed in the USA
LVHW062039110722
723183LV00019B/34

9 783030 936419